HANDBOOK OF JAPAN–UNITED STATES ENVIRONMENT–BEHAVIOR RESEARCH

Toward a Transactional Approach

HANDBOOK OF JAPAN–UNITED STATES ENVIRONMENT–BEHAVIOR RESEARCH

Toward a Transactional Approach

Edited by

Seymour Wapner
Clark University
Worcester, Massachusetts

Jack Demick
Suffolk University
Boston, Massachusetts

Takiji Yamamoto
The Japanese Institute of Health Psychology
Tokyo, Japan

and

Takashi Takahashi
University of Tokyo
Tokyo, Japan

PLENUM PRESS • NEW YORK AND LONDON

Library of Congress Cataloging-in-Publication Data

Handbook of Japan-United States environment-behavior research : toward
  a transactional approach / edited by Seymour Wapner ... [et al.].
        p.   cm.
    Includes bibliographical references and indexes.
    ISBN 0-306-45340-1
    1. Environmental psychology--Japan. 2. Environmental psychology-
  -United States. 3. Environmental psychology--Cross-cultural
  studies.   I. Wapner, Seymour, 1917-   .
  BF353.H27 1996
  155.9'0952--dc21   96-48409
                                                              CIP

ISBN 0-306-45340-1

© 1997 Plenum Press, New York
A Division of Plenum Publishing Corporation
233 Spring Street, New York, N. Y. 10013

10 9 8 7 6 5 4 3 2 1

Printed in the United States of America

Contributors

Kei Adachi, Department of Architecture, Faculty of Engineering, Kansai University, 3-3-35 Yamate-cho, Suita, Osaka, 561, Japan

Irwin Altman, Department of Psychology, University of Utah, Salt Lake City, Utah 84112

Masaaki Asai, Department of Psychology, College of Humanities and Sciences, Nihon University, Tokyo 156, Japan

Robert B. Bechtel, Department of Psychology, University of Arizona, Tucson, Arizona 85721

Kenneth H. Craik, Institute of Personality and Social Research, University of California-Berkeley, Berkeley, California 94720

Jack Demick, Department of Psychology, Suffolk University, Boston, Massachusetts 02114

Junko Fujimoto, 2-13-7 Kitahorie, Nishi-ku, Osaka 550, Japan

Kunio Funahashi, Faculty of Engineering, Department of Architectural Engineering, Osaka University, Yamadaoka, Suita 565, Japan

Toshihiro Hanazato, Institute of Art and Design, University of Tsukuba, Tsukuba, Iberaki 305, Japan

Tomohiro Hata, Department of Built Environment, Tokyo Institute of Technology,Yokohama-shi 226, Japan

Sonomi Hirata, Department of Human Sciences, Graduate School, Waseda University, Saitama 359, Japan

Sandra C. Howell, Department of Architecture, Massachusetts Institute of Technology, Cambridge, Massachusetts 02139

Tomoaki Imamichi, 2-7-13 Shiroganedai Minato-ku, Tokyo 108, Japan

Wataru Inoue, Research Center for Educational Study and Practice, School of Education, Higashihiroshima University, Higashihiroshima 739, Japan

Yuko Inoue, 1-34-20 Nagao Higashimachi, Hirakata, Osaka 573-01, Japan

Shinji Ishii, Department of Psychology, School of Education, Hiroshima University, Hiroshima 739, Japan

Masami Kobayashi, Department of Global Environmental Engineering, Graduate School of Engineering, Kyoto University, Kyoto 606, Japan

Miki Kondo, Department of Built Environment, Tokyo Institute of Technology, Yokohama-shi 226, Japan

Satoshi Kose, Building Research Institute, Ministry of Construction, Tatehara, Tsukuba 305, Japan

Yoichi Kubota, Department of Construction Engineering, Faculty of Engineering, Saitama University, Saitama 338, Japan

Setha Low, Department of Environmental Psychology and Department of Anthropology, Graduate School and University Center of the City University of New York, New York, New York 10036

Taber MacCallum, Paragon Space Development Corporation, Tucson, Arizona 85714

Norio Maki, Department of Global Environmental engineering, Graduate School of Engineering, Kyoto University, Kyoto 606, Japan

William Michelson, Department of Sociology and Centre for Urban and Community Studies, University of Toronto, Toronto, Ontario M5S 1A1, Canada

Hirofumi Minami, Faculty of Education, Kyushu University, Fukuoka 812, Japan

Ken Miura, Department of Global Environmental Engineering, Graduate School of Engineering, Kyoto University, Kyoto 606, Japan

Ryuzo Ohno, Department of Built Environment, Tokyo Institute of Technology, Yokohama-shi 226, Japan

Hisao Osada, Tokyo Metropolitan College of Allied Medical Sciences, Tokyo 116, Japan

Jane Poynter, Paragon Space Development Corporation, Tucson, Arizona 85714

George Rand, Department of Architecture and Urban Design, School of Arts and Architecture, University of California-Los Angeles, Los Angeles, California 90024

Amos Rapoport, Department of Architecture, University of Wisconsin-Milwaukee, Milwaukee, Wisconsin 53201

Susan Saegert, Department of Environmental Psychology, Graduate School and University Center of the City University of New York, New York, New York 10036

Miho Saito, School of Human Sciences, Waseda University, Tokorozawa, Saitama 359, Japan

Toshihiko Sako, Department of Human Health Sciences, School of Human Sciences, Waseda University, Saitama 359, Japan

Ichiro Soma, School of Human Sciences, Waseda University 359, Japan

Daniel Stokols, School of Social Ecology, University of California, Irvine, Irvine, California 92697-7050

Takeshi Suzuki, Department of Architecture, Faculty of Engineering, University of Tokyo, Tokyo 113, Japan

Takashi Takahashi, Department of Architecture, Faculty of Engineering, University of Tokyo, Tokyo 113, Japan

Kunio Tanaka, 1579-4 Mikage-cho, Kishimoto, Higashinada-ku, Kobe 658, Japan

Kanae Tanigawa, Bell Road Rokko #204, 14-14 Teraguchi-cho, Nada-ku, Kobe 657, Japan

Karl Toews, 1558 Stonemill Road, Lancaster, Pennsylvania 17603

Seymour Wapner, Heinz Werner Institute, Clark University, Worcester, Massachusetts 01610

Takiji Yamamoto, The Japanese Institute of Health Psychology, 5F Mikuni Building, 1-3-5 Kojimachi, Chizoda-Ku, Tokyo 102, Japan

Ervin Zube, School of Renewable Natural Resources, University of Arizona, Tucson, Arizona 85721

Preface

This volume is an outgrowth of research on the relations between human beings and their environments, which has developed internationally. This development is evident in environment–behavior research studies conducted in countries other than the United States. See Stokols and Altman (1987) for examples of such work in Australia, Japan, France, Germany, the Netherlands, Sweden, the United Kingdom, the former Soviet Union, and Latin and North America. The international development of this research area is also evident in the establishment of professional organizations in different countries such as the Environment–Behavior Design Research Association (EDRA) in the United States, the Man–Environment Research Association (MERA) in Japan, the International Association for People–Environment Studies (IAPS) in Great Britain, and the People and Physical Environment Research Association (PAPER) in Australia.

This volume focuses on environment–behavior research within Japan and the United States as well as cross-cultural studies involving both countries. As we note in detail in Chapter 1, the conference on which the work presented herein is based was preceded by three Japan–United States conferences on environment–behavior research, the first of which took place in Tokyo in 1980.

As currently conceived, the present volume stands alone as a compendium of a significant proportion of cross-cultural research on environment–behavior relations in Japan and the United States that has been developing over the last 15 years. As such, we envision the volume as a basic interdisciplinary reference for anthropolgists, architects, psychologists, sociologists, urban planners, and environmental geographers. Pedagogically, the book might be appropriate for advanced undergraduate courses in environmental and/or cross-cultural psychology. More definitely, the work can serve as a primary text in a range of graduate courses on environment–behavior research.

We would like to extend our sincere appreciation to the following institutions and individuals who contributed to the Japan–United States seminar on which this volume is based. First, the seminar would not have occurred without the generous support of the United States National Science Foundation, the Japan Society for the Promotion of Science, and both the Heinz Werner Institute for Developmental Analysis and the Frances L. Hiatt School of Psychology of Clark University. Second, there are several individuals who contributed greatly to the actual success of the seminar: Joyce Lee, who organized all details of the seminar from beginning to end (including expert typing and retyping of manuscripts); and Lorraine G. Wapner, who provided an exceptional social context that nurtured the camaraderie among—and fostered the continued collaboration of—conference participants.

Contents

1. **Introduction**   1
   *Seymour Wapner, Jack Demick, Takiji Yamamoto, and
   Takashi Takahashi*

I. PHYSICAL ASPECTS OF THE PERSON

2. **Visual Perception in Elderly Persons with Dementia**   15
   Kei Adachi

3. **Dwelling Design Guidelines for Accessibility in the
   Aging Society: A New Era in Japan?**   25
   Satoshi Kose

4. **Urban Housing and the Elderly**   43
   Takashi Takahashi

5. **Quality of Life in Japanese Older Adults**   51
   Hisao Osada

6. **A Study on a Relocation of a Nursing Home for
   Blind Older Adults**   59
   Takiji Yamamoto and Toshihiro Hanazato

II. PSYCHOLOGICAL ASPECTS OF THE PERSON

7. **Body and Self Experience: Japan versus USA**   83
   Jack Demick, Shinji Ishii, and Wataru Inoue

8. **Cross-Cultural Survey on Color Preferences in Three
   Asian Cities: Comparisons among Tokyo, Taipei,
   and Tianjin**   101
   Ichiro Soma and Miho Saito

9. **Mode of Being in Places: A Case Study in Urban Public
   Space**   113
   Takeshi Suzuki

III. SOCIOCULTURAL ASPECTS OF THE PERSON

10. **Urban Renewal and the Elderly: An Ethnographic Approach**   133
    Hirofumi Minami

11. **Integrating Environmental Factors into Multidimensional
    Analysis**   149
    William Michelson

IV. PHYSICAL ASPECTS OF THE ENVIRONMENT

12. **Experiencing Japanese Gardens: Sensory Information and
    Behavior**   161
    Ryuzo Ohno, Tomohiro Hata, and Miki Kondo

13. **Preference for Trees on Urban Riverfronts**   181
    Yoichi Kubota

14. **Landscape Values: Congruence and Conflict in Desert
    Riparian Areas**   199
    Ervin H. Zube

15. **Environmental Transition and Natural Disaster: Restoration
    Housing for the Mt. Unzen Volcanic Eruption**   209
    Masami Kobayashi, Ken Miura, and Norio Maki

16. **Environmental Psychology and Biosphere 2**   235
    Robert B. Bechtel, Taber MacCallum, and Jane Poynter

V. INTERPERSONAL ASPECTS OF THE ENVIRONMENT

17. **Rethinking Stereotypes of Family Housing in Japan and
    the USA**   247
    Sandra C. Howell

VI. SOCIOCULTURAL ASPECTS OF THE ENVIRONMENT

18. **Sociopsychological Environments of Japanese Schools as
    Perceived by School Students**   261
    Masaaki Asai and Sonomi Hirata

19. **"Big School, Small School" Revisited: A Case Study of
    a Large-Scale Comprehensive High School Based on
    the Campus Plan**   273
    Toshihiko Sako

20. **Sojourn in a New Culture: Japanese Students in American
    Universities and American Students in Japanese Universities**   283
    *Seymour Wapner, Junko Fujimoto, Tomoaki Imamichi,
    Yuko Inoue, and Karl J. Toews*

21. **Public Space as Art and Commodity: The Spanish American Plaza**............................ 313
   Setha M. Low

22. **Urbanization and Quality of Life in Asia** 325
   Kanae Tanigawa and Kunio Tanaka

VII. FUTURE THEORETICAL AND EMPIRICAL DIRECTIONS FOR ENVIRONMENT–BEHAVIOR RESEARCH

23. **Directions of Environmental Psychology in the Twenty-First Century** 333
   Daniel Stokols

24. **Transactional Perspective, Design, and "Architectural Planning Research" in Japan** 335
   Kunio Funahashi

25. **Some Arguments for a Comparative Developmental Environmental Psychology with a Long-Term View of History and Cultural Psychology** 365
   George Rand

26. **Prospects for Environmental Psychology in the Third Millennium**... 377
   Kenneth H. Craik

27. **What Is the Situation?: A Comment on the Fourth Japan–USA Seminar on Environment–Behavior Research**................. 385
   Susan Saegert

28. **Theory in Environment–Behavior Studies: Transcending Times, Settings, and Groups** 399
   Amos Rapoport

29. **Environment and Behavior Studies: A Discipline? Not a Discipline? Becoming a Discipline?**........................ 423
   Irwin Altman

Appendix: List of Presentations of Previous Japan–U.S. Seminars on Environment-Behavior Research.................... 435

Name Index 439

Subject Index 447

HANDBOOK OF JAPAN–UNITED STATES ENVIRONMENT–BEHAVIOR RESEARCH

Toward a Transactional Approach

Chapter **1**

INTRODUCTION

Seymour Wapner, Jack Demick, Takiji Yamamoto, and Takashi Takahashi

This volume is based on an ongoing collaboration of scientists from Japan and the United States through a series of seminars over the past 15 years. The First Japan United States Seminar, "Interaction Processes Between Human Behavior and Environment," took place in Tokyo on September 24–27, 1980. In a preface to the report on the proceedings of the seminar, Hagino (1981) described its origins as follows:

> Usually, in order to accomplish anything, there are many preliminary conditions which must first be met and there are many arrangements which must be made. In other words, something must exist to set things in motion. However, there are special unexpected instances when an unusual opportunity can trigger a big event without the usual long years of required pre-planning.
>
> The conditions which led to our Japan–U.S. Seminar on Interactive Processes Between Human Behavior and the Environment seem to come under the second category. When we look back, we can say that the time was surely ripe for holding such a seminar, but a trigger would be needed to set everything into motion. The starting point for all of this "motion" was in August of 1979. The trigger was Professor Takiji Yamamoto, who happened to be at Clark University lecturing just at the time when Professors William Ittelson and Seymour Wapner were thinking to organize a joint Japan–U.S. Environmental Psychology seminar on Japanese soil. They proposed the idea to Professor Yamamoto and soon afterwards he returned to Japan and began to discuss this plan with me and a few of our other colleagues. On the Japanese side, however, there was very little organization among researchers in Environmental Psychology. In fact, in Japan environmental psychological studies have been done from many different disciplines and workers have presented and published their research only in their respective associations and fields

Seymour Wapner • Heinz Werner Institute, Clark University, Worcester, Massachusetts 01610. Jack Demick • Department of Psychology, Suffolk University, Boston, Massachusetts 02114. Takiji Yamamoto • The Japanese Institute of Health Psychology, 5F Mikuni Building, 1-3-5 Kojimachi, Chiyoda-Ku, Tokyo, 102, Japan. Takashi Takahashi • Department of Architecture, University of Tokyo, Tokyo 113, Japan.

Handbook of Japan–United States Environment–Behavior Research: Toward a Transactional Approach, edited by Seymour Wapner, Jack Demick, Takiji Yamamoto, and Takashi Takahashi. Plenum Press, New York, 1997.

and thus communication between them was insufficient. Research exchange could only be done on an individual basis at both national and international levels. Because of this we were naturally taken by surprise at the suddenness of the proposal of such a seminar, but since the necessity for more communication and organization in Environmental Psychology has been strongly felt for a long time, we decided to hold the seminar in Japan with the co-operation of Psychologists, Architects, Urban Planners, Urban Engineers and Social Engineers.

Fortunately, we had the kind support of both the Japan Society for the Promotion of Science and the National Science Foundation and our seminar was made possible. It was held beginning on the 24th and lasting until the 27th of September 1980 in Tokyo. In spite of our early apprehensions, the congress was a great success. The rich content of the lectures, the stimulating discussions and the active interchange were such an intense experience that the four days allowed us just didn't seem to be enough time. An expected benefit arose from this in that participants from both countries enthusiastically requested that a future such congress be held to take up where the first one left off.

In regard to the meaning of this seminar and possible future seminars, I would like to add the following comments. Research on interactive processes between the environment and human behavior requires basic research from each separate discipline and of course requires interdisciplinary and joint cross-cultural studies. Furthermore, I think that more academic communication between researchers from different cultures can help to find some needed general laws for Environmental Psychology. It is a very positive "sign" on the road to progress in our discipline that international communication and joint research is increasing. I fervently hope that the Second Japan–U.S. Seminar will be held in the near future on American soil. In Japan, an important impact of this seminar was the increasing communication and commitment of researchers working on environment and human behavioral problems. This increase in communication will add meaning to this type of seminar.

It has finally become possible for us to publish the collected papers and a summary of the discussions held during the seminar. It is our sincere desire that this volume will be a milestone for both future American-Japanese research co-operation and will stimulate the organization and unification of Japanese researchers.

Finally, I must acknowledge the efforts of Professors Ittelson and Wapner and the other American participants who come such a very far distance, and I must express my thanks for the excellent co-operation of all the Japanese participants. I am deeply grateful to the Japan Society for the Promotion of Science and the National Science Foundation for their generous financial support. Without it, this seminar certainly could not have been held. I should like to offer my thanks to the College of Humanity and Sciences, Nihon University which provided rooms for this conference, and in particular to Dean and Professor Shotaro Tsumakura who gave the congratulatory address. I must also mention that I am deeply indebted to Professor Yamamoto, Professor Masaaki Asai, Professor Takashi Takahashi and Mr. Joe Hicks for all of their co-operation from the very beginning. (pp. v–vii)

In the same preface, Ittelson (1981) viewed the First Japan–U.S. Seminar as:

the beginning of what we hope will be a continuing process of cooperation between Japanese and American scholars working on a range of problems involved in the inter-relationships between people and their physical settings. While this area of study is perhaps commonly referred to as environmental psychology, it is also called by various other names including man-environment systems, environment and behavior studies, human geography, environmental sociology and others. It is an interdisciplinary field that includes methods and findings from anthropology, sociology, geography, architecture, landscape architecture and planning as well as psychology.

Environmental psychology is a relatively new field having its most immediate beginnings as a discipline in the 1950s. The following twenty years saw a period of remarkable growth until by the time of this conference the field was firmly established as an internationally recognized area of teaching, research and application.

The growth probably reflects a general concern over the human consequences of both alteration of our natural environment as well as our increasingly complex social and technological environments. The resulting problems are not defined by disciplines nor do they recognize international boundaries. It is our belief that multi-disciplinary, international efforts of which this conference is but a small sample, will be needed to provide a comprehensive scientific understanding of the complex issues involved in the interrelationships between human behavior and the environment. (p. viii)

It is of significance that the First Japan–U.S. Seminar, involving 25 Japanese and 8 American participants, served as a significant step in the establishment of the Japanese Man-Environment Research Association (MERA). As reported by Wapner (1983):

A group of the original participants at the USA/Japan Seminar met several times and discussed the possibility of establishing a Japanese organization to promote such cooperation among the various fields in the hope of bettering the understanding of Man-Environment related problems. This core group consisted of three senior committee members, M. Mochizuki (Psychologist), Y. Yoshitake, past president Architectural Institute of Japan (Architect), and G. Hagino, president of the Japanese Psychological Association (Psychologist), and six acting members, T. Yamamoto (Psychologist), M. Inui (Architect), K. Hiroshima (Population Research), T. Takahashi (Architect), M. Asai (Psychologist), and I. Soma (Psychologist) who serves as Secretary-General. This group presently operates the steering committee for MERA. MERA was formally launched on the 19th of August, 1982. The first meeting was held at Waseda University in Tokyo, Japan. The main speakers for the inaugural meeting were Dr. Seymour Wapner of Clark University, who spoke on "Some Trends and Issues in Environmental Psychology" . . . and Professor Mamoru Mochizuki who spoke on "Specific Problems Facing Japanese Environmental Psychology." During this meeting members contributed various ideas as to how to further Man-Environment Relations research in Japan and how to work together to promote such research. In addition to promoting Man-Environment research in Japan, another important task facing the group was seen in its role as a bridge between Japan and other countries. (pp. 4–5)

In keeping with the hopes expressed by Professors Ittelson, Hagino, and others, the Second Japan–U.S. Seminar was held at the University of Arizona, Tucson, on October 6–9, 1985. Ittelson (1986) noted:

As in most successful conferences, the formal agenda ran in parallel with an informal one, which, unplanned and unanticipated, emerged from the spontaneous contacts among ever-changing groups in usually unscheduled social settings. As the conference developed, in addition to the sharing of ideas and plans, intellectual confrontations and collaborations and other substantive developments, a general consensus gradually emerged that became formalized on the last day. In brief this seminar was initially viewed by all participants as a follow-up to the initial seminar held in Tokyo in October 1980 . . . . After four days of interaction, the view had subtly changed, and participants began to talk of this seminar as the one between the first and the third . . .

Such an outcome was by no means certain when the first seminar was planned five years ago. At that time, personal contacts between individuals in the two countries suggested the existence of a body of common interests. The first seminar was exploratory in orientation, seeking out those common themes and asking whether further collaboration was desirable and feasible. The success of that meeting can be assessed by what followed. In Japan it served as a catalyst, bringing a number of individuals and institutions into a professional organization [MERA]. Some of these events are described by Professor Hagino in his message to the seminar. In the US the first seminar crystallized the interest of a number of researchers in cross-national studies. The second seminar was originally conceived as an opportunity to look back over five years and to assess what has been accomplished. This was achieved but as suggested above, the spirit of the group quickly changed from looking backward to looking forward. A committee is already actively engaged in planning for the third seminar to be held in Japan. (pp. iii–iv)

The Third Japan–U.S. Seminar on environment behavior relationships was held in Kyoto, on July 19–20, 1990. There were 13 presenters from Japan, 10 from the United States, and 27 observers from various parts of the world. In his preface to the volume containing the papers presented at the seminar, Takahashi (1990) described the papers as dealing

> mainly with three kinds of problems which reflect recent interest in this field: 1) the problem of consistent methods for comparing different environmental behaviors in different cultures. 2) the problem of generality and specialty of environmental behaviors in Japan and the uniqueness of Japanese architecture. 3) the problem of describing a broader model of "man as a people attached to a place" either in positivistic studies or in more normative studies. It is remarkable that the studies in this volume, especially the ones dealing with the third problem, show the very recent development of the model of man-environment relationships into a more culture dependent one. (p. xii)

In his message for the opening of the seminar, Hagino (1990) made the following relevant observation:

> It is our responsibility to prevent destruction of the human environment and, with scientific substantiations, to participate actively even in, say, political and social areas in order to improve our environment. (p. xiii)

For a complete listing of the papers presented at each of the first three seminars, the interested reader is referred to the appendix of this volume.

The editors of this volume served as organizers of the Fourth Japan–U.S. Seminar on Environment–Behavior Research that took place at Clark University on August 8–9, 1995. In our message for the opening of the seminar, we noted the following:

> The series of published seminars had a marked effect in advancing the development of environment-behavior research in both countries resulting, for example, in the development of the Man-Environment Research Association (MERA) and its journal in Japan as well as of cross-cultural research involving the joint efforts of investigators within the two countries. Given current concerns about the serious environmental problems facing humanity at the end of the 20th century (e.g., air pollution, resource shortages, overpopulation, urban decay), there is a powerful need to find solutions, both intra- and inter-culturally, to develop more optimal relations between humans and their environments. This seminar is directed toward that end. This will be accomplished by discussions in each session and a general discussion at the end of the conference devoted to identifying and making arrangements for cross-cultural research between Japanese and American investigators.
>
> As we see it, expected results of the Fourth Seminar include: 1. Uncovering and initiating relevant, specific research problems—for example, impact of technology on behavior—that require solution in keeping with the goals of the Seminar; 2. Developing a synergistic, transactional relationship between Japanese and American researchers; 3. Conducting investigations that are practically directed toward changing human behavior with regard to the environment; 4. Proposing global environmental policy regarding means of optimizing human-environment relations; 5. In general, changing the current

state of person–environment relations so that a more optimal relation exists in the 21st century.

As will be seen from the nature of the chapters in this volume, which are based on the presentations at the seminar, a number of these goals have already been accomplished.

A brief synopsis of the chapters contained in this volume follows to help orient the reader. Organizationally, we have divided the volume into six interrelated categories generated against the backdrop of a holistic, developmental, systems-oriented approach to person-in-environment functioning (e.g., Wapner, 1981, 1987; Wapner and Demick, 1990, 1991).

Specifically, person-in-environment processes are treated in relation to *physical aspects of the person* in Part I (Chapters 2–6), in relation to *psychological aspects of the person* in Part II (Chapters 7–9), in relation to *sociocultural aspects of the person* in Part III (Chapters 10–11), in relation to *physical aspects of the environment* in Part IV (Chapters (12–16), in relation to *interpersonal aspects of the environment* in Part V (Chapter 17), and in relation to *sociocultural aspects of the environment* in Part VI (Chapters 18–22). Although these chapters have been organized within one of these six categories for the sake of convenience, it should be emphasized that, in reality, these aspects of persons and environments mutually define one another (transactionalism) and operate contemporaneously (holism) within the functioning of the person-in-environment system.

Part VI is devoted to future theoretical and empirical directions for the field of environment–behavior research (Chapters 23–28). Finally, all of the chapters in the volume, with the exception of that by the participant-observer Rand (Chapter 25), were presented at the conference.

Part I, dealing with physical aspects of the person, focuses exclusively on the transactions (experience and action) of older adults, an increasingly complex social problem for both Japan and the United States. It begins with Kei Adachi's (Chapter 2) presentation of some data concerning the visual perception of older adults with dementia. He has shown that the range of eye fixations and scan paths in the older adult with dementia is limited and biased toward the center of presented stimuli to a significantly greater extent than in mentally retarded and normal groups. His studies have relevance for the study of environmental perception and cognition over the course of the human life span.

Satoshi Kose (Chapter 3) discusses the development of building facilities suited to the needs of contemporary Japanese older adults (emphasis on accessibility). Consideration is also given to how this development impacts past, present, and future governmental policy.

In line with this discussion, Takashi Takahashi (Chapter 4) presents findings from his ongoing research on urban housing for Japanese older adults. Here his focus is twofold: (1) assessment of older adults' adaptation as a function of the physical layout of their home settings and (2) examination of older adults' use of surrounding outdoor facilities (e.g., association with friends and neighbors). The meaning of these studies for both public policy and future environment–behavior research is stressed.

Hisao Osada (Chapter 5) presents the findings from a study on some relations between Japanese older adults' subjective well-being and various organismic and environmental factors. Specifically, he found that subjective well-being was greater in males than in females and that, irrespective of gender, subjective well-being was related to subjective health status, chewing ability, visual capacity, falls, pain, and aspects of social support. This research has clear implications for optimal environment–behavior relations in older adult populations.

Takiji Yamamoto and Toshihiro Hanazato (Chapter 6) present the results of two studies on the physical relocation of a nursing home for blind older adults in Japan. Study 1 demonstrates the microgenetic development of older adults' spatial representation (cognitive sketch maps of the new environment) over the course of relocation. Study 2 documents that the requirements of "behavioral focal points" (places where users can get together naturally and interact) for blind older adults differ from those of sighted individuals. These studies contribute to the burgeoning literature, both here and abroad, on the effects of environmental relocations and transitions more generally.

Part II, which focuses primarily on psychological aspects of the person, begins with the work of Demick, Ishii, and Inoue (Chapter 7) on similarities and differences between aspects of Japanese and American individuals' body and self experience. Following from an elaborated organismic-developmental perspective (e.g., Wapner, 1987), findings indicated that the Japanese and Americans exhibited differences in *sensorimotor* (e.g., Americans moved their bodies through space faster than the Japanese), *perceptual* (e.g., relative to the Japanese, Americans exhibited greater degrees of body boundary diffuseness and personal space), and *conceptual* experience (e.g., relative to the Japanese, Americans provided less detail on self-drawings but reported more satisfaction with body and self and higher self-esteem). Implications are discussed for those interested in environment–behavior relations as well as those concerned with unified theory.

Ichiro Soma and Miho Saito (Chapter 8) present findings from their long-term research program on cross-cultural similarities and differences in color preference (Tokyo, Taipei, and Tianjin). Their findings indicated that each area preferred different hues and tones; however, a strong preference for white was common to all areas (possibly related to religion, mythology, a sense of purity, etc.). Implications are discussed for environmental design as well as for cross-cultural psychology.

Takeshi Suzuki (Chapter 9) presents the results of case studies of urban public space in Tokyo, Taipei, and Paris with a focus on typologies for the "mode of being in places." He argues that, when someone is in a space (e.g., reading a book in the park), he or she either consciously or unconsciously chooses a mode or state of being that includes perception of surroundings, vistas of the city, social contact, how others can recognize him or her, and so on. Though being in places is a fundamental experience in our normal daily life, Suzuki has identified a new problem area by arguing that we have limited language (both in Japanese and English) to describe this phenomenon.

Part III, which deals primarily with sociocultural aspects of the person, begins with the work of Hirofumi Minami (Chapter 10). Using an ethnographic approach,

Minami reports changes in the roles of two samples of older adults during the urban renewal of their neighborhoods. In general, he describes how changes in aspects of the physical environment facilitated generational changes in the social organizational dynamics of the community (e.g., in one context, older adults gradually disengaged, whereas, in the other, older adults adopted roles as leaders). This research adds significantly to our bodies of knowledge on both urban renewal and environmental transitions.

William Michelson (Chapter 11) also considers the sociocultural aspect of the person. Specifically, by focusing on the daily travel of employed versus unemployed women, he demonstrates the efficacy of the time use approach (multidimensional analysis), which integrates environmental considerations with human activity, social interaction, and subjective interpretation. The notion of significant relations between problem, theory, and method in environmental psychological research is highlighted.

Part IV, which deals primarily with physical aspects of the environment, can be divided into those chapters that treat the *natural environment* and those that consider the *built environment*. Ryuzo Ohno, Tomohiro Hata, and Miki Kondo (Chapter 12) discuss a study on the experience of Japanese gardens. Participants were asked to stroll a circuit-style garden in Kobe city at will; each participant's motion and walking pace were recorded at every 0.5-meter point along the path. Findings indicated that people's behavior commonly changes at certain places in the Japanese garden and can be explained by sensory information in the environment. The use of such novel techniques is discussed for environmental psychological research more generally.

Yoichi Kubota (Chapter 13) reconsiders the meaning of trees planted on the waterfront from an environment–behavior perspective. Traditionally in Japan, trees—particularly cherry and willow trees—had animistic significance. Results from a computer-generated experiment revealed that contemporary Japanese individuals still prefer cherry and willow trees; however, trees in a row along the river channel were preferred over traditional random arrangements.

Ervin H. Zube (Chapter 14) concludes the subsection on the natural environment by describing a 15-year research program addressing riparian area values in the desert landscapes of Arizona. Three stages of the program are presented: exploratory interviews to identify a broad spectrum of values; a statewide mail survey to assess the relative importance of those values; and, based on data from that survey, a series of values-oriented mail surveys targeted on specific riparian areas. Cases of congruence and conflict of values—for example, among groups of respondents, between respondents and the sustainability of riparian areas—are then examined. The study highlights the importance of considering the often overlooked area of values and valuative functioning.

Beginning the subsection on the built environment, Masami Kobayashi, Ken Miura, and Norio Maki (Chapter 15) report the findings of a study on the effects of environmental transition related to a natural disaster (the Mt. Unzen volcanic eruption in Japan). Specifically, they found that successful relocation depended not only on the physical reconstruction of the house but, perhaps more importantly, on the victims' active participation in the process of rebuilding their homes.

Robert Bechtel (Chapter 16) focuses on one particular built environment, namely, Biosphere 2 (a materially closed ecosystem), and reports some findings from an assessment of its feasibility. Specifically, he reported that the isolation of the environment was overcome by use of electronic media and that the biospherians were very similar in MMPI profiles to astronauts. Implications for the future of such projects are also discussed.

Part V, which deals with interpersonal aspects of the environment, is represented by a single contribution, namely, that of Sandra Howell (Chapter 17). Howell attempts to link selected built characteristics of Japanese and North American dwellings with observed and reported data on intrafamily behaviors. She also illustrates the ways in which changes in the fixed and semifixed features of the dwelling, made by homeowners, interacted with and reflected developmental changes in the household, particularly in parent–child and sibling relationships. The need for more research in this general area is clearly indicated.

Part VI, which explores sociocultural aspects of the environment most generally, can be divided into those chapters that treat the sociocultural aspect of the *school* and *urban* environments, respectively. Masaaki Asai and Sonomi Hirata (Chapter 18), for example, focus on the sociocultural environment of Japanese schools as perceived by high school students. Comparing these perceptions with those of Japanese delinquents in correctional institutions and with U.S. normative data, Asai and Hirata suggest that similar perceptual patterns associated with the sociocultural environment of the classroom may elicit similar behaviors despite cultural and physical milieu differences.

Toshihiko Sako (Chapter 19) explores the long-standing issue of big versus small schools by presenting a case study of a Japanese large-scale comprehensive high school based on the campus plan (schools-within-a-school). He reports that smaller settings (e.g., home rooms, clubs) were described as "familiar," whereas bigger settings (e.g., overall campus) showing anonymous properties were described as "distant" and "shallow." This leads him to conclude that, in the ideal setting such as that described herein, the home room is an effective small school setting supporting a big school.

In contrast to the prior two chapters that address the high school environment, Seymour Wapner, Junko Fujimoto, Tomoaki D. Imamichi, Yuko Inoue, and Karl J. Toews (Chapter 20) report on a series of studies that assessed the experience of Japanese undergraduates in American universities and of American undergraduates in Japanese universities. The holistic, developmental, systems-oriented perspective is utilized in analyzing the (cognitive, affective, and valuative) experience of these students. Many Japanese students found the United States to be big, inexpensive, ineffective in transportation, dangerous, and free as compared to Japan; American students found Japan to be small, expensive, effective in transportation, safe and regulated. Both groups experienced an awareness of being a foreigner and language barriers. Japanese students in the United States were more satisfied with academics than their American counterparts in Japan. Relations among problem, theory, and method in environment–behavior research are again highlighted.

With respect to the urban environment, Setha M. Low (Chapter 21) examines public space as art and commodity to understand the contradictions between the artistic and often idealized representational purposes of the urban plaza and its political and economic base (sociocultural aspects of environment). Examples are drawn from 5 years of ethnographic fieldwork on two plazas in urban Costa Rica. The history, design, social use, and environmental meaning of these two plazas are compared as distinct ways of producing and socially constructing urban public space. Imagings of the producers and concerns of the users are contrasted with the intentions of designers and government officials in all countries.

Concluding this section, Kunio Tanaka and Kanae Tanigawa (Chapter 22) examine relations between urbanization and quality of life. Specifically, from assessing urban administrators' perceptions of urban problems in 128 Asian cities, they report that, while rapid growth poses serious problems (e.g., weaknesses in public transportation, housing, and pollution), some conditions (e.g., high quality health and family planning services) seem to mitigate these growth problems. The need to complement subjective measures with more objective ones is also discussed.

Part VII, devoted to future theoretical and empirical directions for the field of environment–behavior research, begins with Daniel Stokols's (Chapter 23) discussion of directions for environmental psychology in the 21st century. Directions for research and theory development are considered in light of several current societal concerns, including global environmental change, the spread of violence at regional and international levels, the impact of new information technologies on work and family life, rising costs of health care delivery, and processes of societal aging. Such community concerns are expected to create new opportunities for cross-paradigm research within psychology and between psychology and other disciplines. Examples of these directions include theoretical analyses of individual and subgroup differences in people's reactions to natural and built environments; research on the role of cultural, geographic, and technological factors in creativity and theory development; and the formation of contextually broader theories and community problem-solving strategies.

Kunio Funahashi (Chapter 24) contributes to this section with his discussion of the history of architectural planning research (ARP) in Japan. He demonstrates how this field compares to the field of environment–behavior studies (EBS) by focusing on the characteristic theories and methods of each.

George Rand (Chapter 25) presents some arguments for a comparative, historical, and developmental environmental psychology. Specifically, he advocates the reconstruction of person–environment relationships in accord with new understandings of historical determinism, dynamics of race and class, perception of social rules that reinforce existent inequities, consideration of the developmental status of person environment systems, and the like. Though Rand's approach is consonant with that of Wapner et al. (Chapter 20) and Demick et al. (chapter 7), he focuses on sociocultural processes, while they concentrate on the interplay of biological, psychological, and sociocultural processes.

Kenneth Craik (Chapter 26) takes the opportunity to organize his comments about the conference papers along a temporal dimension. Specifically, he forecasts trends in environment–behavior relations (e.g., technology as an engine of discontinuity in environment–behavior relations) as well as in environmental psychology as a field of inquiry (e.g., the need to promote and teach person–environment theorizing). The realization of such trends, he argues, will lead environmental psychology to become an essential core paradigmatic field of contemporary scientific psychology (e.g., similar to the ways in which notions such as life span development and the individual in social interaction have become key within the field).

Susan Saegert (Chapter 27) comments on the seminar's range of participants. Using Bruner's argument that the subject matter of psychology concerns the narratives that make sense in a culture, she delineates the ways in which the conference papers contribute to a cross-cultural understanding of these narratives and concludes with specific suggestions for future research based on such an understanding.

Amos Rapoport (Chapter 28) identifies and discusses a number of problems with the field of environment–behavior studies. A number of suggestions are made, in outline form, for possible responses to these problems with an emphasis on synthesis and theory development. Among topics discussed are the attributes of a good explanatory theory, the need for knowing the literature, the need for maximum linear and lateral conditions within and outside EBS, and the need for minimal assumptions and maximum generality. It is finally argued that explanatory theory is also needed for research application (design).

Irwin Altman (Chapter 29) engages in a discussion of the ways in which both current and future scholars in the field of environment–behavior relations need to confront fundamental philosophical assumptions about their research and action. These assumptions include those related to unit of analysis, stability versus change, philosophy of science, research methodology, and conceptual and theoretical approaches. In this regard, his work resonates with that of Wapner et al. (Chapter 20) and Demick et al. (Chapter 7) insofar as they strongly acknowledge the interrelations between problem, theory, and method and hence the need for all researchers to make hidden assumptions overt.

We hope that these chapters will stimulate readers of this volume to think about and/or engage in collaborative cross-cultural research on environment–behavior relations. As we see it, such research has both theoretical and practical significance for the field of psychology and related disciplines as well as for the world in which we live.

REFERENCES

Hagino, G. (1990). Message for the opening of the Third Japan-U.S. Seminar. In Y. Yoshitake, R. B. Bechtel, T. Takahashi, & M. Asai (Eds.), *Current issues in environemnt behavior research* (p. xiii). Tokyo: University of Tokyo.

Hagino, G., & Ittelson, W. H. (1981). Preface. In G. Hagino & W. H. Ittelson (Eds.), *Interaction processes between human behavior and environment.* (pp. v–vii). Proceedings of the Japan–United States Seminar, Tokyo, Japan, September 24–27, 1980. Tokyo: Bunsei Printing Company.

Ittelson, W. H. (1986). Preface. In W. H. Ittelson, M. Asai, & M. Ker (Eds.), *Cross cultural research in environment and behavior* (pp. iii–iv). Tucson, Arizona: University of Arizona Press.

Ittelson, W. H., Asai, M, & Ker, M. (Eds.). (1981). *Cross cultural research in environment and behavior.* Tucson, Arizona: University of Arizona Press.

Takahashi, T. (1990). Preface. In Y. Yoshitake, R. B. Bechtel, T. Takahashi, M. Asai (Eds.), *Current issues in environment–behavior research* (p. xii). Tokyo, Japan: University of Tokyo.

Wapner, S. (1981). Transactions of persons-in-environments: Some critical transitions. *Journal of Environmental Psychology, 1, 223–239.*

Wapner, S. (1983). Japan's newly formed Man–Environment Research Association (MERA). *Design Research News, XIV*(1), 4–5.

Wapner, S. (1987). A holistic, developmental, systems-oriented environmental psychology: Some beginnings. In D. Stokols & I. Altman (Eds.), *Handbook of environmental psychology* (pp. 1433–1465). New York: Wiley.

Wapner, S., & Demick, J. (1990). Development of experience and action: Levels of integration in human functioning. In G. Greenberg & E. Tobach (Eds.), *Theories of the evolution of knowing: The T. C. Schneirla Conference Series*, (Vol. 4, pp. 47–68). Hillsdale, NJ: Erlbaum.

Wapner, S., & Demick, J. (1991). Some relations between developmental and environmental psychology: An organismic-developmental systems perspective. In R. M. Downs, L. S. Liben, & D. S. Palermo (Eds.), *Visions of aesthetics, the environment & development: The legacy of Joachim F. Wohlwill* (pp. 181–211). Hillsdale, NJ: Erlbaum.

Yoshitake, Y., Bechtel, R. B., Takahashi, T., & Asai, M. (Eds.). (1990). *Current issues in environment behavior research*. Tokyo, Japan: University of Tokyo.

Part **I**
PHYSICAL ASPECTS OF THE PERSON

Chapter **2**

VISUAL PERCEPTION IN ELDERLY PERSONS WITH DEMENTIA

Kei Adachi

INTRODUCTION

To investigate visual perception, eye fixation behavior was examined in this study. Eye fixation behavior in the normal adult has been widely studied since early in this century. Recent improvements in recording apparatuses have made it possible to develop further research (Mackworth and Otto, 1970).

Very little is known about eye fixation behavior in mentally disabled persons such as psychiatric and mentally retarded patients (Holzman et al.,1974; Shagass, Amadeo, & Overton, 1974). This is mainly because these persons have difficulties in maintaining stability and in understanding verbal instructions during experimental procedures. Relatively stable data can be obtained, however, if experiments are done with repeatable instructions and with the subject's headset supported by a familiar person (Osaka, 1973).

Specifically, the purpose of this study is to examine eye fixation behavior in demented elderly persons for comparison with that of mentally retarded and normal persons.

Kei Adachi • Department of Environmental Systems, Wakayama University, Wakayama 640, Japan.

Handbook of Japan–United States Environment–Behavior Research: Toward a Transactional Approach, edited by Seymour Wapner, Jack Demick, Takiji Yamamoto, and Takashi Takahashi. Plenum Press, New York, 1997.

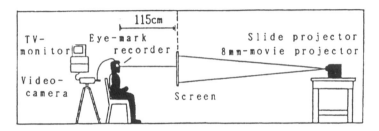

FIGURE 2.1. Apparatus.

METHOD

Subjects

Three groups were selected for the experiment: a demented group, a retarded group, and a normal group. The demented group was drawn from the nursing home residents whose Hasegawa Dementia Scores (HDS) were not more than 10. This group contained one male and six females with an average age of 74 years. The retarded group was drawn from colony residents whose average IQs were 42.1. It contained five males and five females with an average age of 28 years. The normal group consisted of university students. It was composed of five males and five females with an average age of 21 years.

Apparatus

The experiment was conducted in similar-sized rooms at each institution. Subjects were seated in chairs facing a screen, as seen in Figure 2.1. The stimuli were transmitted by a slide projector and an 8-mm movie projector through rear-screen projection at a distance of 115 cm from the subject. While they were looking at a series of slide and 8-mm movie presentations, the subjects' right eye movements were recorded on videotape at the rate of 30 frames per second through a NAC Eye-Mark Recorder type #5, a corneal reflection method.

Stimuli

The stimuli were given by abstract figures symbolizing way-finding factors about the pedestrian environment. As Figure 2.2 shows, slides A, B, C, and D were static figures represented by refractive boundary, contrast, continuity, and direction,

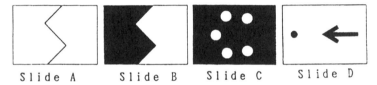

FIGURE 2.2. Slide presentation.

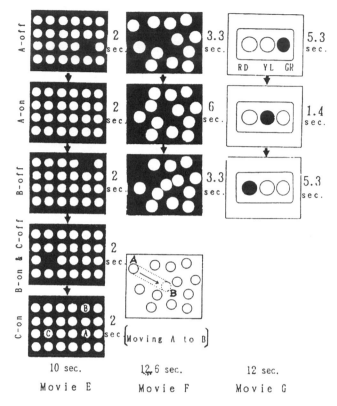

FIGURE 2.3. 8-mm movie presentation.

respectively. Movies E, F, and G of 8-mm film were represented by flashing, moving, and signal, as shown in Figure 2.3.

The presentation time of each slide was 8 s. The presentation times for movies E, F, and G were for 10 s, 12.6 s, and 12 s, respectively. The projection size was 42 cm × 60 cm (21° × 30° in visual angle) for slides and 48 cm × 60 cm (24° × 30°) for movies. The projection size was within the maximum visual angle (43° × 60°) of the eye-mark recorder.

Analytical Procedure

Eye fixation behavior was analyzed with fixation time, range, and scan path as parameters. Eye movement of less than 1° in visual angle was defined to be a fixation in the experiment, and was traced on an 8 in. × 10 in. modular grid sheet (3° per grid).

RESULTS

The ratio of summed fixation times per presentation times was 76.3% in the demented group, 80.5% in the retarded group, and 81.1% in the normal group. The ratio of the demented group was significantly lower than that of the other groups. The ratio

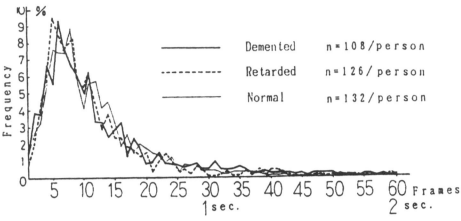

FIGURE 2.4. Frequency distribution of fixation time.

of summed fixation times in the case of normal persons after training is generally over 90% (Watanabe & Hiwatari, 1965). Our data were slightly lower than this figure because no subjects had received training; therefore, this difference can be considered negligible.

Fixation Time

Figure 2.4 shows the frequency distribution of fixation time. The distribution generally looked similar for each group. The most frequent value of fixation was 0.20 s in the demented group, 0.17 s in the retarded group, and 0.27 s in the normal group. The fixation time of 0.20 s ~ 2.0 s is generally enough to recognize the meaning of stimuli. The percentage of fixation times over 2.0 s was 2.3% in the demented group, 1.2% in the retarded group, and 0.4% in the normal group. The maximum considered too much for tropistic setting, was highest in the demented group, at 4.73 s, followed by 4.06 s in the retarded group, and 2.36 s in the normal group.

As Table 2.1 shows, mean fixation time was 0.47 s in the demented group, 0.43 s in the retarded group, and 0.41 s in the normal group. There were significant differences among the three groups ($p < .01$). There were also significant differences between the demented group and the other two groups ($p < .05, p < .01$).

Mean fixation time on each presentation is also shown in Table 2.1. In the case of normal persons, the mean fixation time on static pictures and photographs is in the range of 0.26 to 0.33 s, and on moving pictures, 0.33 to 0.49 s (Watanabe & Hiwatari,1965). In our experiment, there are no significant differences between slides and movies and a series of presentations.

Eye Fixation Behavior on Stimuli

Eye fixation behavior on stimuli was examined (see Figures 2.5 and 2.6) with the following findings.

TABLE 2.1. Mean Fixation Time for Each Presentation[a]

Group	Total	Slide A	Slide B	Slide C	Slide D	Movie E	Movie F	Movie G
Demented	0.47±0.52	0.42±0.32	0.43±0.37	0.47±0.57	0.54±0.69	0.51±0.58	0.45±0.50	0.47±0.48
Retarded	0.43±0.45	0.46±0.46	0.42±0.40	0.48±0.58	0.45±0.44	0.38±0.39	0.43±0.44	0.40±0.35
Normal	0.41±0.32	0.37±0.27	0.36±0.29	0.43±0.34	0.39±0.30	0.42±0.33	0.46±0.38	0.40±0.32

[a]M ± SD; SD = Standard deviation.

Slide A

Sixty-two percent of fixations by the demented group were centered on only two grids at the center of the slide. Accordingly, they lacked adequate coverage of the stimuli on refractive boundary. The retarded group covered most of the stimulus grids and the normal group covered all of the stimulus grids on refractive boundary.

Slide B

The refractive boundary is emphasized by contrast between black and white on Slide B. The demented group centered fixation on only three stimulus grids at the center, however, increased fixation from 11% on slide A to 24% on slide B on grids of refractive portions. The retarded group increased fixation on those portions in comparison with coverage on Slide A. The normal group extended the fixation range, perhaps because of distinguishing black from white area.

Slide C

Sequential fixation is expected on the stimuli of five circles. The demented group centered little fixation on the stimuli. The percentage of fixation on the circles was 13%. Four of seven subjects in the demented group did not fixate on any of the five circles, and no sequential scan paths were found on those . The retarded group fixated 57% on the circles. Only one subject in the group covered all circles and made sequential scan paths. The normal group fixated 42% on the circles. Three of ten subjects in the group fixated on all of them and five subjects made sequential scan paths.

Slide D

The stimuli, a point and an arrow, are expected to induce fixations and round-trip scan paths between the point and the arrow. The demented group fixated 68% on the arrow and 0% on the point. The mean round-trip scan path per person was 0.14 times for the group. The retarded group fixated 47% on the arrow and 8% on the point. The mean round-trip scan path per person was 1.7 times. The normal group fixated 50% on the arrow and 10% on the point, similar to the retarded group, but had 2.7 times for the mean round-trip. Thus, the stimulus of the arrow was ineffective for the demented group.

Movie E

The stimuli were a group of circles in a regular arrangement that flashed on and off irregularly, one after another. The demented group failed to fixate on the

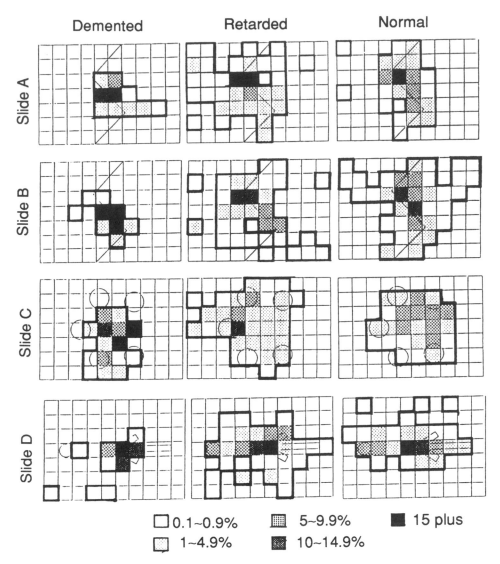

FIGURE 2.5. Percentage of total fixations in each modular grid for slides A, B, C, and D.

on-off pattern of circles A and B. They were unable to fixate on circle C sufficiently. Both the retarded and the normal groups were able to fixate on all on-off patterns.

Table 2.2 shows the percentage of subjects who were able to pursue the stimulus under changing scenes for all subjects tested. The percentage was 0% in the demented group, 25% in the retarded group, and 72% in the normal group. Under the scene of one circle on and one circle off simultaneously, both the demented and retarded groups missed the stimuli.

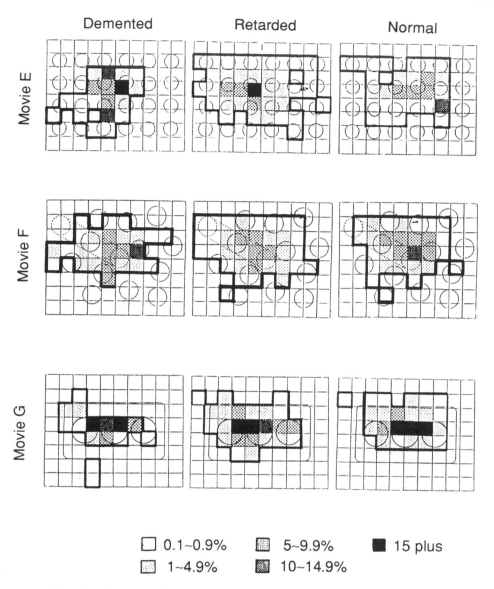

FIGURE 2.6. Percentage of total fixations in each modular grid for Movies E, F, and G.

Movie F

One of twelve circles moved at a speed of 2.2° s for 6 s in the movie lasting 12.6 s. Before a circle started moving, fixations were centered in the demented group and spread in the other two groups.

Table 2.3 shows the percentage of subjects who were able to pursue the moving circle for Movie F. The pursuit percentages were 71% in the demented group and 100% in the other two groups. The percentages obtained in Movie F

TABLE 2.2. Percentage and Number of Subjects Who
Pursued Stimulus under Changing Scenes for Movie E

Group	Pursuit percentage	Number pursuing stimulus[a]					
		A	A'	B	B'	C	C'
Demented	0	0	0	0	0	0	0
Retarded	25	3	2	2	0	4	4
Normal	72	5	10	8	7	7	6

[a]A: A off; A': A on; B: B off; B': B on; C: C off; C': c on; B' & C occur simultaneously.

were significantly higher than those in Movie E. Table 2.3 also shows a time lag for beginning eye pursuit after the circle started in Movie F. The mean time lag in the demented group was significantly longer than that in the retarded group ($p < .05$) and the normal group ($p < .01$).

Movie G

The signal lights of green, yellow, and red were turned on sequentially in this movie. During the duration of the green light, the percentage of fixation on the stimulus was only 3% in the demented group, 18% in the retarded group, and 35% in the normal group. During the duration of the yellow light, fixation on the stimulus was 50% in the demented group and 48% in other two groups. The high rate in the demented group for the yellow light was mainly due to fewer eye movements in its location at the center of the screen.

During the duration of the red light, the fixation on the stimulus was 33% in the demented group as well as in the retarded group and 25% in the normal group. The percentage of fixation on the red signal in the demented group increased from 22% during the duration of the yellow to 33% during the duration of the red light.

Table 2.4 shows the percentage of the subjects who pursued signal changes for Movie G. The mean was 57% in the demented group, 83% in the retarded group, and 100% in the normal group. In the case of the demented group, the percentage of eye pursuit from green to yellow, as well as that from yellow to red, was 71%.

TABLE 2.3. Percentage of Subjects Who Pursued the
Moving Circle in Movie F and Time Lag for Beginning Eye
Pursuit after a Circle Moved

Group	Pursuit percentage	M ± SD[a]	Maximum	Minimum
Demented	71	2.96 ± 1.83	5.87	1.17
Retarded	100	1.00 ± 0.57	2.10	0.25
Normal	100	0.50 ± 0.16	0.83	0.30

[a]SD = Standard deviation.

TABLE 2.4. Percentage of Subjects Who
Pursued Signal Changes for Movie G

Group	Total	Green	Yellow	Red
Demented	57	29	71	71
Retarded	83	80	80	90
Normal	100	100	100	100

CONCLUSION

The results are summarized as follows:

1. The range of fixation points and scan paths in the demented group was significantly limited and more biased toward the center of stimuli than in the other two groups.
2. The demented group and the retarded group missed more stimuli on static figures than the normal group.
3. Eye fixations in the dementia group increased on stimuli where clear contrast and flashing were provided.
4. It took longer for the demented group than for the other two groups to fixate on stimuli.
5. The normal group and the retarded group were able to fixate on and pursue moving stimuli. The demented group was also able to do so, although with the lowest degree of efficiency among the three groups.

We, therefore, infer that the level of visual perception in the demented group can be raised to some extent by visual stimuli that are emphasized by clear contrast, moving, and flashing.

REFERENCES

Adachi, K., Araki, H. (1988). Zukei tokushitsu ni taisuru chushikeikou [Eye fixation behavior to figure characteristics]. *Journal of Architectural Planning & Environmental Engineering, 399,* 52–59.
Holzman, P.S., Proctor, L., Levy, D., Yasillo, N., Meltzer, H., & Hurt, S. (1974). Eye-tracking disfunctions in schizophrenic patients and relatives. *Archives of General Psychiatry, 31,* 143–151.
Mackworth, N., & Otto, D. (1970). Habituation of the visual orienting response in young children. *Perception & Psychophysics, 7,* 173–178.
Osaka, R. (1973). Shinrigaku Jikken-ho [Experimental methods in psychology], vol. 3. Tokyo: Tokyo University Press.
Shagass, C., Amadeo, M., & Overton, D. (1974). Eye-tracking performance in psychiatric patients. *Biological Psychiatry, 9,* 245–260.
Watanabe, A., & Hiwatari, K. (1965). Gazo to chushiten bunpu [Fixations on TV-figures]. *NHK Technical Report, 17,* 4–20.

Chapter **3**

DWELLING DESIGN GUIDELINES FOR ACCESSIBILITY IN THE AGING SOCIETY

A New Era in Japan?

Satoshi Kose

INTRODUCTION

Japan will be one of the most aged societies in the years to come (Campbell, 1992). It is crucial to prepare for this change since the speed of aging of Japan's population is unparalleled in the world. It took only 24 years for the aged population in Japan to double from just 7% (1970) to 14% (1994); in 2020, the percentage of aged will be over 25% (National Institute of Population Problems, 1991). A rather pessimistic, but still realistic, estimate is that the aged will exceed 30% because birthrate, which is just around 1.5 in recent years, will not reverse its declining trend.

Recognizing the importance of preparing for a highly aged society, the Ministry of Construction started a project on aging and dwelling design in 1987 (Kose & Nakaohji, 1991b). The project, titled "Development of Technology for the Enhancement of Residential Environment in the Aging Society," aimed at solving the problems of housing for the aged with respect to both policy on housing supply and guidelines on dwelling design. It also aimed at improving the quality of the environment at the community level. This chapter reports only on the developments in housing.

Satoshi Kose • Building Research Institute, Ministry of Construction, Tatehara, Tsukuba 305, Japan.

Handbook of Japan–United States Environment–Behavior Research: Toward a Transactional Approach, edited by Seymour Wapner, Jack Demick, Takiji Yamamoto, and Takashi Takahashi. Plenum Press, New York, 1997.

THE ISSUE

At the beginning of the project, many members of the project committee (most are researchers and professionals in architectural design and building science) thought that special housing forms would suffice. This was coupled with the notion that the silver-housing scheme, a joint policy measure between the Welfare Ministry and Construction Ministry to provide age-specific housing started in 1987, would solve the housing problems of frail, aged persons.

Reality, however, was different. It gradually became evident that almost all dwellings have to be prepared for the time when the residents got older. Special housing can never be a solution when one person in four is 65 years of age or over. There are two major difficulties: First, the combined number of constructed special housing units for the aged and nursing homes of any kind can never catch up with the increase in the aged population. Second, the ultimate ratio of the aged population is unbelievably high (one in four!) so it is much more sensible to assume that almost all dwelling units will ultimately be resided in by aged persons.

A recent survey by the Management Agency revealed that the majority (68%) of aged people prefer to continue to live in their present dwellings as they grow older; further, 15% expressed their wishes to rehabilitate their dwellings adaptively to meet their needs (Asahi Evening News Editorial, 1995). This clearly demonstrates that dwellings have to be designed to accommodate the needs of aging residents; that is, they have to be universally designed to match future needs in addition to current requirements.

Even though the aged persons wish to do so, they are unlikely to live with their children in an extended family, a popular form of household in the traditional Japanese way of life. This is partly because children's job opportunities are usually in larger cities, away from where the parents have lived, and partly because the aged parents will choose to live by themselves as a couple or as a single and will "age in place" as long as they can. (The granny annex is one solution that has traditionally been adopted in some regions of Japan. It was basically built in one of the corners of the land-lot of the family, being able to accommodate the needs of frail, aged persons. The move to a granny annex meant that the aged person was to retire to the backstage of life, but still keep his or her role as an adviser in case of need. Such a tradition is almost lost in most parts of Japan.) Appropriate dwelling design will become vital so as not to hinder the aged from living a comfortable life. Accessible dwelling in its broadest terms (or universal design) will become a necessity. At least, it is certain that care of the aged in institutions or even in special housing for the aged will never be a feasible solution when nearly 30% of the entire population will be 65 and older.

SURVEY OF AGED PERSONS' CAPABILITIES

The crucial problem was to identify the characteristics of the aged persons for whom the dwellings would be designed. Should we expect all aged persons to become "wheelchair-bound," "mentally disordered (by Alzheimer's)," and/or "bedfast" (Netakiri)? Those three words are most commonly cited by the mass media when raising the issue of aging problems. Or, could they be more active and have more positive attitudes to life? For many, the report by the mass media seemed too distorted.

Three surveys were conducted, therefore, to clarify the actual situation of aged persons who live in an extended family, those who live in special housing for the aged, and those who live independently.

Survey of Aged Persons Living in an Extended Family

The first survey was intended to explain the relationship between the physical capabilities of aged persons and the design features of their dwellings, that is, how healthy and active the aged persons are, and to what extent considerations are given to improving the quality of living in the dwellings. The survey was administered to 900 aged and 700 nonaged (below 65 years of age) persons (Kose, 1992).

The findings were as follows:

1. Participants were relatively healthy, much healthier than normally assumed. About a third of the aged subjects could run; about half could walk without any assistive devices. Only 2% of the subjects were wheelchair-bound or bedfast. As they grew older, however, their physical capability deteriorated and about half of those 85 years and older needed some assistive devices such as handrails and walking sticks.

2. The subjective evaluation of the design features in dwellings varied between aged and nonaged persons with respect to aspects such as step and level differences in entrances and at doors, that is, door sills and upstands (serving as water barriers). The aged persons were experiencing a lot of trouble in those places, whereas the nonaged did not note them as problematic. Stairs were an exception because the aged persons normally gave up climbing up and down.

3. The aged persons expressed their wishes for improvement in design only when their physical capabilities deteriorated and some supportive devices were needed.

4. In contrast, the accident incidence clearly indicated that the aged suffered from falls on level ground (tripping over small level differences, slipping on the floor, etc.) even before they needed some devices. Those who could walk without any assistive devices had a much higher incidence of accidents than nonaged subjects.

The above findings suggest that design considerations for the aged have to give priority to safety rather than to usability for specific needs.

Survey of Aged Persons Living In Special Housing for the Aged

The second survey was conducted with those residents who live in special housing called "silver-housing" (Kose, Ohta, Tanaka, & Watanabe, 1992). The scheme is basically similar to British sheltered housing, with a resident warden living in the same block as residents. Two samples, both with 40 residents, were chosen for the survey. The survey was composed of two parts, one on physical capability and the other on the effectiveness of design details. The management of such schemes naturally assumed that details designed by architects gave due consideration to the needs of the aged residents. But were they truly age-conscious?

The residents were independent when they were admitted to the housing, and as only a couple of years had past since the housing was built, no residents needed wheelchairs or were bedfast. Some of the design details were, however, already inappropriate for the aged residents because the designers did not correctly realize the needs of the aged or they just misunderstood the requirements. Examples included wrong placement of handrails, level differences that had no meaning, inappropriate choice of light bulbs, and the like. Another notable problem was inadequate social services to cover the needs of residents, a problem that also occurs in the UK, possibly because wardens are not necessarily qualified to provide social services.

Survey of Aged Persons in a Local Community

To verify the validity of the first survey, another survey was administered to the aged residents in a suburban area of Tokyo. Subjects this time were not limited to the aged persons living with extended family, but were randomly chosen from the residents registered in the city. In spite of the differences among subjects, the results were quite similar to those obtained in the first survey and common design problems were pointed out. The most important finding was that the need to live an independent life seems to keep aged persons relatively healthy and active, suggesting the positive effect of aging in place.

PROPOSAL OF DESIGN GUIDELINES

To avoid the above-mentioned architectural design mistakes and to establish a minimum accessibility/usability level for future public rental housing, the Housing Construction Division of the Ministry of Construction requested new design guidelines. These guidelines were basically a revision of draft guidelines originally made in 1987 (Kose, 1987; Kose & Nakaohji, 1991a).

The new "Design Guidelines of Public Multifamily Housing" were prepared in March 1991 (Kose, Suzuki, Yamada, & Tanaka, 1991) to be used in the construction of public housing in 1991 and onward (the sixth 5-year Plan of Housing Construction). This was the first step toward providing accessible dwellings for most of the population without need of major rehabilitation work in the years to come.

The concept of the design guidelines can be summarized as follows:

1. Elimination of level differences within the dwelling unit.
2. Standard requirements for handrail installation (or preparation for installation).
3. Standard detailed design for the benefit of aged dwellers such as door hardware and other operating devices.

4. Requirements for equipment to support the life of the aged, including bathroom units.

The design guidelines were targeted to become universal; that is, they were intended to be based on a universal design concept. Everyone who lives in the dwellings will benefit from the upgraded design standards. It must be noted, however, that the dwellings designed and built within the above guidelines will not be sufficient to accommodate persons with wheelchairs. There is still a wide gap, at least in Japan, between the concept of wheelchair housing and the concept of universal design housing. Some gaps stem from cultural reasons and some from climatic and practical difficulties.

It is perhaps the Ministry's next job to integrate the complexities and eliminate this crucial gap. There are several issues needing immediate solution such as a shift from sleeping on the floor on a futon mattress to sleeping in bed (this will make it substantially easier for the caregiver to assist the frail elderly up and about), and a change of bathing behavior from immersing in a bathtub to having a shower (this will reduce the risk of falls for the resident as well as the risk of back pain for the caregiver).

Public multifamily housing is for rent to the lower-income population; it is constructed by local governments with subsidies from the Ministry of Construction. For middle-class people, there are other public housing programs that are provided by the Housing and Urban Development Corporation and local government housing corporations. The government has some control over them and can impose accessibility to design features that are in line with those in public rental housing.

In 1992, design guidelines for detached houses were also established because the majority of housing construction in Japan is in the form of detached houses and some design features are different from those for multistory housing. Prototype houses were built by private housing manufacturers according to the guidelines, and they proved to be economically feasible to build.

EXAMPLES OF DETAILED DESIGN THAT COMPLIES WITH THE GUIDELINES

The following gives some idea of what the design guidelines require in real situations. Their change from existing traditional detailing is substantial.

First, in spite of the accessibility requirements, we had to allow level differences between indoors and outdoors for main entrance and balcony. Because the risk of driving rain is much higher in Japan than in most western countries, it is important to have the floor level raised to prevent flooding. You will see these points in Figures 3.1 and 3.2.

Second, we required flat indoor floor finishing between different spaces, including elimination of door sill level differences. This change is vital for the prevention of falls from tripping, the major cause of accidental injury of the aged. It is a clear departure from previous requirements for wheelchair accessibility, which in

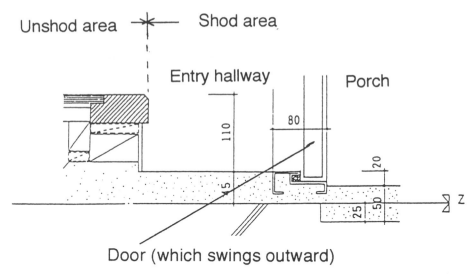

FIGURE 3.1. Level difference at the entrance. This example is for multistory buildings. For detached houses, the height difference between shod and unshod areas will be around 180 mm instead of 110 mm.

reality allowed 3 cm step differences in some cases. The requirement also applies between the tatami room and other spaces (with some exceptions). Please refer to Figures 3.3 and 3.4. (One possible alternative to a flat floor between a wooden floor and the tatami room is raising the tatami room about 40 cm. In this case, the tatami

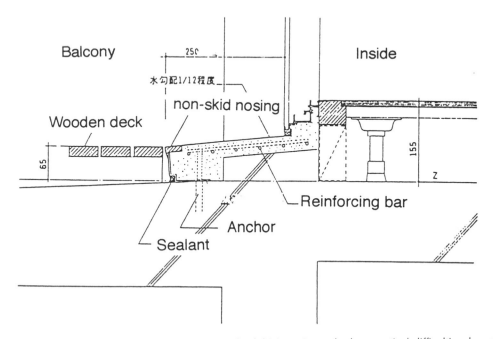

FIGURE 3.2. Level difference at the balcony. Risk of driving rain, and other practical difficulties do not make it feasible to flatten here.

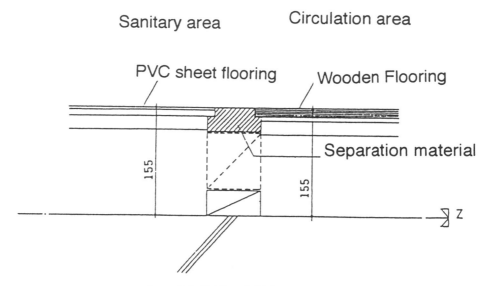

FIGURE 3.3. Flat floor finish between spaces.

area could be used as a kind of bed. It reminds us of the traditional space arrangement where the daily activity space was on the ground level and shod, while the tatami space was reserved for sleeping and formal activities.)

Third, we proposed that the bathroom unit be basically made flat. The difficulty comes from the fact that the Japanese wash the body outside the bathtub during

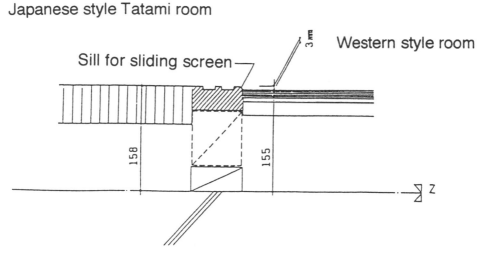

FIGURE 3.4. Flat floor finish between Tatami room and other space. Tatami has been the typical floor finish in Japanese spaces, but because of its thickness the floor was several centimeters higher than in other areas. By adjusting the dimensions of materials, the floor finish can be made flat.

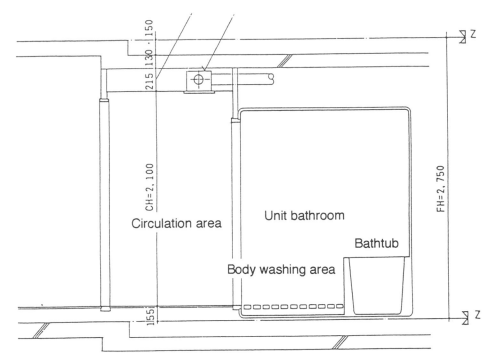

FIGURE 3.5. Conceptual drawing of Bathroom Unit. The concrete slab must be lowered under the bathtub a minimum of 150 mm compared to previous practice, which increases the floor-to-floor height in many cases because ceiling height of Japanese dwellings is very low compared to Western dwellings. The dimension was based on the assumption that people sit directly on the floor instead of using chairs.

bathing and so the washing area will be completely wet. The normal custom has been to have a raised water barrier as the boundary, which was also a barrier against mobility for the aged and disabled. This problem was easy to solve for detached houses by placing the bathroom on the ground floor. To make the bathroom unit basically flat for upper floors of multistory housing, however, the floor slab must be lowered to allow for bathtub placement because the Japanese bathtub is deeper than the western bathtub. See Figures 3.5–3.7.

Fourth, handrails were required or advised in toilets and bathrooms because keeping good balance in these places is a most difficult task for the aged. The handrail installation has to be stable. The recent professional practice of wall finishes has, however, used plaster or gypsum boards, which cannot stand the force applied to handrails. Thus, some new ideas need to be introduced. You will find these ideas in Figures 3.8–3.11.

In Figures 3.12–3.15, you will see the above design features integrated in the demonstration houses built by the private sector as a standardized model.

These designs clearly mark a departure from the traditional design concepts for young and able-bodied adults as dwellers, that have resulted in many difficulties for aged persons.

The design guidelines will be used particularly to improve the quality of housing design with public financial support, not only publicly constructed housing but

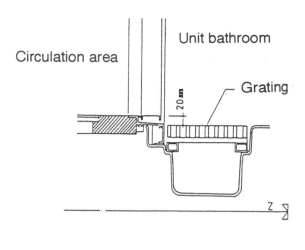

FIGURE 3.6. An example of finishing at the doorstep to the bathroom washing area. Water will be trapped at the drain so the floor can be basically flat.

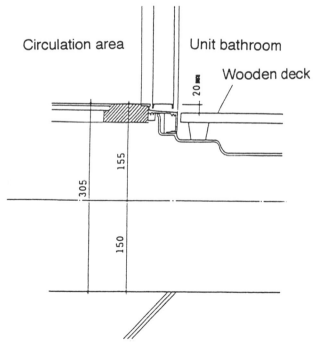

FIGURE 3.7. Another example. The wooden deck will allow water to stay under it. Maximum 20-mm level difference between dry area and washing area in the bathroom is permitted here.

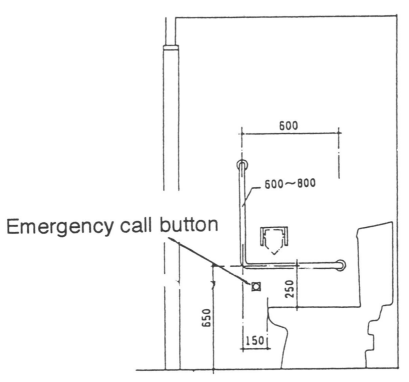

Figure 3.8. Recommended handrail layout for toilet stool. The handrail's distance from the stool is crucial to allow for effective force application.

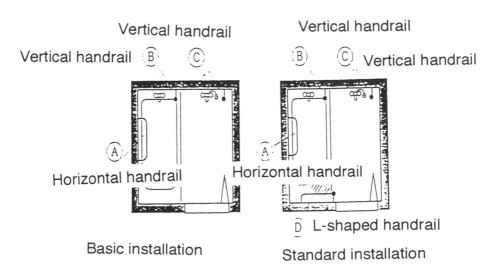

Figure 3.9. Required arrangement for handrails within the bathroom. One to assist in and out of the bathtub; another to assist sitting/standing in the washing area; the third to assist secure posture in the bathtub; the fourth one is recommended at the entrance, if there is a level difference between dry area and washing area.

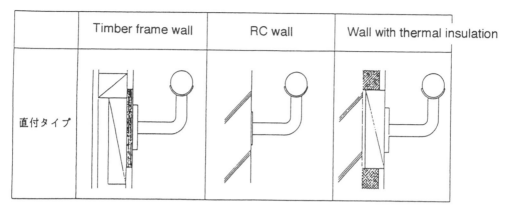

	Timber frame wall	RC wall	Wall with thermal insulation
直付タイプ			

FIGURE 3.10. Handrail installation for walls—examples of standard practice.

also housing mortgage assisted by the government, because the mainstream of Japanese housing construction has relied and will continue to rely on private initiative (Kose, 1993; 1994b; 1996).

A comparison between U.S. Fair Housing Amendment Act accessibility guidelines and our design guidelines reveals some similarities. Recent news from the UK reveals that they are discussing the feasibility of similar accessibility requirements for new dwellings. Of course, there are some differences in guidelines among the countries, but the trends seem to demonstrate that the universal design concept is now becoming worldwide, whether it is called life span dwelling, lifetime home, universal design house, or the like.

LESSONS FROM THE GREAT HANSHIN EARTHQUAKE

The most recent earthquake disaster in Kobe (January 17, 1995) clearly demonstrated that the accessibility concept in dwellings must be at the heart of housing policy.

First, the dwellings that collapsed were mostly very old and badly maintained ones. In the past, such dwellings had been visited and inspected by the carpenters who built them, but recent social change has swept away such tradition. The residents, however, never realized the importance of inspection and maintenance. The dwellings, thus, could not withstand the overwhelming force of the earthquake. If the dwellings had been given any kind of maintenance, it is quite likely that barrier-free modifications were also done at the same time.

Second, the temporary housing that the Ministry of Welfare arranged lacked the most important issue of accessibility. What was ordered was temporary housing with the minimum quality standards for healthy robust workers in the construction site. The dwellings were full of barriers, and many refugees, children, aged persons, and persons with disabilities had difficulties living in them. Only after the troubles became evident were remodelings done. If the accessibility issues had been widely known, most of the problems could have been avoided from the beginning.

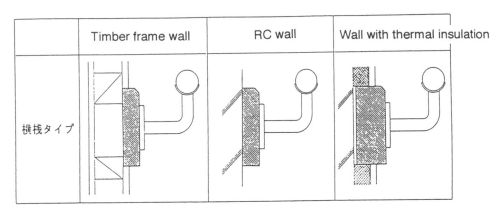

FIGURE **3.11.** Reinforcement for future handrail installation.

Third, public housing now planned for construction by the local government to accommodate those who lost their dwellings is following the basic barrier-free concept of the guidelines. However, desirable design features not covered in the design guidelines are unlikely to be introduced in the construction specification. The requirements closely connected to structures were easier to adopt because they have to be done at the stage of initial construction, but desirable design features that can be introduced afterward tended to be ignored since they are not urgent enough to justify the increased cost, or simply because such design features are not realizable due to nonexistence of proper building products. As to housing construction in the private sector, the only control could be through the Housing Loan Corporation; however, the accessibility requirements were only partly obligatory for special housing for aged persons and they were introduced only in 1995 for extended family dwelling loan schemes. The Ministry's policy is yet to be determined on this issue.

Fourth, in the regions where the majority of dwellings collapsed or burned down, reestablishing the existing communities seems quite difficult. It will perhaps be necessary to have support from the local government to build dwellings, whether they are owner occuped or accommodations to let. Aged persons will not easily find funds to borrow for owner-occupied housing construction or apartment units. It may be that aging and disabled residents will have to find accommodations provided by the public sector. Lack of funds will surely lessen their alternatives, and it may force them to leave their previous communities. If almost all the dwellings were designed and built barrier-free, the difficulty of finding suitable dewellings to live would be only social or economical, but if the dwellings in the private sector continue to be built without due regard to the aging society, the possible choice of dwellings would be discriminatory.

CONCLUSION: THE REMAINING ISSUES

The publication of the new design guidelines from the Ministry of Construction on June 23, 1995, and the Housing Loan Corporation's step toward requiring

FIGURE 3.12. Level difference at the entrance is supplemented with a vertical handrail.

FIGURE 3.13. Level difference between a tatami room and a carpeted room next to it is eliminated.

basic accessibility (choosing some items among the guidelines) to extended family living loans (which qualify larger loan sums) are welcome moves. It is argued that a substantial portion of houses recently built are complying with some of the basic requirements of the guidelines, flat floor level in particular.

Taking these circumstances into consideration, the Housing Loan Corporation moved forward a step to introduce "energy conservation design" or "barrier-free design" as minimum requirements for its standard loan schemes beginning fiscal year 1996. Such policy-based requirements will surely become a driving force in instituting change. Hopefully, the majority of housing stock will be replaced by universal design dwellings within 20 to 30 years, in time for the later years of the Japanese baby-boomer generation.

The remaining problem is whether the design should allow wheelchairs into the dwelling itself. Current design requests that the wheelchair for outdoor use be left in the entrance hall and exchanged for a specially designed indoor wheelchair if it is necessary to use one indoors as well. This scheme comes from the traditional custom of taking one's shoes off at the entrance hall in Japanese houses. (A point worth mentioning is that although many persons take off (muddy) shoes at the entrance hall in Sweden, they normally do not change wheelchairs nor do they have step differences there (although there may be door sills).An option would be to leave the electric wheelchair at the entrance hall for charging the battery and use a

FIGURE 3.14. This bathroom floor is flat from the outside to the inside. Water problems were solved with a water channel covered with grating.

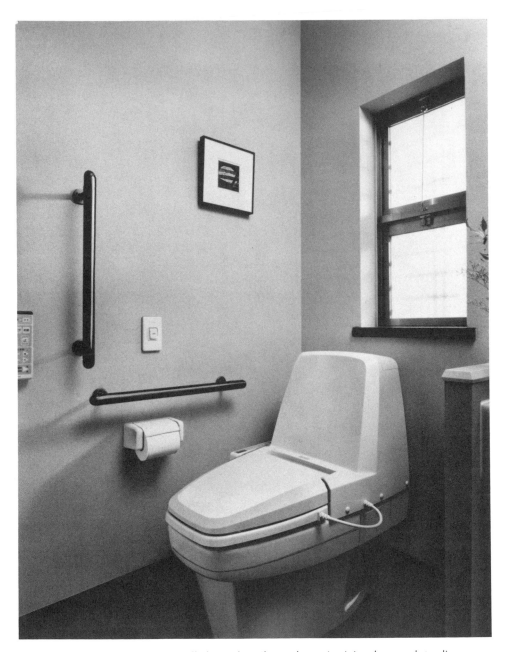

FIGURE 3.15. Handrails were installed near the toilet stool to assist sitting down and standing up.

compact wheelchair indoors as conditions are more favorable there compared to the outside.) A step difference at the entrance hall keeps dust and mud away from the unshod floor and prevents wetting from driving rain. Should this step be eliminated as well? This will be an issue for the baby-boomer generation, not the present aged, to decide. It is expected to take 5 to 10 years more until the direction becomes evident.

NOTES

An original version of this chapter was submitted to CIB-W84 seminar and workshop in Budapest, September 1991, and its abridged version (Kose, 1994a) is printed in the report published by the Royal Institute of Technology (Ratzka & Walmsley, 1994). The present version is further updated with more information on recent policy developments, including figures and photos.

The author would like to thank Asahi Chemical Co. for allowing use of the photos in this chapter.

REFERENCES

Asahi Evening News. (1995, July 2). Editorial: Belated guidelines for homes are inadequate, 07, 02

Campbell, J. C. (1992). *How policies change: The Japanese government and the ageing society.* Princeton: Princeton University Press.

Kose, S. (1987). Aging population and the impact on buildings (Kourcisha to Kenchiku). *Kenchiku-Gijutsu, 427,* 155–165 (in Japanese).

Kose, S. (1992). Capability of daily living of the elderly and their accident experiences: Implication to design of safer, easier-to-use dwellings. In E. G. Arias and M. D. Gross (Eds.), *Equitable and Sustainable Habitats. Proceedings of EDRA 23* (pp. 158–166).

Kose, S. (1993, August). *Design guidelines of dwellings for the ageing society: How to implement them into the policy from a long-range perspective.* Paper presented to an International Conference on Ageing and Housing Policy, Kobe, Japan.

Kose, S. (1994a). *Design guidelines of public collective housing for the aging society: The move toward a new era in Japan.* In A. D. Ratzka & K. Walmsley (Eds.), Report of the Fourth International Expert Seminar on Building Non-Handicapping Environments: Access Legislation and Design Solutions (pp. 91–96). Stockholm: Building Function Analysis of the Royal Institute of Technology.

Kose, S. (1994b). Housing for the aging society: Meaning of barrier-free design in Japan. *MERA Journal, 2*(1), 59–64.

Kose, S. (1996). Possibilities for change toward universal design: Japanese housing policy for seniors at the crossroads. *Journal of Aging and Social Policy.* Manuscript submitted for publication.

Kose, S., & Nakaohji, M. (1991a). Design guidelines for dwellings for an aging society: A Japanese perspective. *Building Research and Information, 19*(1), 24–30.

Kose, S., & Nakaohji M. (1991b). Housing the aged: Past, present and future; policy development by the Ministry of Construction of Japan. *The Journal of Architectural and Planning Research, 8*(4), 296–306.

Kose, S., Ohta, A., Tanaka, Y., & Watanabe, K. (1992). Examination of design effectiveness of special housing for the aged: Is Japanese "Silver-housing" a success? *IAPS12 Proceedings, 3,* 161–166.

Kose, S., Suzuki, A., Yamada, S., & Tanaka, K. (Eds.). (1991). *Design specifications of the publicly operated collective rental housing construction* (in Japanese). Tokyo: Ministry of Construction.

National Institute of Population Problems (1991). *Population estimate* (in Japanese). Tokyo: Author.

Ratzka, A. D., & Walmsley, K. (Eds.). (1994). Report of the Fourth International Expert Seminar on Building Non-Handicapping Environments: Access Legislation and Design Solutions. Stockholm: Building Function Analysis of the Royal Institute of Technology.

Chapter **4**
URBAN HOUSING AND THE ELDERLY

Takashi Takahashi

INTRODUCTION

To cope with today's aging population, there exist many types of elderly housing and facilities for the elderly. Historically, there have been no special building types for elderly housing with the exception of *inkyo-beya* (a small hut or a room for the aged) in Japan. This is mainly because the elderly have relied upon family or relatives.

The public sectors in Japan have now begun to provide housing for the elderly such as "silver-housing." The designs of these houses are based on the "barrier-free" concept, especially in building elements and facilities. The units are aimed to serve as a home for an elderly person or couple. The only remarkable feature of this design, from the point of view of the elderly, is that it incorporates the barrier-free concept.

This chapter deals with an appraisal of urban housing for the self-supportive aged. Toward this end, our research group visited four housing sites and recorded the actual living conditions of the elderly.

CURRENT TRENDS OF PLANNING THEORY IN ROOM LAYOUT OF A HOUSE

Under the prevailing principles of housing design, rooms are allocated to such activities as sleeping, dining, and living. As a result, each room has its own name, such as bedroom, dining room, living room, and so on.

Takashi Takahashi • Department of Architecture, Faculty of Engineering, The University of Tokyo, To-kyo 113, Japan.

Handbook of Japan–United States Environment–Behavior Research: Toward a Transactional Approach, edited by Seymour Wapner, Jack Demick, Takiji Yamamoto, and Takashi Takahashi. Plenum Press, New York, 1997.

However, younger architects are now proposing a new principle for the spatial layout of a house. This principle reflects the variety of recent lifestyles that stress individual activity rather than family or group activity, as well as the notion that the use or name of a room or space should be formed in the process of the occupants' adaptation to the house.

Figures 1 and 2 show recent examples designed by young Japanese architects. In these houses, spatial configuration and naming do not follow the usual method.

In the case of Figure 1, two floors are allocated to a wife and a husband and inner walls have been reduced to a minimum. The lower floor is called the wife's domain and the upper one is called the husband's domain. In the process of living in the new house, the husband and wife find their appropriate positions and patterns of movement within the large open space.

Figure 2 shows a house for a family of four: parents and two children. The living space is divided into three zones—south, middle, and north. In the middle of the house, there are small, enclosed rooms. These rooms are used for personal activities such as studying, sleeping, and relaxing, but none of the rooms have traditional names.

The above two examples indicate that traditional separation of space within a house into public and private zones need not be so rigid. Parts of a "public zone" may be used for personal activities. This sequential process of appropriating space would be applicable to housing for the elderly because people, including the elderly, tend to find their most convenient positions for daily living, consciously or unconsciously. Moreover, as they get older, people have a tendency to restrict their activities to a few places that are used not only for their personal and private activities, but also for interpersonal and social interactions such as chatting with a friend.

In this research, we observed and interviewed the occupants of public housing units for the elderly and tried to establish how they appropriated space for their daily living activities. The result of this research will, we hope contribute to improving the quality of future housing for the elderly.

METHOD

Sites and Interviewees

We studied four sites: (1) N (Nezu) site: part of the old town area filled with detached houses (457 questionnaires, 26 interviews); (2) A (Akabanedai) site: fifties-style public housing (433, 23); (3) H (Hikarigaoka) site: public housing for the aged recently built (16, 15); (4) K (Kibougaoka) site: public housing for the aged recently built (19, 18). In every site, we studied the units of the elderly above 65 years old.

Methods of Analysis

Based on questionnaires, interviews, and detailed activity maps (a floor plan of a unit including furniture, tools, miscellaneous goods, and seating or occupying position

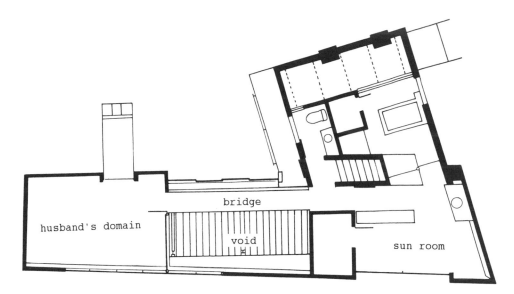

first floor plan

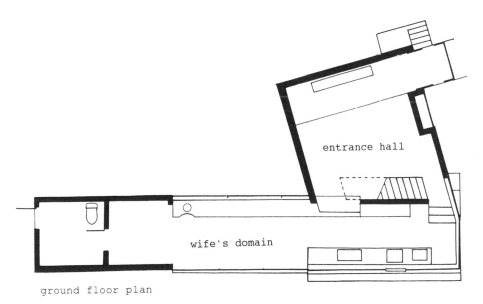

ground floor plan

FIGURE 4.1. Floor plan for two-person dwelling. (Architect: J. Akoi)

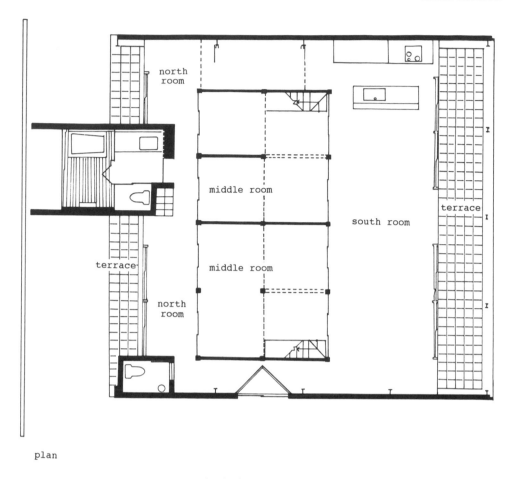

plan

FIGURE 4.2. House plan for family of four. (Architect: S. Yoshino)

for daily activities), issues of analysis are as follows: (1) appraisals of units of formation of fixed seating and occupying position within rooms; (2) appraisals of neighbourhood environment; and (3) proposals for future planning of housing for the elderly and the surrounding neighbourhood. This chapter mainly refers to the first issue.

ANALYSIS OF FORMATION OF FIXED SEATING AND OCCUPYING POSITION

Relation of Dining, Sleeping, and Appropriating Position

In English, the word "appropriating' is used to express a general meaning such as "to lay claim to." In this chapter "appropriate" has the meaning of adopting a fixed position for daily activities, such as reading, watching TV, relaxing, and so on. The fixed position or positions may be used for dining and sometimes even for sleeping as well. The latter is particularly true for the elderly living in a Tatami-style room.

Situations have been observed where these activities overlap, as well as where these activities are separated. The five types of situations found are as follows:

1. Dining, sleeping, and appropriating take place in the same room   Type (D, S, A). This is the most popular situation when the number of rooms in a house is limited. However, this tendency does not necessarily correlate with a resident's health or physical condition. Even in the case of the K site of recently built public housing (where the separation of dining is possible), the three activities tend to center in a comfortable room facing the south. In many cases, a Futon is used for sleeping and is put in a large cupboard during the day.

2. Dining is separated from a room for sleeping and appropriating   Type (D+S, A). This is a rare case even though the separation of dining and sleeping was the most important planning principle of modern public housing. However, in a house for the elderly (especially a single person), the separation of dining and appropriating position is very rare.

3. Sleeping is separated from a room for dining and appropriating   Type (S+D, A). This is an atypical type in the elderly couple's unit, but the tendency of sleeping and appropriating in separate rooms is increasing.

4. Appropriating is separated from a room for dining and sleeping   Type (A+D, S). In general, this type is very rare. Just after the death of a husband/wife or in case of some professions, this living style may appear.

5. Dining, sleeping, and appropriating have separated positions in different rooms   Type(D+S+A). If the number of rooms is restricted, this type of living would not be possible. Some of the older people who are endowed with vitality could continue this type of living.

USES OF ROOMS IN THE OLD COUPLE'S DWELLING

Modern functional planning of a house allocates a function to a room based on the concept of public space and private space. Actually, each room does not afford just one function. The use of a room changes from time to time. In the case of a single person's dwelling, the allocation of the position of an activity is the most important concern, but in a couple's dwelling, it is necessary to consider whether a room is shared or used exclusively.

Increase in the Number of Rooms and Allocation of Activities

All activities inevitably take place in the same area in a one-room dwelling. In the case of two rooms, there are two types of separation of activities. Each room is shared but the sleeping position is separated from dining and appropriating positions; each person has an exclusive appropriating position but dining and sleeping positions are shared. In the case of three rooms, each person has his or her own room for sleeping and appropriating activities.

It is difficult to draw firm conclusions from these results. I think the results suggest that couples wish to limit the amount of daily interaction. This would be a serious social problem in the highly aged society. However, the younger generation might be expected to adopt a different attitude.

THE LAYOUT AND THE AMOUNT OF STORAGE

The results showed that residents were not satisfied with the layout and the amount of storage space. Because of the paucity of storage space, household goods tended to accumulate around an appropriating position. In a narrow dwelling unit such as the H site, dwelling space is like a corridor because of furniture or household goods stacked against the walls.

It has been observed that many goods tend to be put near the sitting position because of senility. This is not always true, however, because some active residents who are not senile do not mind disordered goods or a very dull living style named Mannen-doko (a sole living position around a Futon that has never been folded up).

Where there is enough storage space for household goods, whether these goods are put away or arranged around appropriating positions depends mainly on the personality or living customs of the residents.

OVERLAPPING AND SEQUENTIAL SEPARATION OF ACTIVITIES

Overlapping and sequential separation of activities are influenced by various conditions.

Number and Size of Rooms

If there is just one room and the size is small, the positions of dining, sleeping, and appropriating activities overlap. Where the number and size of rooms are not so restricted, residents tend to choose an appropriating position based on environmental factors, for example, whether the room receives direct sunlight or whether it receives a cooling wind in summer.

Physical Conditions

If one cannot put the Futon away onto a shelf, appropriating and sleeping positions tend to overlap.

Living Custom and Personal Preference

To watch different TV programs, each member of a couple has different appropriating positions. When a Tatami area is found in the middle of the room, positions of various activities overlap. In a dwelling with two rooms, if one of the rooms is full of household goods, positions of activities tend to overlap in the other room.

COMPARISON OF HOUSING UNITS AND NEIGHBORHOOD IN FOUR SITES

Distribution of Activities and Variability of Settings

In the H site, the area and the number of rooms are limited, so activities are focused in one place, and the settings of the place are changed sequentially. In the K site, activities are distributed in a large room. However, in a small unit, the positions of activities tend to follow the original functional design of the room. Compared with the H and the K sites, the A site has more variations. Some units strictly follow the principle of functional distribution of activities and others have focused positions and sequential changes of settings. The greatest variety of living styles exists in the N site because it is located in the traditional downtown area of Tokyo and there remain many kinds of housing types in this area. This variety in the use of place would be helpful for future principles of housing for the elderly.

Pattern of the Use of Neighborhoods

The four sites have very different compositions. For example, the number and quality of public and commercial facilities are different in each site. Modes of social interaction also are very different. Each site has merits and demerits. The A site follows the modern design principle, which divides the area according to the use of the land. The many public facilities used by the elderly are located at some distance from residential areas. Rearrangement of neighbourhood facilities should be considered.

QUALITY OF SPACE IN HOUSING FOR THE ELDERLY

Based on the above analyses, the desirable quality of space in housing for the elderly is summarized as follows.

Physical Settings Should Be Easy to Use and Should Promote Relaxation

Neither too much functional nor automatic settings or devices are suitable for the elderly even though accessible and universal design have to be taken into account. The concept of competence (Lang, 1987) and a relaxing environment are the most important requirements of housing for the elderly.

Process of Negotiation on the Way a Room Is Furnished

In order not to cause unpleasantness to others, to avoid conflicting tastes, and to promote harmonious interaction, the way a room is furnished needs to be based on a process of negotiation between the residents.

Place for Keeping Goods and Memories of a Family or a Person

A house is not only a place for daily living but also a storehouse of personal memories. People can trace their personal history in every part of the house. Layers of memories kept stably in the house are an important source of support to cope with crises which we, and especially the elderly, meet with during the course of our lives.

Environment in Which We Can Recognize the Transition of Time

Everything changes with time. A dwelling must be an environment in which we can notice environmental change. Such an environment would be refreshing for the elderly. Home and house must be a place where "The Fountain of Age" (Friedan, 1993) gushes out and we experience "age as adventure."

REFERENCES

Friedan, B. (1993). *The fountain of age*. New York: Simon & Schuster.
Lang, J. T. (1987). *Creating architectural theory*. New York: Van Nostrand Reinhold.
Takahashi, T. (1992). Housing for elderly and environment transition. Smile on "Housing Forum," No. 23. Tokyo: Housing Research Foundation.
Takahashi, T. et al. (1993). *Report on environment behavior research of housing for elderly*. Tokyo: Housing and Urban Development Corporation.

Chapter **5**

QUALITY OF LIFE IN JAPANESE OLDER ADULTS

Hisao Osada

INTRODUCTION

The population of Japanese people over 65 years of age is estimated at 1,821 as of September 15, 1995, which is 14.5% of the entire population. While the ratio is still lower than that of Sweden (17.6%) or England (15.8%), the number is expected to continue to increase and reach as many as 32.8 million by the year 2021. As of 1994, the average life expectancy of the Japanese was 83 years for females and 76.7 years for males, the highest in the world. The number of adults 100 years old totaled as many as 6,378 as of September 8, 1995. The rapid aging of the Japanese society is said to be the result of both the decline in the birthrate and improved longevity. An aging society also means a long-life society, which is itself a very welcome phenomenon. On the other hand, living longer often means not only having to face more social or psychological experiences of loss (e.g., loss of spouse) but also having to face physical and mental decline or pathological deterioration. The negative experiences tend to increase the risk factors threatening health and comfortable living.

In this long-life society, successful aging is a requisite theme. To attain successful aging, a high quality of life in the community must be maintained. The high quality of life can be interpreted as "the good life" mentioned by Lawton (1991). The constituent elements of the good life are the objective environment, behavioral competence, perceived quality of life, and psychological well-being. Psychological well-being is one of the most studied fields as is subjective well-being in socio-gerontology (Koyano et al., 1990).

Hisao Osada • Tokyo Metropolitan College of Allied Medical Sciences, Tokyo 116, Japan.

Handbook of Japan–United States Environment–Behavior Research: Toward a Transactional Approach, edited by Seymour Wapner, Jack Demick, Takiji Yamamoto, and Takashi Takahashi. Plenum Press, New York, 1997.

It is helpful to measure subjective well-being by questionnaire surveys for estimating the quality of life of older adults in the community. Standard questionnaires on subjective well-being, such as the Philadelphia Geriatric Center Morale Scale (PGC) of Lawton (1975) and the Life Satisfaction Index (LSI) (Neugarten et al., 1961), have been developed. According to these scales, the confirmed factors include: previous agitation; lonely dissatisfaction; attitudes toward one's own aging; moodiness, zest for life, and congruence.

In Japan, Sugiyama (1981), Sugiyama et al. (1981), Koyano et al. (1983), and Koyano et al. (1989, 1990) constructed a Japanese version based on both the PGC and LSI. They reviewed the reliability, validity, and structure of this edition. Three factors were selected by Koyano et al. (1989, 1990) as representing the older adults' subjective well-being: they encompass entire life satisfaction, psychological stability, and one's estimation of aging. Maeda et al. (1989) analyzed the structure using the Japanese version of the PGC and selected four factors: being optimistic and positive, psychological stability, feeling healthy and needed, and attitudes toward aging. Ishihara et al. (1992) have discovered four psychological factors indicating the quality of life: present satisfaction, vitality or zest for life and hobbies, psychological stability, and self-control. Other studies using these scales have highlighted the following factors in subjective well-being: ego strength (Koyano et al., 1984), participation in athletic activities (Sugiyama et al., 1986), relationship between the family and the institutionalized elderly (Niino et al., 1988), the presence or absence of diseases, physical functioning, level of satisfaction with one's social support, level of satisfaction with the assistance one can provides to others, satisfaction with one's economic status, depressive status, self-esteem, socio-human relationship (Maeda et al., 1989), lifestyle (Haga et al., 1994), functional health, prestige in their occupations, marital status, coresidence with married children,and the size of friendship network (Koyano et al., 1995).

This chapter aims to investigate the factors related to the subjective well-being of Japanese older adults in a rural community. Specifically, we focused on physical function, activities of daily living, and social support.

METHOD

The subjects in our investigation are the rural community elderly in the northeastern region of Japan. At the time of the investigation, the number of the residents aged 65 years and over were 940 (377 males with an average age of 72 years; 565 females with an average age of 73 years). The present analysis utilized 739 residents (78.6%: 294 males, 445 females) who answered the LSIK (Life Satisfaction Index-K), is the dependent variable in this study.

The investigation was conducted from July to August, 1992. The items investigated were sex, age, years of education, cohabitation with a partner, stroke, heart disease, hypertension, diabetes mellitus, hospital visits, systolic blood pressure, falls, pain, subjective health status, hearing capacity, visual capacity, grip strength, chewing ability, locomotion, high functional capacity, alcohol drinking, smoking, regular walks and/or exercise, and social support: among the family living with (IF)

TABLE 5.1. Correlation of LSIK and Independent Variables

Items	Male r	Female r
Age	.02	−.01
Years of education	.02	.05
Cohabiting with the partner	−.03	−.04
Stroke	.02	.00
Heart disease	.04	.08
Hypertension	−.00	.02
Diabetes mellitus	.11	.09
Hospital visits	.08	.07
Systolic blood pressure	.11	.02
Fall	−.04	−.15**
Pain	.13*	.15**
Subjective health status	−.29**	−.23**
Hearing capacity	−.05	.00
Visual capacity	−.20**	−.12*
Grip strength	.12*	.11*
Chewing ability	−.30**	−.28**
Locomotion	−.04	−.15**
High functional capacity	.12	.14**
Alcohol drinking	.05	−.06
Smoking	.00	−.03
Regular walks and/or exercise	.10	−.07
Listening-IF	−.18**	−.18**
Helping-IF	−.12*	−.10*
Considerate-IF	−.14*	−.12*
Encouraging-IF	.12*	−.12*
Financing-IF	−.15**	−.15**
Relax-IF	−.09	−.11*
Household-IF	−.11	−.08
Nursing-IF	−.11	−.19**
Listening-OF	−.09	−.06
Helping-OF	−.06	−.10*
Considerate-OF	−.02	−.09
Encouraging-OF	−.03	−.07
Financing-OF	−.09	−.12**
Relax-OF	−.08	−.07
Household-OF	−.12*	−.18**
Nursing-OF	−.10	−.17**

*$p<.05$; **$p<.01$.

items were someone listening to your anxiety (listening-IF), someone helping you in the case of being ill in bed for a few days (helping-IF), someone being considerate to you (considerate-IF), someone encouraging you (encouraging-IF), someone financing you (financing-IF), someone making you relaxed (relax-IF), someone managing household matters for you (household-IF), and someone nursing you in the case of a serious illness (nursing-IF). Among the other than family living with (OF) items were someone listening to your anxiety (listening-OF), someone helping you in the case of being ill in bed for a few days (helping-OF), someone being considerate to you (considerate-OF), someone encouraging you (encouraging-OF), someone financing you (financing-OF), someone making you relaxed (relax-OF), someone managing household matters for you (household-OF), and someone nursing you in the case of a serious illness (nursing-OF)

The LSIK is a questionnaire consisting of nine items, in which the higher total score represents a higher subjective well-being. High functional capacity was assessed with the TMIG Index of Competence (1991) and consisted of 13 items. We analyzed the results using correlation and multiple regression with LSIK as the dependent variable and other variables as the independent variables using the Statistics Package Program SPSS.

RESULTS

Analysis of the LSIK scores indicated significantly higher scores in males than in females (males, 6.0, females, 5.7, $t = 2.16$, $df = 737$, $p < .05$). In light of this finding, we will present the remainder of the findings separately for males and females.

There was no significant correlation between LSIK scores and age. For males, the variables that showed significant correlations were pain, subjective health status, visual capacity, grip strength, chewing ability, listening-IF, helping-IF, considerate-IF, encouraging-IF, financing-IF, household-OF. For females, they were falls, pain, subjective health status, visual capacity, grip strength, chewing ability, locomotion, high functional capacity, listening-IF, helping-IF, considerate-IF, encouraging-IF, financing-IF, relax-IF, nursing-IF, helping-OF, financing-OF, household-OF, nursing-OF (Table 5.1).

To strengthen and clarify the relationship between LSIK and each variable, we employed multiple regression using age and the independent variables with a high correlation coefficients ($p < .05$). As a result, in males, a strong relationship between subjective health status,visual capacity, chewing ability, listening-IF, financing-IF, and LSIK scores was shown (Table 5.2). In females, strong relationships between LSIK scores and falls, pain, subjective health status, chewing ability, listening-IF, and nursing-IF were uncovered (Table 5.3).

DISCUSSION

Among the older adults over 65 years of age in this Japanese rural community, there was no correlation between the subjective well-being as assessed by LSIK scores and age. LSIK scores in males were higher than in females, suggesting that subjective well-being in males was higher.

With respect to the relationships between LSIK and the independent variables, for both males and females, the following was found: high subjective health status, high chewing ability, and availability of someone within the family who would listen to their anxieties related to high subjective well-being. The relationship between having coresidents and high subjective well-being replicated previous findings (Koyano et al., 1995). With respect to subjective well-being in old age the findings

Table 5.2. Multiple Regression in Males

Items	β	r
pain	.09	.13*
Subjective health status	−.22**	−.29**
Visual capacity	−.13*	−.20**
Grip strength	.00	.12*
Chewing ability	−.23**	−.30**
Listening-IF	−.12*	−.18**
Helping-IF	.01	−.12*
Considerate-IF	−.04	−.14*
Encouraging-IF	−.04	−.12*
Financing-IF	−.11*	−.15**
Household-OF	−.06	−.12*
R	.44**	

$*p<.05; **p<.01.$

suggested that what was important was not objective health status such as chewing ability, diseases, and hospital visits, but the subjective health condition such as subjective health status. Our finding that subjective health status was closely related to subjective well-being is also consonant with previous findings (Maeda, 1988).

Sex differences are observed in the relationships between LSIK scores and the independent variables. In males, visual capacity in performing daily activities and a sense of security with respect to financial support had strong relationships to their subjective well-being. In females, having no fall experiences, no physical pain, and the existence of someone living with them who would support them in the event of a serious disease had strong relationships to subjective well-being in old age.

To maintain subjective well-being and a high quality of life in old age, it is important to support the older adults' daily lives, helping them feel healthy and preventing their functional decline, which affect their daily activities such as chewing. At the same time, it is clear that emotional support from family members is important particularly if older adults are in stressful situations. In addition, sex differences were observed in such variables as daily activities, lifestyles, and necessary social resources. For that reason, it is necessary to take into consideration sex differences when viewing the quality of life in old age.

Our results are based on older adults living in suburban areas of Japan. It was pointed out at a recent seminar that there must be a difference between the lifestyles of the Japanese and Americans. An elderly American, for example, might feel that the ability to drive a car by himself or herself is closely related to his or her subjective well-being. Even within Japan, it should be understood that the living environment of the urban areas differs from that of rural areas, which naturally leads to lifestyle differences in both areas. To improve the subjective well-being of older adults in the community, the study suggests the establishment of an appropriate support system to assist older adults to obtain more independence in their daily living and action. This should be done after assessing the living environment of the particular group of older adults being studied. More basic and interdisciplinary studies will be needed in this area.

TABLE 5.3. Multiple Regression in Females

Items	β	r
Fall	−.11*	−.15**
Pain	.10*	.15**
Subjective health status	−.17**	−.23**
Visual capacity	−.07	−.12*
Grip strength	.05	.11*
Chewing ability	−.28**	−.28**
Locomotion	−.06	−.15**
High functional capacity	.05	.14**
Listening-IF	−.11*	−.18**
Helping-IF	−.08	−.10*
Considerate-IF	−.00	−.12*
Encouraging-IF	.03	−.12*
Financing-IF	−.01	−.15**
Relax-IF	−.01	−.11*
Nursing-IF	−.16**	−.19**
Helping-OF	−.01	−.10*
Financing-OF	−.05	−.12**
Household-OF	−.06	−.18**
Nursing-OF	−.09	−.17**
R		.46**

$*p < .05$; $*p < .01$.

NOTES

The present study is part of the Nangai Study, a longitudinal multidisciplinary study on normal aging conducted by the Tokyo Metropolitan Institute of Gerontology. The author is grateful for the collaboration of Vice-Director Shibata H. MD, Ph.D. from the Tokyo Metropolitan Institute of Gerontology.

REFERENCES

Haga, H., Shibata, H., Suzuki, T., Nagai, H., Kumagai, S., Wantabe, S., Amano, H., Yasumura, S., & Sakihara, S. (1994). Lifestyles and quality of life in older people living in the community. *Japanese Journal of Gerontology, 16*(1), 52–58.

Ishihara, O., Naito, K., & Nagashima, K. (1992). Developing a quality of life index. *Japanese Journal of Gerontology, 14*, 43–51.

Koyano, P. W. (1983). Common dimensions and interrelationships of the measures of morale, life satisfaction and happiness of the aged: A re-examination. *Japanese Journal of Gerontology, 5*, 129–142.

Koyano, P. W., Okamura, K., Anda, T., Hasegawa, M., Asakawa, T., Yokayama, H., & Matsud, T. (1995). Correlates of subjective well-being and social relations in middle-aged and older adults living in urban areas. *Japanese Journal of Gerontology, 16*(2), 115–124.

Koyano, P. W., Shibata, H., Haga, H., & Suyama, Y. (1990). Structure of a life satisfaction index-invariability of factorial structure. *Japanese Journal of Gerontology, 12*, 102–116.

Koyano, P.W., Shibata, H., Maeda, D., Shimonaka, Y., Nakazato, K., Haga, H., Suyama, Y., & Matsukai, T. (1984). Indices of successful aging and their correlates: An interdisciplinary analysis. *Japanese Journal of Gerontology, 6*(2), 186–196.

Koyano, P.W., Shibata, H., Haga, H., & Suyama, Y. (1989). Structure of a life satisfaction index: Multidi-
    mensionality of subjective well-being and its measurement. *Japanese Journal of Gerontology, 11*,
    99–115.
Lawton, M. P. (1975). The Philadelphia geriatric center morale scale: A revision. Journal of *Gerontology*,
    30, 85–89.
Lawton, M. P. (1991). A multidimensional view of quality of life in frail elders. In J. E. Birren, (Eds.), *The
    concept and measurement of quality of life in the frail elderly* (pp. 3–27). San Diego, CA: Academic
    Press.
Maeda, D. (1988). "Quality of life" of the elderly: Longitudinal analysis of its socio-behavioral aspects.
    Social Gerontology, 28, 3–18.
Maeda, D., Noguchi, Y., Tamano, K., Nakatani, Y., Sakata, S., & Liang, J. (1989). Structure and possible
    causes of subjective well-being of Japanese elderly. *Social Gerontology, 30*, 3–16.
Neugarten, B.L., Havighurst, R. J., & Tobin, S. S. (1961). The measurement of life satisfaction. *Journal of
    Gerontology, 16*, 134–143.
Niino, N., Koizumi, A., Kawakami, N., & Morimoto, K. (1988). Factors affecting the life satisfaction of
    institutional elderly-stratified by ADL and subjective complaint. *Japanese Journal of Gerontology,
    10*(1), 227–233.
Sugiyama, Y., (1981a). Standardization of the revised Japanese questionnaire of the PGM as a psychologi-
    cal scale for measuring the level of life satisfaction of the elderly- (1) On the reliability and construct
    validity of the scale. *Japanese Journal of Gerontology, 3*, 57–69.
Sugiyama, Y., Takekawa, T., Nakamura, K., Sato, S., Vrasawa, K., & Sato, Y. (1981b). Standardization of the
    revised Japanese questionnaire of the PGM as a psychological scale for measuring the level of life
    satisfaction of the elderly- (2) On the concurrent validity of the scale. *Japanese Journal of Gerontol-
    ogy, 3*, 70–82.
Sugiyama, Y., Nakamura, K., Sato, S., Takekawa, T., & Saito, K. (1986). Physical activities of the elderly in
    relation to life satisfaction. *Japanese Journal of Gerontology, 8*, 161–176.

Chapter 6

A STUDY ON A RELOCATION OF A NURSING HOME FOR BLIND OLDER ADULTS

Takiji Yamamoto and Toshihiro Hanazato

INTRODUCTION

People encounter major environmental transitions during the course of their lives (e.g., entering school, marriage, retirement, relocation, natural disaster, and bereavement). These transitions cause radical changes in an individual's person–environment system (P–E system). Relocation might operate as a critical transition if the processes involved are difficult for the individual (cf. Wapner, 1987). Since visual impairment is a definite barrier in the development of environmental cognition, careful interventions are required in, for example, the case of relocation for the visually impaired.

This chapter analyzes the impact of the relocation of a nursing home for blind older adults. As a consequence of relocation, the residents were forced to reconstruct their P–E system and to develop new environmental representations, namely, cognitive maps of the nursing home environment. The long-term reconstructional process of cognitive maps is analyzed through sketch maps of the physical environment. The communal relations of the blind older adults were studied by focusing on the usage of a so-called behavioral focal point (BFP) or a place where individuals can interact together (Bechtel, 1987, 1990).

Takiji Yamamoto • The Japanese Institute of Health Psychology, 5F Mikuni Building, 1-3-5 Kojimachi, Chizoda-Ku, Tokyo 102, Japan. **Toshihiro Hanazato** • Institute of Art and Design, University of Tsukuba, Tsukuba, Iberaki 305, Japan.

Handbook of Japan–United States Environment–Behavior Research: Toward a Transactional Approach, edited by Seymour Wapner, Jack Demick, Takiji Yamamoto, and Takashi Takahashi. Plenum Press, New York, 1996.

Literature Review

There have been many studies on cognitive mapping since Tolman's (1948) study of cognitive mapping of rats. Relevant to this study are some concepts developed in past studies.

Various research involved in developing the typology of the cognitive map has been undertaken, especially in the 1970s (e.g., Appleyard 1970; Downs & Stea, 1977). These authors discuss how the shape of cognitive maps become transformed as a person becomes accustomed to his or her environment (cf. Schouela, Steinberg, Leveton, & Wapner, 1980). Further, a distinction between two types of environmental representation was made. One is a linear, or sequential, shape, related to a person's movement and decision plan in space. The other is an organized shape, which has a topographical relation with critical elements of the explored environment and represents a person's understanding of the spatial entity. The former is referred to as a route map and the latter as a survey map.

Some have emphasized the distinction between the cognitive map and the figural presentation (Lloyd, 1982; Pinker 1979). Cognitive maps are regarded as frameworks in which environmental experience is organized, and figural presentations are the equivalent of the environmental feature.

Spatial cognition has often been associated with vision. A totally blind person's ability to understand and to represent the physical environment has often been questioned. For example, several studies (e.g., Byrne & Salter, 1983; Casey, 1978) have dealt with the cognitive mapping of blind people. There have also been several theoretical accounts of the environmental cognition of blind individuals. Other studies (e.g., Landau, Gleitman, & Spelke, 1981) suggest that, although the blind person's spatial image is based on the integration of nonvisual information, they can have mental images like sighted people.

According to Passini, Proulx, and Rainville (1990), there are two different theoretical approaches for understanding the spatio-cognitive ability of visually impaired persons, the "deficiency theory" and the "quantitative difference theory" (see also Fletcher, 1980, 1981a, 1981b). Deficiency theory hypothesizes that a visually impaired person lacks the necessary competence to perceive the spatial world. By contrast, quantitative difference theory hypothesizes that the competence of the blind individual is equal to that of the sighted individual's in quality but that there is a difference in behavioral quantity. In other words, it suggests a visually impaired person's spatio-cognitive competence as low in quantity. A majority of recent studies support the latter theoretical position. A central objective of this chapter is to test these two theories.

As Passini, Proulx, and Rainville (1990) have found, the visually impaired person, in general, needs a relatively longer time to improve his other competence in spatio-cognition. To create a cognitive map, blind persons utilize nonvisual sensory cues such as the tactile feeling of a guide rail, echo of sounds, voice of others, scent, wind, etc., which ordinary people tend to miss because they mainly use visual cues in their development of a cognitive map. Passini et al. (1990) explored this by administering eight way-finding tasks to several control groups. Results suggested that the ability of spatio-cognition is obtainable without vision or visual experience, though the visually impaired need a longer time to finish the task than the sighted. The researchers explained that this finding is inconsistent with deficiency theory and supports quantitative difference theory.

The orthogenetic principle (Werner, 1957) assumes that the developmental and learning process of environmental cognition is microgenetic. This principle states that the more differentiated and hierarchically integrated a system is in terms of its parts and of its means and ends, the more highly developed the system is said to be. In keeping with this principle it is assumed that the developmental and learning process of the visually impaired is similar microgenetically (cf. Schouela et al.,1980 and Yamamoto, Stevens, & Wapner, 1980 for a microgenetic analysis of a partial organization).

Yamamoto (1991) administered sketch maps to a totally blind group and to a sighted blindfolded group. The developmental process of the spatio-behavioral ability was compared by spatio-organization scores. Though there was a quantitative difference, the scores of both groups increased gradually. The scores of the blind group became lower than the sighted blindfolded group over time. This suggested that the developmental process is microgenetic, that it is a relatively difficult task for the blind person to construct a spatial organization by relating spatial components. The blind group could reach a nonintegrated level of spatial organization because it was based on using nonvisual information, but they could not reach the higher integrated level possibly because of the lack of visual information.

The study of the relocation of members of a psychiatric therapeutic community by Demick and Wapner (1980) is another example. The study follows the developmental process in a 2-month time period. They measured various aspects of self, of environment, and of self–environment relations (e.g., hours spent in specific locales, intensity and permanence of relationship with others) over the course of relocation. Long-term developmental processes such as relocations should be viewed from various aspects including self–environment relations because the P-E system has transactional characteristics.

Bechtel (1987, 1990) has discussed the importance and necessity of a place where users of a building can get together naturally, connect, and interact with each other. He has named this a "behavioral focal point" (BFP) and has delineated the main requirements as follows:

1. Central location with maximum accessibility to all.
2. Location at the main crossroads of walk paths in daily activities.
3. Minimum social commitment.
4. Visual access to see others and to be seen.
5. Places for sitting, preferably connected with eating.

Since these conditions include several visual aspects, requirements for blind older subjects may be different from those suggested by Bechtel. In this study, we attempted to discuss blind adults' requirements for a BFP.

Conceptualization

Since we cannot observe the recognized environment directly, we must treat its indirect spatial manifestation. This chapter utilizes sketch maps and drawings by subjects as their spatial representation.

Though many studies have already been undertaken on the spatio-cognition of the visually impaired person, however, these studies focus on relatively short-term transformations of cognitive maps. Our plan was to administer observations over a 7-month period, which is a much longer time than that used in other studies.

In the process of a person's normalization to the new physical environment, we suppose that there would be guidance to understand the spatial structure of the building. Therefore, the cognitive process we observe is not necessarily ideal in terms of experimental setting. It is important to note that the subjects learned and developed their spatio-cognition with the help of instruction from the caretaker or their friends' advice. We do not, however, regard such instruction or advice as irrelevant to this research because it is a matter of course in daily life, where the life-long development happens. Thus, we are taking an ecological approach toward the spatio-cognition of blind older adults.

Since we hypothesize that the developmental process of spatio-cognition is microgenetic, the tests and observations should be undertaken over trial to ascertain the gradual differences that take place.

It was also expected that there might be some difference in terms of behavior or movement as spatio-cognition proceeded.

The communal and spontaneous interaction of blind older adults is significant to ease their relations with others as well as new human–environment relations. Such new human relations are expected to be developed especially in the case of relocation. To understand the behavioral aspects of the new setting, two studies were undertaken; the first employed behavioral mapping and the second employed time sampling in observing lobbies and lounges. Changes of behavior can be expected as blind older adults appropriate their own physical environment. A simple questionnaire was also utilized to learn about blind older adults' ideas concerning mutual communication. Through use of these methods, the study is expected to shed light on the developmental process underlying the behavioral characteristics of the visually impaired following relocation.

The purposes of the study can be summarized as follows:

1. To clarify the long-term developmental process of spatial representation of blind older adults in a new environment through cognitive mapping (Study 1).
2. To test whether the BFPs requirements are applicable to blind older adults (Study 2).

STUDY 1. SPATIAL REPRESENTATION BY BLIND OLDER ADULTS OF A NEW ENVIRONMENT

Method

The visually impaired who are situated in a new environment have to construct their cognitive maps without the help of visual sensory cues. Because the adaptability of blind older adults is relatively low, it is assumed that the process of developing

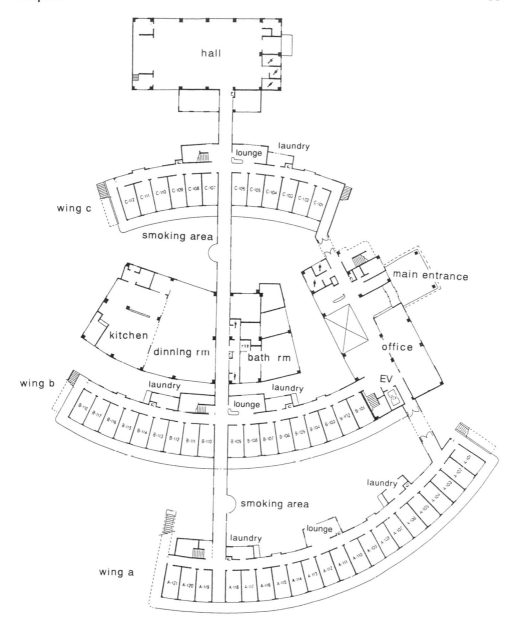

Figure 6.1. Plan of the nursing home.

one's cognitive map is slow, is not an easy task, and takes longer for blind adults than for sighted adults to achieve.

The cognitive mapping test was administered to 17 blind older adults (7 male and 10 female subjects, 7 totally blind, and 10 with visual residues), who resided in a nursing home in the suburbs of Tokyo (Figure 6.1). The average age was 72.2 years. The first test administration was undertaken 1 week after relocation to the

FIGURE 6.2. Walking the right-hand side of the corridor.

newly built nursing home, the second 3 months after, and the third 7 months after. Figure 6.2 shows a scene in which the blind older adults walk in the central corridor on the right-hand side, following the traffic rule of the nursing home.

First, the subjects were given a pen and a sheet of 15 in. × 21 in. (38 × 54 cm) drawing paper and were asked to confirm the size of the paper. Then, they were instructed as follows: "You have moved to this new house. Please draw the internal structure and places as detailed as you know." We used paper that could be easily engraved by a pen in order for subjects to touch and to trace their engraved pictures. On each test occasion, two examiners helped a subject to draw if he or she looked confused and recorded the verbal interaction between them. The examiners wrote the names of places or rooms on the map as the subject expressed them.

Results

Figure 6.3 (a, b, and c) shows the sketch maps of a female subject with visual residue. The maps illustrate the typical changes of the spatial representation over the 7-month period of testing.

To describe the major qualitative changes observed in the sketch maps, we examined changes in the following aspects: (1) number of places on drawings, (2) errors of relational positions of places, (3) curves of the corridors, and (4) overall structure.

The development of each aspect is described in the following data.

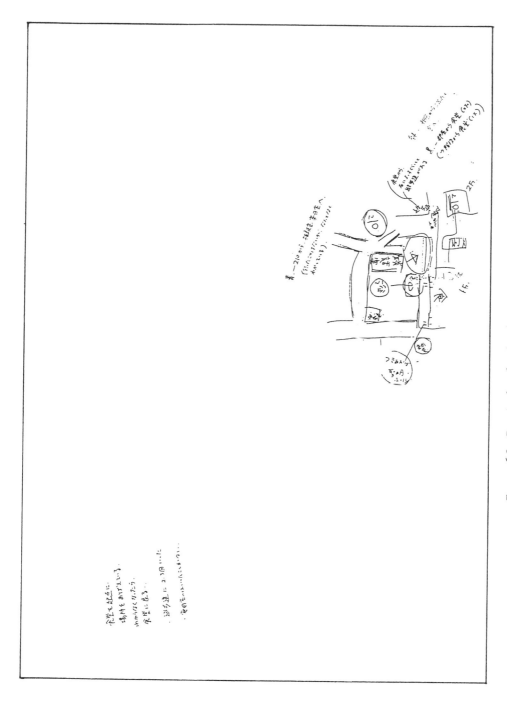

FIGURE. 6.3a. Drawing by a female subject 1 week after relocation.

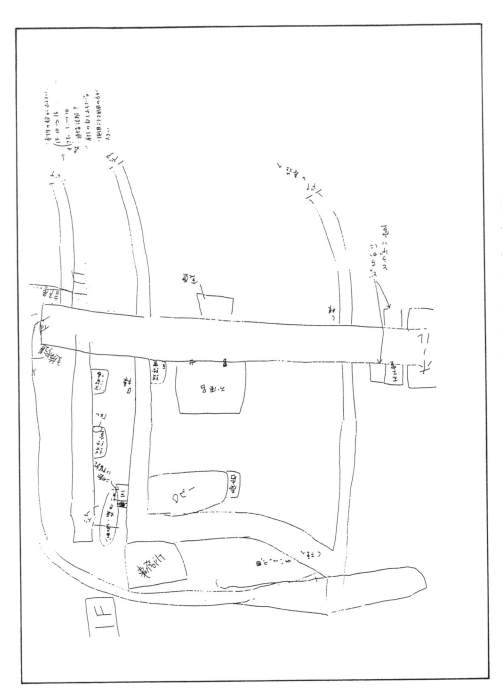

FIGURE. 6.3b. Drawing by a female subject 3 months after relocation.

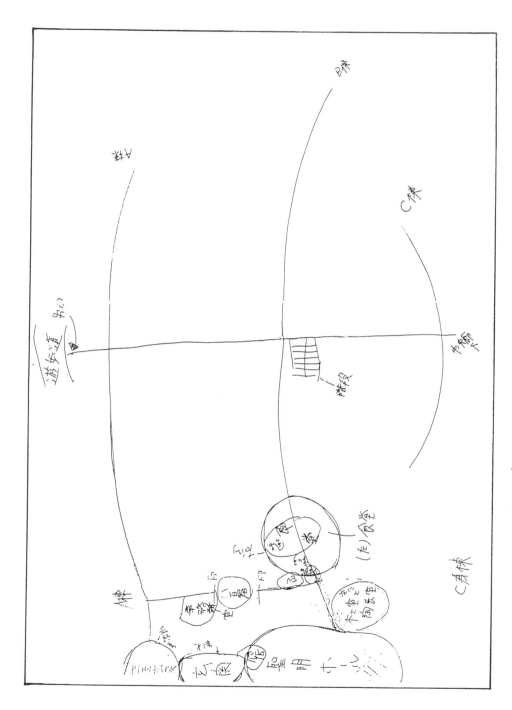

FIGURE 6.3c. Drawing by a female subject 7 months after relocation.

TABLE 6.1. Number of Places on the Drawings

	First	Second	Third
Average	14.29	25.18	37.94
Totally blind	10.14	17.60	31.80
Visual residue	17.20	30.40	41.90

Number of Places on Maps

One of the major changes over 7 months was the increase in the number of places drawn on the map. The average number increased from 14.3 to 25.2, to 37.9 (Table 6.1, Figure 6.4). This result suggests that the spatial knowledge of subjects increased substantially in 7 months. There is, of course, a difference between the totally blind group and those who had some visual capacities available. The latter performed better. There were similar differences in other measurements.

Table 6.2 lists the places drawn by the subjects, and the places drawn by at least 50% and by at least 90% of the subjects are boxed. The numbers of places on each list became four or five times larger in 7 months (Figure 6.5). The main communal places such as the dining room, bathroom, and caretaker's room were listed from the beginning. Thus, the frequently used main communal spaces were recognized first by the subjects, then the less used spaces were recognized gradually.

In addition, the drawn figures themselves became larger in three tests, which may also suggest the feeling of expansion of the living and moving area during the course of habitation.

Errors of Relational Positions of Places

To evaluate the accuracy of the drawings, relational position of the places in the maps were checked. We quantified from two different measures, namely, the number of mistakes per person, and the mistake ratio per number of places. The

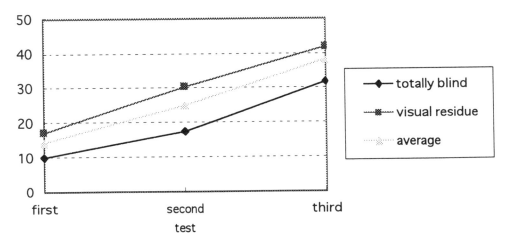

FIGURE 6.4. Number of places on the drawings.

TABLE 6.2. Places Drawn and Percentage of Subjects That Drew Them

Place	%	Place	%	Place	%
Dining rm	94	Dining rm	94	Dining rm	100
Bath rm	94	Caretaker's rm	94	Bath rm	100
Own rm	88	Own rm	94	Hall	100
Caretaker's rm	76	Meeting rm	88	Caretaker's rm	100
Stair-A	59	Bath rm	82	Meeting rm	100
Doctor's office	53	Consultants' rm	82	Main entrance	94
Consultants' rm	53	Kiosk	76	Office	94
Hall	41	Stair-ELV	76	Kiosk	94
Stair-B	35	Hall	71	Stair-C	94
Stair-C	35	Main entrance	71	Lounge-B2F	94
Lounge-A2F	35	Doctor's Office	71	Stair-ELV	88
Main entrance	29	Barber	71	Doctor's office	88
Kiosk	29	Stair-A	65	Consultants' rm	88
Stair-ELV	29	Elevator	65	Stair-A	82
Dispensary	29	Lounge-B2F	65	Elevator	82
Laundry-A2Fa	29	Wood-work rm	59	Laundry-A2Fa	82
Office	24	Office	59	Lounge-C2F	82
Elevator	24	Dispensary	59	Own rm	82
Barber	24	Stair-B	53	Stair-B	76
Lounge-B2F	24	Exit-AB	53	Barber	76
Lounge-C2F	24	Entrance lobby	47	Lounge-A2F	76
Smoking-AB2F	24	Stair-C	47	Exit-AB	76
Promenade	24	Promenade	47	Wood-work rm	71
Entrance lobby	18	Pottery rm	41	Pottery rm	71
Locker rm	18	Lounge-B1F	41	Entrance lobby	71
Lounge-A1F	18	Smoking-BC2F	35	Dispensary	71
Laundry-A2b	18	Exit-BC	35	Lounge-B1F	71
Laundry-C2F	18	Lounge-C1F	29	Lounge-C1F	59
Meeting rm	18	Laundry-C2F	29	Laundry-B2Fa	59
Smoking-AB1	18	Lounge-C2F	29	Exit-A2Fa	59
Smoking-BC2F	18	Smoking-BC1F	29	Smoking-BC2F	53
Lounge-B1F	12	Smoking-AB2F	29	Exit-BC	53
Lounge-C1F	12	Lounge-A2F	24	Lounge-A1F	47
Laundry-B2Fa	12	Laundry-B2Fa	24	Smoking-AB1	47

average number of mistakes per person did not change drastically (Figure 6.6). However, the mistake ratio per number of places constantly decreased over the 7 months from 9.05%, to 6.54%, to 3.26%. The latter result suggests the cognitive map became more accurate over time.

We interpret this result as follows: On the first test, the area in which the subjects moved was limited to the minimum required; on the second test, they had moved around the building but they made mistakes because of the lack of experience; and on the third test, they came to recognize the relational positions of the places they usually used.

Curves of the Corridors

As is shown in the plan (Figure 6.1), there are three corridors in the building that are curved in different directions. We rated the understanding of these curves

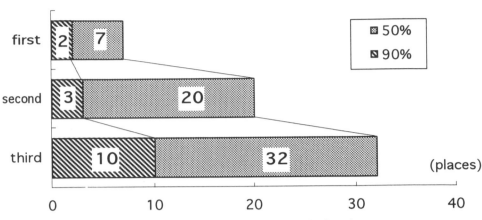

FIGURE 6.5. Number of places mentioned by the subjects.

by using a range from 0, meaning no correctly drawn curved corridor, to 3, meaning a perfectly drawn corridor (Table 6.3). The average rating increased constantly from 0.4 to 1.7 in 7 months, though at that time the score is still relatively low. As shown in Figure 6.7, 0 scores at the first test occurred for 88.2%, while at the third test, 17.7% of the subjects did not as yet understand the curve at all. Even at the third test, 7 months after relocation, only 23.5% could understand the curve thoroughly. This shows that the understanding of a curve is a difficult concept for blind older adults.

Overall Structure

To assess blind adults' understanding of the overall spatial structure, portrayal of the five main corridors was rated from 0 to 5, 5 meaning perfect understanding

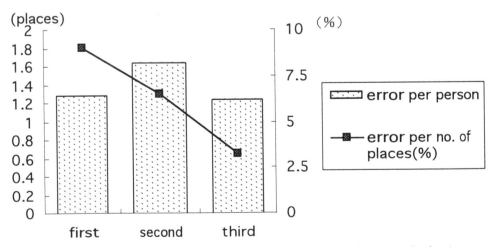

FIGURE 6.6. Average error per person and error ratio per the number of places on the drawings.

TABLE 6.3. Average Score Understanding the
Curves of the Corridors

	First	Second	Third
Average scores (0-3)	0.35	1.41	1.65

of the overall structure. Over 7 months, the average score increased constantly from 1.6, to 4.1, to 4.9 (Figure 6.8). The subjects achieved relatively higher scores by the second test administration. The percentage of subjects who understood the overall structure perfectly reached 88.2% at the last test. Thus, most subjects came to understand the overall structure.

We categorized maps as follows: route map (RM), survey map (SM), and intermediate map (IM) (Figure 6.8). The number of route-map drawings decreased and that of survey-map drawings increased in 7 months. On the first assessment, some subjects could draw survey maps of main functional places such as the dining room and the bathroom. However, they could only draw route maps to show ways to less used places. On the second and third tests, however, the maps tended to be more structured, and the places and rooms were drawn in relatively correct positions, which suggests that with time, the blind older adults came to understand the spatial structure of the building more accurately. Eventually, at the third test, there were no route maps, and survey maps increased.

According to our categorization, the percentage of route maps decreased from 76.7% to 0%, while that of survey maps increased from 11.7% to 41.2%. If intermediate maps are considered as partial survey maps, all maps have survey-map char-

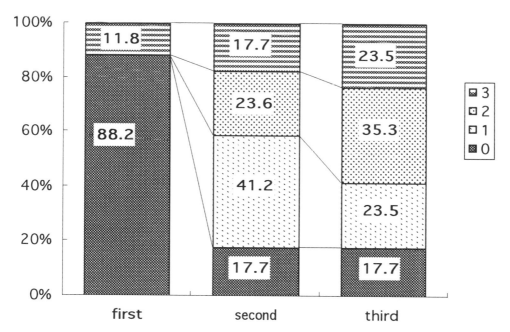

FIGURE 6.7. Scores on understanding the curves of the corridors.

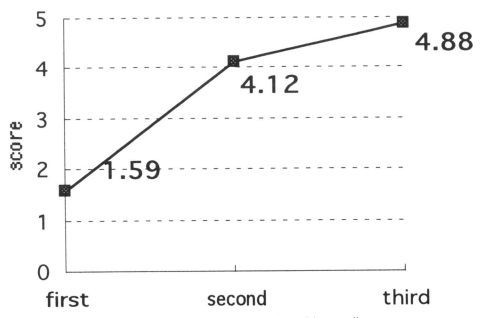

FIGURE 6.8. Average score of the understanding of the overall structure.

acteristics at the last assessment. In other words, on the first examination, the subjects have only fragmental representations of experienced routes (i.e., sequential or one dimensional); at the second and third examination, *as they experience the building*, they develop spatial representations that show the relations of optional places (i.e., topographical or two dimensional).

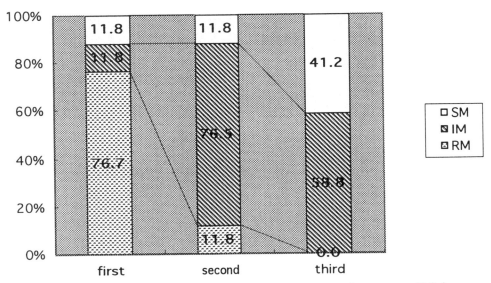

FIGURE 6.9. Percentage of route maps (RM), intermediate maps (IM), and survey maps (SM) drawn.

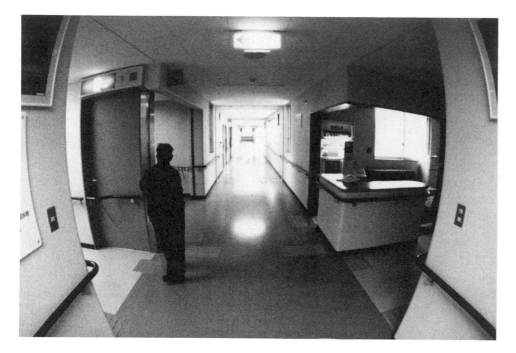

FIGURE 6.10. Intersection of the corridors. On the left is a lounge, a BFP.

Conclusion

With regard to numbers of places on maps, errors of relational positions of places, and overall structure, the results indicate that the subjects' spatial cognition developed within the 7 months. However, understanding of the curves of the corridors was not achieved very well by the time of the third test. It may be that the curve is too subtle for the blind individual to understand. Although an overall understanding was obtained, certain aspects of curve detail may be difficult for blind adults to understand.

Except for the results shown in Figure 6.5, in the error per person on the drawing, most of the evidence shows gradual increase in understanding as the subjects got accustomed to the new physical environment following relocation. Thus, there is evidence that the development of spatio-cognition is microgenetic (cf. Schouela et al., 1980).

In sum, our sketch map analysis documented how blind adults' understanding of spatial structure developed and their movement behaviors expanded. For a sighted person, the spatio-cognitive process we examined could be performed at least within a day. The extent of visual competence, however, seems to influence the speed of understanding spatial structure.

Though we did not attempt to treat the difference between the totally blind group and those with some visual capacity specifically, the result was consistent with Passini et al. (1990). That is, the group of totally blind people needed more time to perceive the structure than the weak-sighted.

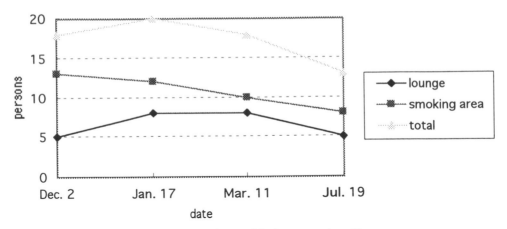

FIGURE 6.11. Observed users of the lounges and smoking areas.

As shown in the interpretation of the test of relational position, we presume the development of spatial cognition has some relationship to the sensory-motor activity of the subjects. To discuss this problem—that there might be some difference in terms of behavior or movement as spatial cognition proceeded—another study was conducted.

STUDY 2. APPLICABILITY OF BEHAVIORAL FOCAL POINT (BFP) REQUIREMENT TO BLIND OLDER ADULTS

Method

As mentioned earlier, the six lounges in this building fulfill the BFPs requirement, if they were used by sighted adults. To assess the applicability of the BFPs requirement for blind adults, we utilized several approaches.

Behavioral mapping observation in the building (Figure 6.1) was the first approach. We observed six lounges and four smoking areas through time-sampling procedures. The observations of the number of users and the mode of their behavior were conducted four times: after 1 week, 1 month, 3 months, and 7 months. Four observations were recorded at 11:00 a.m., 1:30 p.m., 3:30 p.m., and 4:30 p.m..

Observation of some subjects' movement were conducted for a certain period of time per day. If they moved from their original place, the observers followed them to record their movement for a total of 10.5 hours.

We also interviewed and administered a simple questionnaire on the occasion of the third sketch map test to clarify the interactions between the blind older adults.

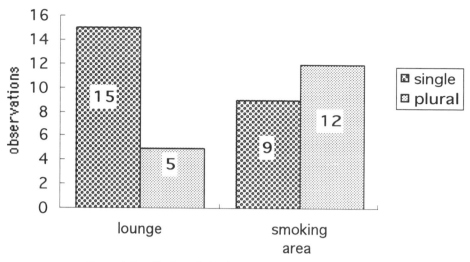

FIGURE 6.12. Single and plural use of lounges and smoking areas.

Results

Behavioral Mapping Observation

Although the lounges in this building fulfill the BFPs requirements, the results of behavioral mapping observation showed that the use of the lounges was relatively lower than that of the smoking areas. It was also observed that the number of users did not change radically over 7 months (Figure 6.11).

Interaction between inhabitants was not often observed in the lounges (Figure 6.10). Three-quarters (75%) of all the observed behaviors were of a single use (one person is using) of the lounge, while plural use (more than one person is using) was observed one-quarter of the time (25%). On the other hand, the plural use of the smoking areas was observed more than half the time (57.1%) compared to single usage (42.9%). Plural use does not necessarily mean interaction between users but only that there is the possibility of interaction. From this result we concluded that the lounges are not functioning as BFPs although they fulfill the essential requirements proposed by Bechtel (1987, 1990).

Observation of Movement

The traces of movement shown in Figure 6.13 are typical of the subjects. The traces show two subjects' movements over a period of 10.5 hours during the daytime. Focused around their own rooms their activities were limited to essential daily tasks such as getting hot water, washing clothes, and going to the dining room for meals even if they used the communal places. During the few interactions between the subjects we observed in this way, communication occurred in their individual rooms. Subjects' activities were relatively low: they tended to stay in their own rooms and did not use the common areas frequently.

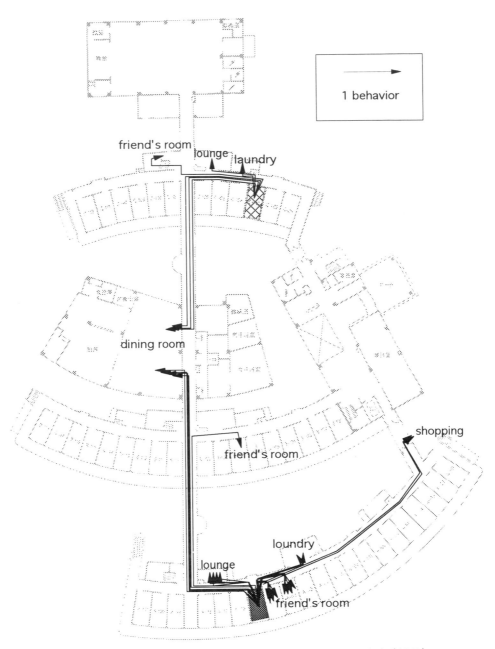

FIGURE 6.13. Traces of two subjects' daytime movements over a period of 10.5 hours.

Simple Questionnaire Test

We have obtained similar results with the use of three simple questions. The first is "Where do you have conversations with your most familiar friends?." As

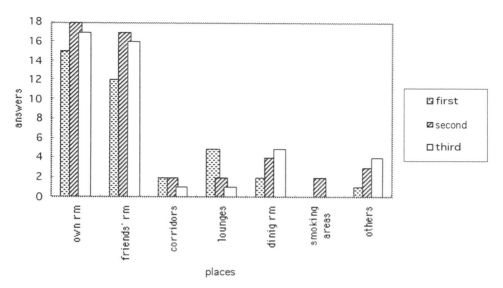

FIGURE 6.14. Places the subjects use for communication with intimate friends.

shown in Figure 6.12, subjects answer that they have conversations mainly in their own rooms or their friends' rooms. The most frequent response (89.7%) to the second question, "Where would you like to have conversation?" was in their own rooms. For the third question, "For what purpose do you use the lounges?" we found a similar result, namely, that 79.3% of the subjects say they use the lounges getting hot water and depositing hauling trash, but not for conversation. Thus, communication is mainly carried on in individual rooms.

Conclusion

As a result of behavioral mapping, we can conclude that the requirements of a BFP for blind older adults are different from those of ordinary people. Possible reasons for the low use of a BFP in this nursing home are as follows: (1) the activity of the blind older adults is relatively low; (2) there are not enough cues to get together, making the purpose of staying in these lounges is unclear; (3) the private rooms are comfortable enough for being with friends; and (4) the inhabitants may feel that those open spaces are under the surveillance of others since they cannot see whether others are listening to their conversation.

We can say, therefore, that some conditions of a BFP for the sighted are not necessarily applicable to blind individuals. The results show that their communication is mainly undertaken in their own rooms. Because they cannot rely on visual communication, the blind usually use verbal communication, which tends to be interpersonal. This, however, does not necessarily mean that the blind older adults neither need a BFP nor that the requirement for the blind person's BFP is totally different from that of the sighted subject. We should observe more actual settings before discussing the requirements.

FIGURE 6.15. A private room.

DISCUSSION

Through the examination of sketch maps, we have documented the developmental process of environmental cognition of blind adults. Most results show a gradual increase in cognition although there are differences of achievement depending on the tasks. Therefore, we can consider the developmental process as microgenetic.

Most of the drawings on the first test administration were route maps, while most on the second were intermediate maps, and most on the third had survey map characteristics. If we follow the microgenetic development assumption, the initial spatial representation is at an undifferentiated state, the second one is a differenti-

ated state, or quasi-differentiated state, and the third is a differentiated and integrated state. This means that the subjects' spatio-cognition changed from an ambiguous to a differentiated or integrated state over time.

Since we evaluated the accuracy of sketch maps by several quantitative methods, a variety of achievements was observed. While most subjects understood the
overall spatial structure of the building, they did not perform very well in understanding the curves of the corridors. Although one may presume that some blind
adults are lacking in the ability to understand the curve, and thus are, inferior to
the sighted (deficiency theory), we think differently. We think they did not perform
very well on the sketch map test because they did not need to understand the curves
practically; that is, they could move around the building without precisely knowing
the curves of the corridors.

These results also show that the spatio-cognitive ability of the visually impaired
is similar to that of the sighted, and thereby they support the quantitative difference
theory. As reviewed earlier, recent studies support the quantitative difference theory
though the speed of development is slow. We obtained a similar result to Passini et
al. (1990) although to a nonsignificant degree. The totally blind took more time in
obtaining structured spatial representation than the weak-sighted.

There were several pieces of evidence that suggested that blind adults construct their behavioral environment by using tactile or auditory sensory cues often
missed by the sighted, but this was not demonstrated quantitatively. We assume that
the blind elderly adults are slow compared to the sighted in obtaining spatial
organization because blind people have not learned an adequate frame of reference
for the spatial concept and also because they lack visual sensory information.

We have also focused on the BFP, where any changes of behavior can be
expected as blind older adults appropriate their own physical environment. Unfortunately, we could not find substantial differences in behavior as the development
of spatio-cognition proceeded.

The results of the second study have indicated that Bechtel's requirements of
BFP do not necessarily apply to blind adults. The requirements of a BFP for the
visually impaired seem to be different from those for the sighted because the spatial
behavior and cognition of blind people is based on nonvisual sensory information.
The behavioral and cognitive world of blind adults differs significantly from seeing
adults, so that, when designing buildings, it is important to consider the characteristics of a BFP for the visually impaired.

Blind adults' communications tend to be undertaken in individual rooms. It is,
however, inaccurate to conclude that they do not need any communal space like a
BFP, and it could be suggested that the internal configuration of the building may
deleteriously affect usage of the communal space.

If there are more concrete incentives for blind people to be in lounges, as well
as a more accurate understanding of spatial representations by the blind in general,
these places might be utilized more actively and effectively. It is obvious that further
study is essential. To observe more examples might be an initial step in finding
answers.

ACKNOWLEDGMENT

This study was supported by the Grant in Aids for Scientific Researches (Reference no.; 1994–0601013) from the Japanese Ministry of Education.

REFERENCES

Appleyard, D. (1970). Style and methods of structuring a city. *Environment and Behavior, 2*(1), 100–188.

Byrne, R.W ., & Salter, E., (1983). Distances and direction in the cognitive map of the blind. *Canadian Journal of Psychology, 37*, 293–299.

Bechtel, R. B. (1987). Behavioral focal points. In R. B. Bechtel, R. B. Marans, & W. Michelson (Eds.), *Methods in environmental and behavioral research.* (pp. 143 153). New York: Van Nostrand Reinhold.

Bechtel, R. B. (1990). Psychological aspects of outer space community design. In B. Wilpert, H. Motoaki & J. Misumi (Eds.), *Proceedings of the 22nd International Congress of Applied Psychology, Vol. 2* (pp. 375–380). Hillsdale, NJ: Erlbaum.

Casey, M. W. (1978). Cognitive mapping by the blind. *Journal of Visual Impairment and Blindness, 72*, 297–301.

Demick, J., & Wapner, S. (1980). Effect of environmental relocation upon members of a psychiatric therapeutic community. *Journal of Abnormal Psychology, 89*, 444–452.

Downs, R., & Stea, D. (1977). *Maps in minds.* New York: Harper & Row.

Fletcher, J. F. (1980). Spatial representation in blind children, 1: Development compared to sighted children. *Journal of Visual Impairment and Blindness, 74*(12), 381–385.

Fletcher, J. F. (1981a). Spatial representation in blind children, 2: Effects of task variations. *Journal of Visual Impairment and Blindness, 75*(1), 1–3.

Fletcher, J. F. (1981b). Spatial representation in blind children, 3: Effects of individual differences. *Journal of Visual Impairment and Blindness, 75*(2), 46–49.

Landau, B., Gleitman, H., & Spelke, E. (1981). Spatial knowledge and geometric representation in a child blind from birth. *Science, 213*, 1275–1278.

Lloyd, R. (1982). A look at images. *Annals of the Association of American Geographers, 72*(4), 532–548

Passini, R., Proulx, G., & Rainville, C. (1990). The spatio-cognitive abilities of the visually impaired population. *Environment and Behavior, 22*, 91–118.

Passini, R. (1992). *Wayfinding in architecture.* Van Nostrand Reinhold.

Pinker, S. (1979). Mental maps, mental images, and intuitions about space. *Behavioral and Brain Sciences, 2*, 513.

Schouela, D. A., Steinberg, L. M., Leveton, L. B., & Wapner, S. (1980). Development of the cognitive organization of an environment. *Canadian Journal of Behavioural Science, 12*, 1–16.

Tolman, E. C. (1948). Cognitive maps in rats and men. *Psychological Review, 55*(4), 189–208.

Wapner, S. (1987). A holistic, developmental, systems-oriented environmental psychology: Some beginnings. In D. Stokols & I. Altman (Eds.), *Handbook of environmental psychology* (pp. 1433–1465). New York: Wiley.

Werner, H. (1957). *The concept of development.* Minneapolis, MN: University of Minnesota Press.

Yamamoto, T., & Wapner, S. (1991). *Developmental psychology of life transition.* Kyoto: Kitaoji-shobo.

Yamamoto, T., Stevens, D. A., & Wapner, S. (1980). Exploration and learning of topographical relationships by the rat. *Bulletin of the Psychonomic Society, 15*, 99–102.

Part II
PSYCHOLOGICAL ASPECTS OF THE PERSON

Chapter 7

BODY AND SELF EXPERIENCE
Japan versus USA

Jack Demick, Shinji Ishii, and Wataru Inoue

INTRODUCTION

The present study examines differences between Japanese and American individuals in aspects of body and self experience. The research is based on previous work in the United States (e.g., Demick & Rivers, 1996; Demick & Wapner, 1987, 1996) that has attempted to provide a holistic, developmental, systems approach to body and self experience most generally. That body and self experience are of significant concern to environmental psychologists as well as to others interested in environment–behavior relations (e.g., architects, planners, sociologists) gains support from the notion that aspects of body and self experience encompass relevant general processes such as spatial relations (e.g., between the individual and his or her physical environment or between the individual and others in his or her interpersonal environment).

In light of these notions, the present investigation will first review literature relevant to our holistic, developmental, systems approach to body and self experience. We will then discuss relevant literature on cross-cultural differences (with particular reference to Japanese and American cultures) in aspects of body experience, aspects of self experience, and aspects of more general processes (e.g., privacy, territoriality, spatial relations; cf. Wapner, McFarland, & Werner, 1963) relevant to both the problem at hand and to environmental psychology more generally.

Jack Demick • Department of Psychology, Suffolk University, Boston, Massachusetts 02114.   **Shinji Ishii** • Department of Psychology, School of Education, Hiroshima University, Higashihiroshima 739, Japan.   **Wataru Inoue** • Research Center for Educational Study and Practice, School of Education, Hiroshima University, Higashihiroshima 739, Japan.

Handbook of Japan–United States Environment–Behavior Research: Toward a Transactional Approach, edited by Seymour Wapner, Jack Demick, Takiji Yamamoto, and Takashi Takahashi. Plenum Press, New York, 1997.

83

Body and Self Experience

Based on a recent elaboration (e.g., Wapner, 1981, 1987; Wapner & Demick, 1990) of Werner's (1957) organismic-developmental theory, aspects of body experience are assumed to develop first within the *sensorimotor* level (body action), then within the *perceptual* level (body perception), and finally within the *conceptual* level of cognitive functioning. As one moves from less advanced to more advanced levels, there is a corresponding increase in self-world distancing. It is further assumed that there is a contingency relationship between these most general levels, that is, conceptual functioning depends on perceptual functioning that in turn depends on sensorimotor functioning. The levels differ qualitatively, and functioning at one level is not reducible to functioning on the prior, less complex level since we assume that higher level functioning does not substitute for but rather integrates and transforms lower level functioning. In line with this, we argue that body experience is a necessary precursor to self experience.

As indicated in Table 1, the first level of *body action* involves the movement of one's body in relation to objects constituting the environment. It encompasses (1) *sensorimotor palpation*, tactile manipulation leading to sensations on the body surface; (2) *body mobility*, movement of body parts and then the body itself through space; (3) *use of body parts in space* (e.g., as tools to represent expressive functions); and (4) *tempo regulation*, movement of the body through space at differentiated speeds.

The second level, *body perception*, involves perception of the body qua object. Here, since the body is taken as an object, the individual may focus on any of the dimensions involved in the perception of objects in general, namely, *geometric-technical* (objective/measurable) and *physiognomic* (expressive/dynamic/functional) properties. Geometric-technical properties include (1) location of the body in space, (2) size of body and body parts, (3) shape of body and body parts, (4) body boundaries, (5) organization of body parts, and (6) judgments of physiological activation (e.g., heart rate). Physiognomic properties include those involved in (1) phantom limb phenomena, (2) active touching versus passively being touched, and (3) pointing.

The third, that is, *conceptual*, level of body experience, divided into *conceptual aspects of body* and *conceptual aspects of self*, encompasses what has most generally been referred to as body- and self-concept (esteem, image). The specific conceptual aspects of body include: (1) *body-concept/ image/representation*, here limited to an individual's systematic cognitive impressions of his or her body including the global-articulated dimension (e.g., as measured by form level, identity or role and sex differentiation, and level of detail); (2) *body fantasy*, that is fantasies, metaphors, and symbolic representations of the body; (3) *body-esteem (cathexis)*, a complex structure with an affective-valuative orientation that can be assessed according to its positive or negative valence. The specific conceptual dimensions of self are organized along parallel dimensions, namely, *self-concept/image/representation*, *self fantasy*, and *self-esteem (cathexis)*. Finally, our approach suggests one last conceptual aspect of self experience, namely, *self-understanding*.

This approach stands in marked contradistinction to those that focus on one or two isolated variables relevant to body and self experience (e.g., body concept, self-esteem) or to those that make beginning attempts at an integrative theory of body and self experience. With respect to the latter, for example, researchers (e.g., Cash, 1990; Cash & Green, 1986; Thompson & Dolce, 1989; Thompson, Penner, & Altabe, 1990) have generally differentiated between two modalities of body experience, namely, *body perception* (e.g., body size estimation) and *body-related attitudes* (e.g., satisfaction/dissatisfaction with overall body and body parts). However, while they have also generally acknowledged that each of these modalities has both cognitive (rational) and affective (emotional) components, they have not been exhaustive (e.g., valuative processes) and/or have not discussed developmental relations between these modalities or components.

In contrast, we feel that sharpening and organizing, rather than downplaying, distinctions between aspects of body/self experience and its development can lead to productivity in both theoretical conceptualization and empirical inquiry. For instance, recent research in our laboratory (e.g., Demick & Rivers, 1996) has provided initial empirical support for our conceptualization. Specifically, from a comprehensive battery of tests assessing various aspects of body/self experience administered to 81 American college students, a confirmatory factor analysis has supported the existence of three factors corresponding to the three levels described here, namely, *sensorimotor*, *perceptual*, and *conceptual* aspects of body/self experience.

Cross-Cultural Differences in Body Experience, Self Experience, and Related Processes

Body Experience

To date, there has been little systematic examination of cross-cultural differences in aspects of body experience (e.g., the index to Fisher's [1986], landmark two-volume treatise on the development and structure of body phenomena contains no listing on cross-cultural or sociocultural issues). An exception to this, however, is the work of Lerner and his associates (e.g., Lerner, Iwawaki, Chihara, & Sorell, 1980). Specifically, these investigators have found that, relative to American adolescents, Japanese adolescents had lower self-esteem (more so for females than males) as well as less favorable views of their bodies' attractiveness and effectiveness. Further, though there were minimal grade differences in both sociocultural contexts, sex differences in self-concept accounted for more variance in the Japanese sample.

Self Experience

Within recent years, there has been renewed interest in the social psychology of the self (e.g., Baumeister, 1991) and, following from this, in cross-cultural differences in aspects of self experience. For instance, Markus and Kitayama (1991) have contrasted two different cultural perspectives on the self, namely, those that empha-

Table 7.1 Aspects of Body Experience[a]

Level of cognitive functioning[b]	Definition	Dimension[c]	Source(s)
Sensorimotor: *Body action*	Movement of body in relation to objects constituting the environment	Body action	
		• Sensorimotor palpation (e.g., tactile)	e.g., Santostefano (1978)
		• Body mobility	e.g., Schilder (1935)
		• Use of body-part-as-object	e.g., Kaplan (1983)
		• Tempo regulation	e.g., Santostefano (1978)
Perceptual: *Body*	Perception of body qua object	Body perception	
		• Geometric-technical properties of objects	
		○ Location of body in space	e.g., Wapner (1968)
		○ Size of body and body parts	e.g., Cash & Green (1986); Shontz (1969); Wapner, Werner, & Comalli (1958)
		○ Shape or form of body	e.g., Fallon & Rozin (1985)
		○ Body boundary	e.g., Fisher & Cleveland (1965)
		○ Organization of body parts (parts vs. whole)	e.g., Barton & Wapner (1965); Cash (1989)
		○ Judgments of physiological activation	e.g., Pennebaker (1982)
		• Physiognomic/expressive/dynamic/functional properties of objects	
		○ Phantom limbs	e.g., Critchley (1965); Simmel (1956)
		○ Active touching vs. passively being touched	e.g., Merleau-Ponty (1962); Schlater, Baker, & Wapner (1981)
		○ Pointing	e.g., Porzemsky, Wapner, & Glick (1965)

Conceptual: Body	"The picture of our own body which we form in our mind, that is to say the way in which the body appears to ourselves" (Schilder, 1935)	Conceptual aspects of body	
		• Body concept/image/representation (cognitive aspects)[a]	e.g., Crider (1981); Witkin (1965)
		○ Form level	
		○ Identity or role and sex differentiation	
		○ Level of detail	
		• Body fantasy	e.g., Shontz (1969, 1974)
		○ Body-esteem (cathexis), e.g., evaluation of attractiveness of body	e.g., Lerner (1969); Secord & Jourard (1953)
Conceptual: Body	"The self-concept or self-structure may be thought of an organized configuration of perceptions of the self which are admissible to awareness. It is composed of such elements as the perceptions of one's characteristics and abilities; the percepts and concepts of the self in relation to others and to the environment; the value qualities which are perceived as associated with experiences and object; and goals and ideals which are perceived as having positive or negative valence" (Rogers, 1951)	Conceptual aspects of self	
		• Self-concept/image/representation (cognitive aspects)[a]	e.g., Gellert (1975); Sandler & Rosenblatt (1962)
		• Self fantasy	
		• Self-esteem (cathexis)	e.g., Secord & Jourard (1953)
		○ Influence of body percept and body image, e.g., perception of psychological gender based on comparison of behavior with societal sex-roles stereotypes	
		• Self-understanding	e.g., Damon & Hart (1982)

[a] It should be recognized that the body plays a dual role both as an object of perception and as a bodily framework for the experience of objects (animate and inanimate) "out there" (Wapner & Werner, 1965). Our focus here is on the former.

[b] See Werner (1957a), Wapner (1969), and Wapner and Demick (1990).

[c] Our focus is developmental in a broad sense. It encompasses, e.g., ontogenesis, pathogenesis, and (drug-induced) primitive states ordered developmentally with relations among cognition and affect de-differentiated at the sensoriomotor level and differentiated at the perceptual and conceptual level (cf. Critchley, 1965; de Ajuriaguerra, 1965; DesLauriers, 1962; Fisher & Cleveland, 1968; Shontz, 1974; Striegel-Moore, Siberstein, & Rodin, 1993).

[d] Authors often use "concept/esteem/cathexis" interchangeably, connoting cognitive, affective, and valuative aspects. Here, "concept/image/representation" is restricted to cognitive (structural) aspects sof the body and self.

size the *individuality* of the self (e.g., most cultures in North America and Europe) versus those that stress the *connectedness of the self with other people* (e.g., many cultures in Asia, Latin America, and Africa). In the former, individuals generally define themselves in terms of abilities, feelings, goals, and inner traits; in the latter, people define themselves more in terms of their roles in social groups and their social relationships.

Further, researchers have documented that the actual words used to describe the self vary across cultures. For example, Hamaguchi (1985) has noted that the Japanese word for self, *jibun*, is defined as "one's share of the shared life space" and that the Japanese self is viewed not as a fixed set of traits but as something that changes in different relationships (cf. the American view of self that focuses on individuality). Such differences in cultural conceptions of the self (cf. Rosenberger, 1992) clearly make for cultural differences in action. For instance, research has shown that (1) whereas Americans are motivated in work and school by the possibilities of expressing their inner abilities, comparing favorably with others, and exerting personal control and power, the Japanese are motivated by the possibilities of fulfilling interpersonal obligations and pleasing their families and teachers (e.g., Bond, 1986; Yu, 1974) and (2) whereas Americans prefer task-oriented leaders who separate personal matters from work, the Japanese prefer demanding and father-like managers with a collective view of achievement (e.g., Hayashi, 1988; Markus & Kitayama, 1991).

Finally, in our own work on the impact of governmental legislation (automobile safety belts), we (Demick et al., 1992) have found that, whereas usage rates increased and remained constant in Hiroshima (98% compliance within 12 months), they decreased steadily in Massachusetts. Further, a significant number of Massachusetts participants voiced their concern that mandatory safety belt legislation was an invasion of privacy/infringement on human rights, which ultimately resulted in repeal of the legislation and further decrease in safety belt usage; no parallel phenomenon was manifest among the Japanese.

Taken together, these studies have supported the view that American culture illustrates those cultures that emphasize the *individuality* of the self, while Japanese culture is representative of those that highlight interpersonal *interdependence*. In our previously mentioned study, for example, American participants spoke of automobile safety belt usage (or nonusage) in terms of personal freedom, autonomy, self-reliance, being in control, and productivity, while their Japanese counterparts spoke of notions concerning *amae*, the concern for approval by others and by groups to which one belongs as well as "a worry that they may not be doing the right thing and thus are opening themselves to criticism or ridicule by others" (Reischauer, 1978, p. 143).

In contrast to this work that speaks to Japanese versus American differences in perception of self, Cousins (1989) has provided an alternative interpretation. Specifically, he asked Japanese and American college students to complete a noncontextualized Twenty Statements Test as well as a contextualized version asking for self-descriptions in various situations. Consistent with prior research, relative to Americans, the Japanese listed fewer abstract psychological attributes and more social roles on the noncontextualized version; however, they also provided more highly abstract global self-references (e.g., "I am a human being," "I am an organic form"). On the contextualized version, this finding was reversed: the Japanese scored higher on abstract psychological attributes than did the Americans, who qualified their self-descriptions. In other words, the Americans were significantly more likely than the Japanese to describe themselves in terms of internal traits that generalized across social settings. Cousins (1989) has interpreted these findings in light of the distinction between individualism (lacking in the Japanese responses by U.S. standards) and individuality (expressed by the Japanese within, rather than beyond, the social context). This has led him to argue for divergent cultural conceptions of the person rather than differential perception of self in the two cultures.

Paralleling our views on the multiple dimensions of body experience, it is also our contention that advances in theory and research will follow from consideration of the multiple dimensions of self experience. For example, cross-cultural research such as that of Cousins (1989; see also Trafimow, Triandis, & Goto, 1991) has the clear potential to sharpen distinctions among aspects of self experience (conceptions of self versus of persons) and its development.

Related Processes

Social psychologists have long considered cross-cultural issues in processes related to human interaction. That is, researchers (e.g., Birdwhistell, 1952, 1970; Hall, 1959, 1966, 1976) have suggested cross-cultural differences in general space usage (e.g., personal space) and in nonverbal communication (e.g., cognitive strategies). For example, Hall (1966, 1976) has observed greater personal space usage in Americans versus Italians (cultural intimacy) and the Japanese (architectural intimacy). Another recent study from our own laboratories (Toshima, Demick, Miyatani, Ishii, & Wapner, 1996) has demonstrated that, on a task of sequential cognitive activity (Stroop Color-Word Test), nonverbal strategic behaviors with molar movement (e.g., pointing to each word, rocking of body) were more frequent in Americans, while static nonverbal behaviors (e.g., holding hands at side of body) were more common among the Japanese (cf. Lewis, 1989).

Stemming from the early work in social psychology, environmental psychologists have more recently provided empirical data on cross-cultural differences in processes relevant to environment–behavior relations. Specifically, the processes that have been addressed include architectural form, crowding, privacy, and territoriality. Each will be briefly discussed in turn.

With respect to architectural design, Rapoport (1982) has documented that, whereas cultural diversity may characterize some features of communities, other features seem to possess near-universal meanings (e.g., religious buildings using height to express holiness and/or the idea of reaching toward the heavens). Whereas studies comparing the crowding experiences of different cultures are sorely lacking, Nasar and Min (1984) found that Mediterranean college students residing in a single North American dormitory room reported more crowding than Asian students, who reported more crowding than American students. Research on privacy (e.g., Altman, 1975; Patterson & Chiswick, 1981; Yoors, 1967) has argued that cultures are not different in the amount of privacy they desire but vary considerably in how they achieve that privacy (e.g., in Gypsy society, one does not wash his or her face in the morning if one does not wish to be social). Finally, research on territoriality (e.g., Campbell, Munce, & Galea, 1982; Edney & Jordan-Edney, 1974; Greenbaum & Greenbaum, 1981; Smith, 1981) has indicated that, whereas the question of whether some cultures are more territorial than others has not been answered clearly, cultures do differ in their expression of territoriality (e.g., Worchel & Lollis [1982] found that Americans removed litter placed on the sidewalks in front of their homes faster than homeowners in Greece).

Hypotheses

Based on the above literature reviews, it is hypothesized that there will be differences between Japanese and American individuals in aspects of body and self experience. These include sensorimotor aspects such as body action (e.g., Lewis, 1989; Toshima et al., 1996), perceptual aspects such as body perception (e.g., Hall, 1966; Nasar & Min, 1984), and conceptual aspects of both body and self (e.g., Lerner et al., 1980). Since sex differences have been implicated in the body and self experience particularly of the Japanese (e.g., Lerner et al., 1980), the present study also assesses whether sex and/or gender roles modify these potential cross-cultural relationships.

These and other relationships are assessed in the present investigation.

METHOD

Participants

The American sample consisted of 47 upper-level undergraduates (27 women and 20 men, mean age = 20.3 years), who participated in all aspects of the research. The Japanese sample was composed of 175 upper-level undergraduates. All 175 (119 women and 56 men, mean age = 20.1 years) filled out the questionnaires; of these 175, 41 (21 women and 20 men, mean age = 20.2 years) also completed the experimental tasks. There were no significant age differences among the three subsamples.

Materials

The battery of instruments consisted of the following:

1. *Body action: Tempo regulation.* As one exemplar of the sensorimotor aspect of body and self experience (body action), participants completed Santostefano's (1978) Tempo Regulation Test, which has been shown to possess adequate reliability and validity. Specifically, participants were asked to walk through a 25-foot-long maze at regular, slow, and fast speeds. These three individual measures have also been combined into a composite measure: slow-fast/regular.

2. *Body perception I: Body boundaries*. As a first example of body perception, participants were administered Wapner and Werner's (1965) apparent head size estimation task, which asked each participant, with eyes closed, to estimate how wide his or her head is as projected onto a mounted meter stick (two trials). Measurements were then subjected to the following formula: (mean apparent head size – mean objective head size/mean objective head size) × 100. Wapner and Werner have provided normative, developmental data on this measure; further, Demick and Wapner (1980) have demonstrated changes in psychiatric patients' apparent head size estimation as a function of the environmental relocation of their therapeutic community. The measure has been presumed to reflect the individual's perception of his or her body qua object or definiteness of body boundaries more generally.

3. *Body perception II: Subject–object distance*. As a second example of body perception, each participant completed Horowitz, Duff, and Stratton's (1964) body buffer zone task, which assessed the individual's experience of comfortable space surrounding his or her body in relation to the experimenter (two trials) and to an inanimate object (two trials). For each participant, two measures of self–object distance (mean distances between subject and experimenter and between subject and inanimate object) were computed. This task has been used in reliable and valid ways in much social psychological research (see Fisher, 1986, for a comprehensive review).

4. *Conceptual aspect of body I: Body representation*. Here, participants were asked simply to provide a self-drawing. Based on the original system proposed by Gellert (1968), drawings were scored for symmetry (0–17), proportionality (0–18), and total (0–33). Naglieri's (1988) quantitative system, which incorporates all of Gellert's aspects, was also employed (0–64).

5. *Conceptual aspect of body II: Body evaluation*. Each participant completed Secord and Jourard's (1953) Body-Cathexis Scale, assumed indicative of the individual's evaluation of his or her body with respect to other people and societal norms. Specifically, each participant was asked to provide a rating of satisfaction (from 1 = satisfied, to 5 = dissatisfied) with 44 different body parts. Ratings of the individual items as well as a mean overall rating served as the measures. This task has also been used extensively in social psychological research (see Fisher, 1986).

6. *Conceptual aspect of self I: Self-concept*. Participants were administered Wylie's (1974) Who Am I Test. This task required each individual to list responses to the brief question, "Who am I?" Measures included total number of responses; number of responses describing *private* aspects of the self (e.g., "I am assertive," "I am friendly); and number of responses describing *collective, social* aspects of the self (e.g., "I am a mother," "I am Roman Catholic"). This coding system, based on the work of Triandis (1989), has been used extensively in research (cf. Cousins, 1989).

7. *Conceptual aspect of self II: Self-esteem*. Much research (e.g., Jahoda, 1958; Rosenberg, 1965; Wylie, 1974) has persuasively demonstrated that the variable of self-esteem is a significant index of mental health and thus of general

adjustment. Here, participants' self-esteem was measured with Coopersmith's (1981) Self-Esteem Inventory (Adult Form), a 25-item scale with each item rated as "like me" or "unlike me." The sum of the number of self-esteem items answered in a manner keeping with positive self-esteem (i.e., negative items such as "I get upset easily at home" scored "unlike me"; positive items such as "If I have something to say, I usually say it" scored as "like me"), multiplied by 4, results in a global self-esteem score of 100. Coopersmith (1981) has reported extensive evidence for the reliability and validity of the (original) School Form; he has also documented that the total score correlation of the School Form with the Adult Form exceeds .80 for three samples of high school and college students ($N = 647$).

8. *Conceptual aspect of self III: Self evaluation.* Each participant also completed Secord and Jourard's (1953) Self-Cathexis Scale. Here, instead of rating satisfaction with body parts (Body-Cathexis Scale), the participant rated satisfaction with 29 aspects of self (e.g., first name, personality). Again, ratings of individual items and a mean overall rating served as the measures.

9. *Background information.* Finally, a brief questionnaire asked each participant for demographic and related information (e.g., age, sex, year in college, major).

Finally, participants were also administered E. Kaplan's (1983) Body-Part-As-Object Test (body action), Higgins' (1987) Test of Self-Understanding (conceptual aspect of self), and Bem's (1974) Sex-Role Inventory. These measures were not included in the present analysis because, respectively (1) there was a ceiling effect on the Body-Part-As-Object Test in both sociocultural contexts, (2) scoring for the Test of Self-Understanding was extremely complex and not easily translated for the Japanese experimenters, and (3) there were not enough participants in the American sample to obtain meaningful groupings of masculine, feminine, androgynous, and undifferentiated sex role types. These problems will be addressed in future research.

Statistical Analyses

The major dependent variables were treated by two-way analyses of variance (ANOVA) with sex (male, female) and sociocultural group (Japan, US) as between-subjects variables. Further, post hoc comparisons were employed to assess specific differences between groups.

RESULTS

Representative findings, organized around our conceptual framework, include the following.

Sensorimotor action. Analyses of tempo regulation test scores revealed that, relative to the Japanese, the Americans were significantly faster under all three conditions: slow—Japanese = 182.5 s, M Americans = 16.0 s, $F(1, 85) = 70.7, p <$

.001; fast—Japanese = 113.9 s, M Americans = 5.2 s, $F(1, 85) = 44.3, p < .001$; and normal—Japanese = 59.4 s, M Americans = 8.7 s, $F(1, 85) = 692.2, p < .001$.

There were no significant effects for sex of participant either alone or in interaction with the variable of sociocultural context. There was, however, a trend on the slow condition ($p < .09$) with males ($M = 117.9$ sec) tending to walk slower than females ($M = 71.2$ sec.).

Body perception. The analyses concerning subject–object distance were as follows. First, relative to Americans ($M = 169.7$ mm), the Japanese ($M = 39.7$ mm) stood significantly closer to the object, $F(1, 85) = 12.3, p < .001$. Second, the Japanese ($M = 67.5$ mm) also exhibited less personal space when approaching the experimenter than did Americans ($M = 196.7$ mm). Sex of participant played no role whatsoever.

With respect to body boundary diffuseness, analyses of head size estimation scores revealed that, relative to Americans ($M = 420.5$), the Japanese were significantly more accurate ($M = 42.7$), $F(1, 79) = 164.1, p < .001$. Again, there was no effect involving sex of participant.

Conceptual aspects of body. On the self-drawing, there was no difference between the Japanese and Americans in terms of the symmetry of their representations. There was, however, a significant main effect of sociocultural group on the total scores. That is, the Japanese ($M = .97$) included significantly more details in their self-drawings than Americans ($M = .85$), $F(1, 85) = 4.86, p < .05$. There was also a trend toward an interaction of sex of participant and sociocultural context ($p < .09$) with American males providing the fewest details of all four groups.

Further, analysis of the overall Body-Cathexis Scale Scores revealed that the Japanese ($M = 3.0$) were generally less satisfied with their bodies than were the Americans ($M = 3.4$), $F(1, 219) = 23.1, p < .001$. There were also trends for sex of participant (with males exhibiting slightly more satisfaction) and for the interaction between sex of participant and sociocultural context (with Japanese females expressing the least satisfaction).

Conceptual aspects of self. Analysis of the Coopersmith scores revealed that the Japanese ($M = 62.2$) reported significantly lower self-esteem than did the Americans ($M = 67.8$), $F(1, 219) = 4.0, p < .01$. Sex of participant had no effect.

Finally, analysis of the overall Self-Cathexis Scale scores paralleled those obtained for the overall Body-Cathexis Scale scores. Again, the Japanese ($M = 3.2$) were generally less satisfied with their selves than were the Americans ($M = 3.8$), $F(1, 219) = 36.7, p < .001$. In contrast to body-cathexis, here sex of participant played no role whatsoever.

DISCUSSION

The significant findings, in keeping with expectation, speak to the heuristic value of the holistic, developmental, systems-oriented approach (e.g., Wapner, 1981, 1987; Wapner & Demick, 1990) to a general person-in-environment system conceptualization and, more specifically, to the individual's experience of body and of self. That is, unlike many investigators who have focused on one or two isolated

variables relevant to body/self experience, our focus on the sensorimotor, perceptual, and conceptual aspects of body and self experience (after Wapner, 1969, and
Werner, 1957) has shed light on the complexity of the processes involved in this
aspect of experience.

There was strong evidence to support the main hypothesis of differential body
and self experience between the Japanese and Americans. These findings—especially those concerning tempo regulation, apparent head width, and self-drawings—can be explained by differences in the general nature of the participants in
the two sociocultural contexts. That is, relative to Americans, the Japanese may
possess a more articulated body concept (exhibited by lesser body boundary diffuseness on apparent head size estimation and more detail on self-drawing) and a
less hurried conceptual tempo. These findings are consistent with earlier cross-cultural research (e.g., Lewis, 1989; Toshima et al., 1996) documenting that Americans
at all stages of the life span (infants, children, and adults) use molar body movements to a significantly greater extent than the Japanese.

The findings on body evaluation (Body-Cathexis Scale), self evaluation (Self-
Cathexis Scale), and self-esteem (Self-Esteem Inventory) are in keeping with Lerner
et al.'s (1980) earlier work, which has documented that Japanese adolescents have
lower self-esteem and less favorable views of their bodies' attractiveness and effectiveness than American adolescents. Thus, the present study extends these specific
findings to young adulthood. However, one must exercise caution in interpreting
these data. That is, it is entirely possible that the Japanese are less concerned (rather
than more distressed) than Americans with the appearance of their bodies. This
notion gains support in light of other data that, relative to Americans, the Japanese
utilize lesser degrees of personal space and have a more outer (connectedness)
than inner (individuality) orientation. Such possibilities seem worthy of further
empirical inquiry.

The relative lack of significant sex differences—either alone or in interaction
with sociocultural context—is surprising and deserves some speculation. That is,
unlike Lerner et al.'s (1980) study, we found only a trend for Japanese females to
feel less satisfied with their bodies (Body-Cathexis Scale). In the American sample,
we found only tendencies for males to provide less detail in their self-representations and to walk slightly more slowly than females (Tempo Regulation, slow
condition). The lack of significant sex differences with respect to conceptual aspects of body and self, however, perhaps speaks to the changing nature of sex role
stereotypes both here and abroad.

The present study also has methodological implications. That is, our analysis
emphasizes the theoretical importance of recognizing that different representational media optimally tap different aspects of experience (here, body and self
experience) and that a fuller picture of experience emerges when multiple methods
for assessing experience are employed (see also Dandonoli, Demick, & Wapner,
1990; Demick, 1993; Demick, Hoffman, & Wapner, 1985). It is worth emphasizing
that different media or, more generally, different subject–environment conditions
of assessment involve substantive issues that must be handled theoretically. Thus,
rather than assuming that given findings must be robust enough to be present
independent of the mode of assessment, it is necessary to account theoretically for

the presence of findings using one set of testing conditions and/or media and the absence of such findings using another. Indeed, to obtain cross-cultural differences for one kind of medium and not another throws light more generally on the nature of cross-cultural differences in body and self experience. Here, we might have obtained a different picture of Japanese body experience had our study been limited only to the assessment of conceptual aspects of body (cf. Lerner et al., 1980).

In line with this, the present investigation is not without its limitations. For example, future research aimed at further exploring the relationships uncovered here would do well to include additional measures of body and self experience, additional participants from both (and other) sociocultural contexts, assessment of the effects of gender role as well as sex of participant on body and self experience in both cultures, discriminant function analyses to assess which variables best differentiate the groups, and attempts at replication of our confirmatory factor analysis supporting our levels conceptualization of body and self experience.

The present study has implications for the area of environmental design. That is, the findings on personal space and body boundary diffuseness (less in the Japanese vs. Americans) appear to support the claim that the Japanese require significantly less space than Americans in, for example, their public and private spaces. The findings on differences in tempo regulation (with the Japanese exhibiting slower speeds than Americans) might suggest the increased usage of objects of contemplation (e.g., gardens, statues) in Japanese versus American environments. Given the Japanese focus on outer-directedness (connectedness to others), environmental design in Japan might do well to maximize social interaction at least in designated places. That the Japanese exhibited a greater degree of detail in their self-representations, less subject–object distance, less body boundary diffuseness, and hence a more vigilant cognitive style than Americans suggests that the use of detail in environmental design might be better suited to Japanese than to American culture. Thus, the correlates of body and self experience as related to environmental design issues appear worthy of future empirical study.

Finally, taken as a whole, this study also makes a contribution to the fields of both environmental and developmental psychology. For environmental psychology, the present investigation provides a theoretical framework and classificatory system of heuristic value for the study of body and self experience and, perhaps more generally, for the study of environmental experience. For the field of developmental psychology, the findings concerning the differential experience of body and self in the Japanese versus Americans add to the powerful body of literature on cultural relativism; further, the conceptualization of body and self experience provided here provides a developmental framework for use in future research. Finally, the large body of significant findings points to the mutual value of using methodology and conceptualization from the one field to advance the methodology and conceptualization of the other. In this way, we will be one small step closer toward perceiving psychology and having psychology be perceived by other disciplines as a unified science that truly treats the complex character of everyday life functioning.

ACKNOWLEDGMENTS

We would like to extend our sincere thanks to Clif Hughes and Georgia Madway for their expert assistance in data analysis.

REFERENCES

Altman, I. (1975). *The environment and social behavior: Privacy, personal space, territoriality and crowding.* Monterey, CA: Brooks/Cole.

Barton, M. I., & Wapner, S. (1965). Apparent length of body parts attended to separately and in combination. *Perceptual and Motor Skills, 20,* 904.

Baumeister, R. F. (1991). *Escaping the self: Alcoholism, spirituality, masochism, and other flights from the burden of selfhood.* New York: Basic Books.

Bem, S. L. (1974). The measurement of psychological androgyny. *Journal of Consulting and Clinical Psychology, 42,* 155–162.

Birdwhistell, R. L. (1952). *Introduction to kinesics.* Louisville, KY: University of Louisville Press.

Birdwhistell, R. L. (1970). *Kinesics and context.* Philadelphia: University of Philadelphia Press.

Bond, M. H. (1986). *The psychology of the Chinese people.* New York: Oxford University Press.

Campbell, A. C., Munce, S., & Galea, J. (1982). American gangs and British subcultures: A comparison. *International Journal of Offender Therapy and Comparative Criminology, 26,* 76–89.

Cash, T. F. (1989). Body-image affect: Gestalt versus summing the parts. *Perceptual and Motor Skills, 69,* 17–18.

Cash, T. F. (1990). The psychology of physical appearance: Aesthetics, attributes, and images. In T. F. Cash & T. Pruzinsky (Eds.), *Body images: Development, deviance, and change* (pp. 51–79). New York: Guilford Press.

Cash, T. F., & Green, G. K. (1986). Body weight and body image among college women: Perception, cognition, and affect. *Journal of Personality Assessment, 50*(2), 290–301.

Coopersmith, S. (1981). *Self-esteem inventories.* Palo Alto, CA: Consulting Psychologists Press.

Cousins, S. (1989). Culture and selfhood in Japan and the U.S. *Journal of Personality and Social Psychology, 56,* 124–131.

Crider, C. (1981). Children's conception of the inside of the body. In R. Bibace & M. Walsh (Eds.), *Children's conceptions of health, illness, and bodily functions.* San Francisco: Jossey-Bass.

Critchley, M. (1965). Disorders of corporeal awareness in parietal disease. In S. Wapner & H. Werner (Eds.), *The body percept* (pp. 68–81). New York: Random House.

Damon, W., & Hart, D. (1982). The development of self-understanding from infancy through adolescence. *Child Development, 53,* 841–864.

Dandonoli, P., Demick, J., & Wapner, S. (1990). Physical arrangement and age as determinants of environmental representation. *Children's Environments Quarterly, 7*(1), 26–36.

de Ajuriaguerra, J. (1965). Discussion. In S. Wapner & H. Werner (Eds.), *The body percept* (pp. 82–106). New York: Random House.

Demick, J. (1993). Adaptation of marital couples to open versus closed adoption: A preliminary investigation. In J. Demick, K. Bursik, & R. DiBiase (Eds.), *Parental development* (pp. 175–201). Hillsdale, NJ: Erlbaum.

Demick, J., & Rivers, S. (1996). Aspects of body experience: A confirmatory factor analytic study. Manuscript in preparation.

Demick, J., & Wapner, S. (1980). Effects of environmental relocation on members of a psychiatric therapeutic community. *Journal of Abnormal Psychology, 89,* 444–452.

Demick, J., & Wapner, S. (1987, November). *A holistic, developmental approach to body experience.* The Body and Literature: An Interdisciplinary Conference, SUNY-Buffalo, NY.

Demick, J., & Wapner, S. (1996). A holistic, developmental, systems-oriented approach to body experience. Manuscript in preparation.

Demick, J., Hoffman, A., & Wapner, S. (1985). Residential context and environmental change as determinants of urban experience. *Children's Environments Quarterly, 2*(3), 44–54.

Demick, J., Inoue, W., Wapner, S., Ishii, S., Minami, H., Nishiyama, S., & Yamamoto, T. (1992). Cultural differences in impact of governmental legislation: Automobile safety belt usage. *Journal of Cross-Cultural Psychology, 23*(4), 468–487.

DesLauriers, A. (1962). *The experience of reality in childhood schizophrenia.* New York: International University Press.

Edney, J. J., & Jordan-Edney, N. L. (1974). Territorial spacing on a beach. *Sociometry, 37,* 523–527.

Fallon, A. E., & Rozin, P. (1985). Sex differences in perceptions of desirable body shape. *Journal of Abnormal Psychology, 94*(1), 102–105.

Fisher, S. (1986). *Development and structure of the body image* (Vols. 1–2). Hillsdale, NJ: Erlbaum.

Fisher, S., & Cleveland, S. E. (1965). Personality, body perception, and body image boundary. In S. Wapner & H. Werner (Eds.), *The body percept* (pp. 48–67). New York: Random House.

Fisher, S., & Cleveland, S. E. (1968). *Body image and personality.* New York: Dover.

Gellert, E. (1968). Comparison of children's self-drawings with their drawings of other persons. *Perceptual and Motor Skills, 40,* 307–324.

Gellert, E. (1975). Children's constructions of their self-images. *Perceptual and Motor Skills, 40,* 307–324.

Greenbaum, P. E., & Greenbaum, S. D. (1981). Territorial personalization: Group identity and social interaction in a Slavic-American neighborhood. *Environment and Behavior, 13,* 574–589.

Hall, E. T. (1959). *The silent language.* Garden City, NY: Doubleday.

Hall, E. T. (1966). *The hidden dimension.* New York: Doubleday.

Hall, E. T. (1976). *Beyond culture.* Garden City, NY: Doubleday.

Hamaguchi, E. (1985). A contextual model of the Japanese: Toward a methodological innovation in Japan studies. *Journal of Japanese Studies, 11,* 289–321.

Hayashi, C. (1988). *National character of the Japanese.* Tokyo: Statistical Bureau, Japan.

Higgins, E. T. (1987). Self-discrepancy: A theory relating self and affect. *Psychological Review, 94,* 319–340.

Horowitz, M., Duff, D., & Stratton, L. (1964). Body buffer zone: Explorations of personal space. *Archives of General Psychology, 11,* 651–656.

Jahoda, M. (1958). *Current concepts of positive mental health.* New York: Basic Books.

Kaplan, E. (1983). Process and achievement revisited. In S. Wapner & B. Kaplan (Eds.), *Toward a holistic developmental psychology* (pp. 143–156). Hillsdale, NJ: Erlbaum.

Lerner, R. M. (1969). The development of stereotyped expectancies of body build relations. *Child Development, 40,* 137–141.

Lerner, R. M., Iwawaki, S., Chihara, T., & Sorell, G. T. (1980). Self-concept, self-esteem, and body attitudes among Japanese male and female adolescents. *Child Development, 51,* 847–855.

Lewis, M. (1989). Culture and biology: The role of temperament. In P. Zelazo & R. Barr (Eds.), *Challenges to developmental paradigms* (pp. 203–223). Hillsdale, NJ: Erlbaum.

Markus, H., & Kitayama, S. (1991). Culture and the self: Implications for cognition, emotion, and motivation. *Psychological Review, 98,* 224–253.

Merleau-Ponty, M. (1962). *Phenomenology of perception.* London: Routledge & Kegan Paul.

Naglieri, J. A. (1988). *Draw a person: A quantitative scoring.* San Antonio, TX: Psychological Corporation.

Nasar, J. L., & Min, M. S. (1984, August). *Modifiers of perceived spaciousness and crowding: A cross-cultural study.* Paper presented at the annual meeting of the American Psychological Association.

Patterson, A. H., & Chiswick, N. R. (1981). The role of the social and physical environment in privacy maintenance among the Iban of Borneo. *Journal of Environmental Psychology, 1,* 131–139.

Pennebaker, J. W. (1982). *The psychology of physical symptoms.* New York: Springer-Verlag.

Porzemsky, J., Wapner, S., & Glick, J. A. (1965). Effect of experimentally induced self-object cognitive attitudes on body and object perception. *Perceptual and Motor Skills, 21,* 187–195.

Rapoport, A. (1982). *The meaning of the built environment: A nonverbal communication approach.* Beverly Hills, CA: Sage.

Reischauer, E. O. (1978). *The United States and Japan* (3rd ed.). Cambridge, MA: Harvard University Press.

Rogers, C. R. (1951). *Client-centered therapy: Its current practice, implications, and theory*. Boston: Houghton Mifflin.

Rosenberg, M. (1965). *Society and the adolescent self-image*. Princeton, NJ: Princeton University Press.

Rosenberger, N. R. (Ed.). (1992). *Japanese sense of self*. New York: Cambridge University Press.

Sandler, J., & Rosenblatt, B. (1962). The concept of the representational world. *Psychoanalytic Study of the Child, XVII*, 128–145.

Santostefano, S. (1978). *A biodevelopmental approach to clinical child psychology: Cognitive controls and cognitive control therapy*. New York: Wiley.

Schilder, P. (1935). *The image and appearance of the human body*. London: Kegan, Trench, Turbner.

Secord, P., & Jourard, S. (1953). The appraisal of body cathexis: Body cathexis and the self. *Journal of Consulting Psychology, 17*, 343–347.

Schlater, J. A., Baker, A. H., & Wapner, S. (1981). Apparent arm length with active versus passive touch. *Bulletin of the Psychonomic Society, 18*, 151–154.

Shontz, F. C. (1969). *Perceptual and cognitive aspects of body experience*. New York: Academic Press.

Shontz, F. C. (1974). Body image and its disorders. *International Journal of Psychiatry in Medicine, 5*, 461–472.

Simmel, M. L. (1956). Phantoms in patients with leprosy and in elderly digital amputees. *American Journal of Psychology, 69*, 529–545.

Smith, H. W. (1981). Territorial spacing on a beach revisited: A cross-national exploration. *Social Psychology Quarterly, 44*, 132–137.

Striegel-Moore, R. H., Silberstein, L. R., & Rodin, J. (1993). The social self in bulimia nervosa: Public self-consciousness, social anxiety, and perceived fraudulence. *Journal of Abnormal Psychology, 102*(2), 297–303.

Thompson, J. K., & Dolce, J. J. (1989). The discrepancy between emotional versus rational estimates of body size, actual size, and ideal body ratings: Theoretical and clinical implications. *Journal of Clinical Psychology, 43*(3), 473–478.

Thompson, J. K., Penner, L. A., & Altabe, M. N. (1990). Procedures, problems, and progress in the assessment of body images. In T. F. Cash & T. Pruzinsky (Eds.), *Body images: Development, deviance, and change* (pp. 21–48). New York: Guilford Press.

Toshima, T., Demick, J., & Miyatani, M., Ishii, S., & Wapner, S. (1996). Cross-cultural differences in processes underlying sequential cognitive activity. *Japanese Psychological Research*, 38(2), 90 96.

Triandis, H. C. (1989). Cross-cultural studies of individualism and collectivism. *Nebraska Symposium on Motivation, 37*, 41–134.

Trafimow, D., Triandis, H. C., & Goto, S. G. (1991). Some tests of the distinction between the private self and the collective self. *Journal of Personality and Social Psychology, 60*, 649–655.

Wapner, S. (1968). Age changes in perception of verticality and of the longitudinal body axis under body tilt. *Journal of Experimental Child Psychology, 6*, 543–555.

Wapner, S. (1969). Organismic-developmental theory: Some applications to cognition. In J. Langer, P. Mussen, & M. Covington (Eds.), *Trends and issues in developmental psychology* (pp. 38–67). New York: Holt, Rinehart & Winston.

Wapner, S. (1981). Transactions of persons-in-environments: Some critical transitions. *Journal of Environmental Psychology, 1*, 223–239.

Wapner, S. (1987). A holistic-, developmental-, systems-oriented environmental psychology: Some beginnings. In D. Stokols & I. Altman (Eds.), *Handbook of environmental psychology: Vol. 2* (pp. 1433–1465). New York: Wiley.

Wapner, S., & Demick, J. (1990). Development of experience and action: Levels of integration in human functioning. In G. Greenberg & E. Tobach (Eds.), *Theories of the evolution of knowing: The T. C. Schneirla conference series* (pp. 47–68). Hillsdale, NJ: Lawrence Erlbaum Associates.

Wapner, S., & Werner, H. (1965). An experimental approach to body perception from the organismic-developmental point of view. In S. Wapner & H. Werner (Eds.), *The body percept* (pp. 9–25). New York: Random House.

Wapner, S., McFarland, J. H., & Werner, H. (1963). Effect of visual spatial context on perception of one's own body. *British Journal of Psychology, 54*(1), 41–49.

Wapner, S., Werner, H., & Comalli, P. E. (1958). Effect of enhancement of head boundary on head size and shape. *Perceptual and Motor Skills, 8*, 319–325.

Werner, H. (1957). *Comparative psychology of mental development* (rev. ed.). New York: International Universities Press (originally published in 1940).

Witkin, H. A. (1965). Development of the body concept and psychological differentiation. In S. Wapner & H. Werner (Eds.), *The body percept* (pp. 26–47). New York: Random House.

Worchel, S., & Lollis, M. (1982). Reactions to territorial contamination as a function of culture. *Personality and Social Psychology Bulletin, 8,* 370–375.

Wylie, R. (1974). *The self-concept: A review of methodological considerations and measuring instruments* (Vol. 1). Lincoln, NE: University of Nebraska Press.

Yoors, J. (1967). *The gypsies*. New York: Simon and Schuster.

Yu, E. S. H. (1974). *Achievement motive, familism, and hsiao: A replication of McClelland-Winterbottom studies*. Dissertation Abstracts International, 35, 593A. (University Microfilms No. 74–14, 942.

Chapter **8**

CROSS-CULTURAL SURVEY ON COLOR PREFERENCES IN THREE ASIAN CITIES

Comparisons among Tokyo, Taipei, and Tianjin

Ichiro Soma and Miho Saito

INTRODUCTION

Several factors are said to be related to color preference: age, sex, and geographical area of residence. Although there have been numerous studies of age and sex differences in color preference, very few have concentrated on geographical regions, especially from a cross-cultural perspective.

Eysenck (1941) suggested that there was a general order of preference for fully saturated hues as follows: blue, red, green, purple, orange, and yellow. Choungourian (1968) reported preferences for eight Ostwald hues (red, orange, yellow, yellow green, green, blue green, blue, purple) in 160 American, Lebanese, Iranian, and Kuwaiti university students in Beirut and consequently concluded that cultural variables determined color preferences. A recent factor analytic study by Adams and Osgood (1973) found similarities in feelings about colors among 23 cultural groups. Saito (1981) demonstrated cross-cultural similarities and differences in color preference among nine cultural groups: Americans, Germans, Danes, Australians, Papua New Guineans, South Africans, Japanese-Americans living in the United

Ichiro Soma and **Miho Saito** • School of Human Sciences, Waseda University, Saitama 359, Japan.

Handbook of Japan–United States Environment–Behavior Research: Toward a Transactional Approach, edited by Seymour Wapner, Jack Demick, Takiji Yamamoto, and Takashi Takahashi. Plenum Press, New York, 1996.

States, foreigners living in Japan, and Japanese. Specifically, 400 subjects were asked to choose the colors that they liked and disliked among 65 colored chips. Results indicated that vivid blue was the only color that was commonly preferred by all groups. Other color preferences were found to be different among the groups, leading Saito to conclude that cultural variables are indeed involved in color preference. One very significant finding in this study was the distinct Japanese preference for white. Twenty-five percent of the Japanese subjects selected white as their first, second, or third choice, whereas such a preference could not be observed in other countries. This strong preference for white in Japan has been subsequently confirmed by Saito (1981), Saito, Tomita, & Kogo (1991), and Saito, Tomita, & Yamashita (1991). Further, to investigate whether this tendency is unique to Japan or may be observed in other Asian countries, Saito (1992) replicated the above study in Korea and compared the findings with a Japanese control group: preference for white was observed in Korea as well as in Japan. It was suggested that the geographical and cultural proximity of Japan and Korea was the reason for the similarity.

The above comparative study showed that a strong preference for white was common to both Japanese and Korean subjects. If, as suggested, the preference for white is based on similarities in cultural background brought about by geographical proximity, then such a tendency should also be seen among other Asian countries where there has been a long history of cultural exchange. To demonstrate that such environmental aspects as cultural similarity and geographical proximity are important factors affecting the preference for white, the above study was extended to two other Asian areas: Taipei and Tianjin.

The main purpose of this study is to compare general color preference tendencies in these three Asian areas with an emphasis on the preference for white.

METHOD

Subjects

The total number of subjects was 474. Each sample population contained almost an equal number of subjects. The subjects in Tokyo (88 male/72 female) were 160 university students. The average age was 19.6 years. The 156 subjects in Taipei (102 male/54 female) were mostly university students, whose average age was 24.1 years. The 158 subjects in Tianjin (69 male/89 female) were also mostly university students (average age = 22.8 years).

Stimulus

The stimulus was a color chart with 77 colored chips: 70 chromatic colors, 5 achromatic colors, silver, and gold. The colors were arranged horizontally by tone (seven tones: pale, light-grayish, dull, light, vivid, deep, dark) and vertically by hue (10 hues: R, YR, Y, GY, G, BG, B, PB, P, RP). The achromatic colors were white, light gray, medium gray, dark gray, and black. Gold and silver were included as metallic colors.

Procedures

While looking at the color chart, subjects were asked to select their three most highly preferred colors and three least preferred colors in order of preference and nonpreference, respectively, and to state the reasons for their choices. In most of the cases, the inquiries were carried out face-to-face. The color chart was observed under either fluorescent light or daylight from northern windows.

RESULTS

General Order of Preference

Figure 8.1 displays the general orders of preference and nonpreference for colors in each area according to how frequently each color was selected as liked or disliked. As indicated in Figure 8.1, white as well as vivid blue, vivid green, light violet, light blue were commonly preferred in all three areas. Pale colors (such as pale sky, pale green, pale greenish sky) were preferred in Tokyo. Vivid purple ranked as both popular (8.1%) and unpopular (7.5%) among the people in Tokyo. Purplish colors (e.g., light violet, vivid violet, pale lavender, light purple) were highly preferred in Taipei. A particular preference for vivid green was observed in Tianjin. Although white ranked third in Tianjin and second in the other two areas, the relative frequency of selection was the highest among the subjects in Tianjin (Tianjin: 29.7%; Taipei: 19.2%; Tokyo: 17.5%). In Tianjin, the frequency with which the top three colors (vivid green, vivid blue, and white) were chosen accounted for about one-third of the total frequency.

Olive, dark gray, gold, and dark yellowish brown were commonly disliked in all three areas. In general, colors of dark tone were unpopular. As seen in Table 8.1, black was not only popular (17.1%) but also unpopular (19.0%) in Tianjin. Grayish pink and grayish yellow-green were particularly unpopular in Tokyo. Silver was disliked in both Taipei and Tianjin; however, it was a preferred color among the subjects in Tokyo (ranked 10th in preference order).

Comparison of the Relative Frequencies of Color Selection as Grouped by Hue and by Tone

The color chips on the color chart were arranged systematically by hue and by tone, and a comparison was made between the relative frequencies of selection among hue and tone groups. Figure 8.2, for example, compares how frequently colors in each hue group were chosen as preferred. Statistically significant differences were uncovered by chi-square tests. The popularity of BG ($p < .001$) and B among the subjects in Tokyo was mainly related to the preference for pale greenish sky (ranked 7th in preference order), vivid blue-green (10th), pale sky (4th), deep

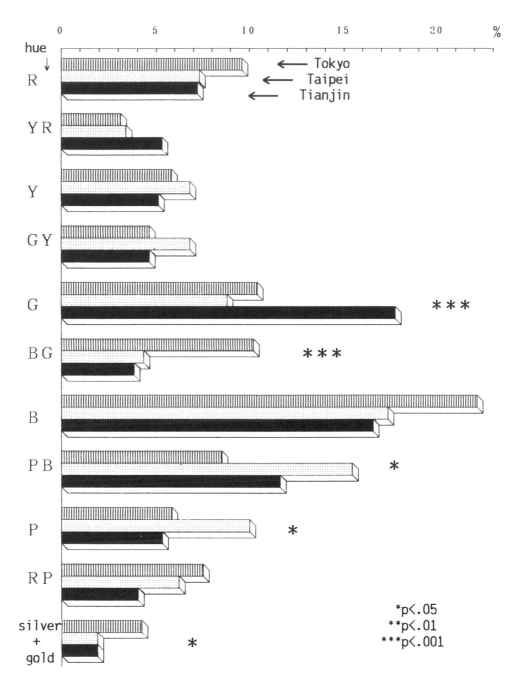

FIGURE 8.1. Order of Preference and Nonpreference.

order	like						dislike					
	TOKYO		TAIPEI		TIANJIN		TOKYO		TAIPEI		TIANJIN	
1st	vivid-blue	33.1%	vivid-blue	23.1%	vivid-green,	34.2%	olive	24.2%	olive	27.6%	dark gray	23.4%
2nd	white	17.5%	white	19.2%	vivid-blue	34.2%	gold	16.9%	gold	23.7%	olive, gold	22.8% 22.8%
3rd	vivid red	10.6%	light-violet	12.8%	white	29.7%	dark red	11.3%	dark gray	22.4%		
4th	vivid green,	10.0%	vivid violet,	12.2%	vivid-violet,	17.1%	dark gray, grayish-pink (ltq24)	10.0%	dark red	19.9%	black	19.0%
5th	pale sky	10.0%	pale lavender	12.2%	black	17.1%		10.0%	dark blue	17.3%	silver	13.9%
6th	deep-blue	9.4%	pale sky, light-blue, black	10.9% 10.9% 10.9%	vivid red	14.6%	grayish-pink (ltq2)	9.4%	dark-greenish-blue, dark-yellowish-brown	13.5% 13.5%	dark blue	13.3%
7th	pale-green, light-violet	8.8% 8.8%			light blue	10.1%	medium-gray	8.8%			olive-green, dark-greenish-blue, dark red-purple	10.8% 10.8% 10.8%
8th	pale-greenish sky	8.8%					dark-yellowish-brown	8.1%	silver	12.2%		
9th			vivid-green	10.3%	light-green, light violet, vivid yellow	9.5% 9.5% 9.5%			dark red-purple	10.3%		
10th	light-green, light-blue, vivid-blue-green, vivid-purple, silver	8.1% 8.1% 8.1% 8.1% 8.1%	light-purple	9.0%			dark blue, grayish-yellow-green, vivid purple, beige	7.5% 7.5% 7.5% 7.5%	olive-green	7.7%	dark-yellowish-brown	10.1%

FIGURE 8.2. Comparison between the three areas of the relative frequencies of preferred color choices by hue.

blue (6th) and light blue (10th). In Taipei, PB ($p<.05$) and P ($p<.05$) were highly preferred. Light violet (placed 3rd in preference ranking), vivid violet (4th), pale lavender (4th), light purple (10th), and deep violet were responsible for this result. PB was the favorite hue for the subjects in Tianjin as well (vivid violet, light violet, and pale lavender were highly preferred). The most preferred hue in Tianjin was G ($p<.001$). The high preferences for vivid green, the most preferred color, light green (8th), and pale green contributed to this finding. Among the metallic colors ($p<.05$), silver was especially preferred by the people in Tokyo (10th).

As indicated in Figure 8.3, the people in Tokyo and Taipei have quite similar preferences with respect to tones. Among the areas that were examined, Tianjin was particularly distinctive in its preference for vivid tones ($p<.001$). For instance, half of the vivid colors, that is, vivid green, vivid blue, vivid red, vivid yellow, and vivid violet, ranked 10th or higher. Pale tone ($p<.01$) was much more popular in Tokyo and Taipei.

In sum, subjects in Tokyo preferred pale colors in bluish and greenish hues. People in Taipei preferred pale and light colors in a purplish hue. Achromatic colors ($p<.001$) were popular in Tianjin and in Taipei: both groups liked both black (4th in Tianjin, 6th in Taipei) and white. Statistically significant differences among the

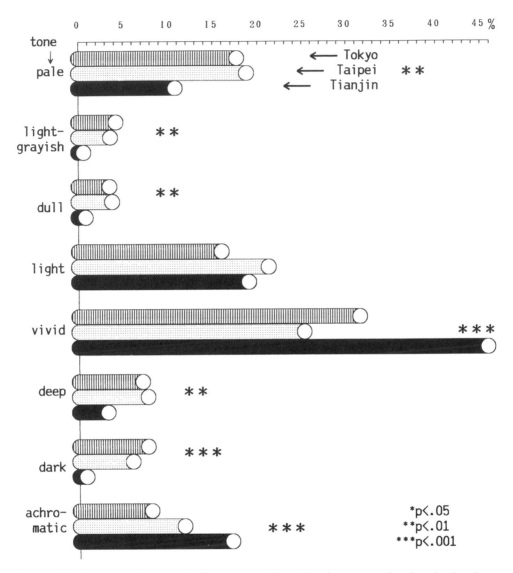

FIGURE 8.3. Comparison between the three areas of the relative frequencies of preferred color choices by tone.

three areas were uncovered by chi-square tests with respect to light-grayish ($p<.01$), dull ($p<.01$), deep ($p<.01$), and dark ($p<.001$) tones.

The results of color selection were also analyzed collectively by the chi-square test to investigate whether the general preference by hue and tone differed among the three areas. The outcomes were statistically significant in all the categories (hue: like, $df=22$, $p<.001$; hue: dislike, $df=22$, $p<.001$; tone: like, $df=16$, $p<.001$; tone: dislike, $df=16$, $p<.001$). This means that differences exist in each area in the tendencies of color preferences by hues and tones.

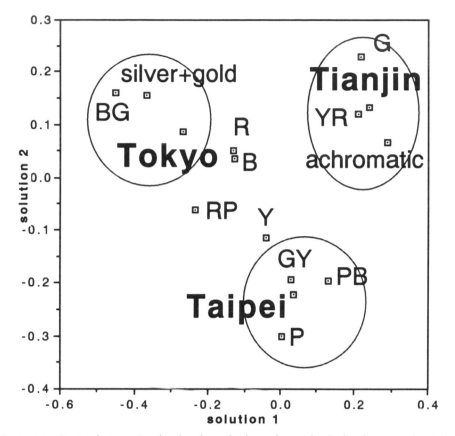

FIGURE 8.4. Scatter diagram. Results of analysis of color preference by dual scaling according to hue.

Mathematical Analysis by Dual Scaling

To understand better the color preferences of each area, dual scaling (Nishisato, 1980) was applied to the results according to hue and tone. Each delta partial (the total variance accounted for) for solution 1 is: hue (like), 64.50%; hue (dislike), 75.60%; tone (like), 87.91%; and tone (dislike), 82.49%. As displayed in Figure 8.4, hue preference in Tokyo was mainly characterized by the popularity of BG and silver; they were grouped relatively close together to Tokyo. Tianjin is distinctive in preferring G, YR, and achromatic colors. Preference for P, PB, and GY (popularity of vivid yellow-green and deep yellow-green were responsible) marked the results in Taipei.

As shown in Figure 8.5, the preference in Tianjin for vivid tone as well as achromatic colors was quite distinctive. The tendency of Taipei for not preferring the colors in dark tone was apparent. The people in Tokyo seemed to be comparatively characterized by the nonpreference for the colors in light-grayish, vivid, pale, and light tones.

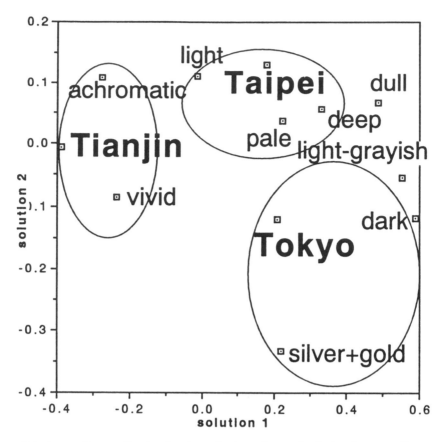

Figure 8.5. Scatter diagram. Results of analysis of color preference by dual scaling according to tone.

DISCUSSION

The results of this study revealed differences in color preferences among three areas when comparisons were carried out according to hue and tone. However, there were also some similarities in color preference besides the preference for white. Vivid blue, vivid green, light violet, and light blue were commonly preferred among the three groups. Vivid blue was also highly preferred in all the countries and groups where surveys have been carried out by Saito (1981, 1990). As a result, one might conclude, as did Birren (1950), that blue is a universal favorite.

In the present study, most of the subjects associated the image of vivid blue with the color of the sky and sea and with spacious images. The people in Tokyo preferred the color also because it was refreshing, beautiful, bright, transparent, and so on. For the people in Taipei, the associative images of vivid blue were light, spacious, lively, free, refreshing, and elegant. For the people in Tianjin, it suggested both spaciousness and open-mindedness. They also thought the color was deep, quiet, and pleasant.

The distinctive preference in Tianjin for the G hue, as mentioned before, was seen in their strong preference for vivid green. For the subjects in Tianjin, vivid green meant the symbol and power of life. The image of this color was also lively, refreshing, natural, and pleasant. Vivid colors were favored by the people in Tianjin in general because the colors looked lively, passionate, and powerful. Purplish colors (e.g., light violet, vivid violet, pale lavender, light purple) were highly preferred in Taipei. The stated reasons for the choice were that these colors were peaceful, gentle, elegant, refined, noble, and romantic. Pale tone colors were preferred in Taipei and Tokyo. For the people in Taipei, they looked beautiful, soft, noble, calm, and pleasant. Tokyo subjects liked the pale tone colors because they were gentle, beautiful, pretty, and refreshing.

Olive, dark gray, gold, and dark yellowish brown were disliked in all three areas. Generally, dark tone colors were unpopular because they seemed dirty, dark, heavy, and muddy. Some subjects in Tianjin stated that they disliked olive because it reminded them of the mud after a big flood. This indicates the involvement of an environmental factor in the subjects' attitudes toward a certain color. Grayish colors like grayish pink, and grayish yellow-green were particularly unpopular in Tokyo because they looked ambiguous, unclear, dirty, and somber. Gold and silver were disliked in Taipei and in Tianjin: they represent the colors of mourning or death since they are used in a wreath that is offered to the dead. These color choices, therefore, reflected the culture and social customs of these areas. Other reasons for the choices were because these colors had the images of being vulgar, cheap, flashy, shiny, and stimulating. In particular, the subjects felt gold implied a sense of vanity or falsehood. Some said that the flashy and showy images of metallic colors had negative implications for the people in Tianjin and in Taipei and, therefore, were not accepted in their areas. By contrast, the subjects in Tokyo felt the colors were popular because of their loud, glittering nature.

Black produced strong feelings of both like and dislike. The positive images for the subjects in Tokyo were clearness, tightness, and sharpness. Reminiscence of death was a negative image along with darkness and heaviness. For the people in Taipei, black was accepted because it had proper dignity with nobleness, whereas it was disapproved for its images of anxiety, fear, and sin. Subjects in Tianjin liked the color because it was solemn, rich, calm, noble, and mysterious. However, black was also associated with the negative images of fear and despair along with the image of death. Other images that the color evoked among the subjects in Tianjin were darkness, oppressiveness, and indecency.

Preference for White

One of the objects of this study was to investigate whether a strong preference for white was common to Asian areas, which are in close proximity to Tokyo both geographically and culturally. The findings showed that the people in Taipei and Tianjin preferred white to an even greater degree than in Tokyo. The results of the present study and the previous study in Korea by Saito (1992), therefore, have shown that four neighboring Asian areas have a strong common preference for white. Geographical and cultural proximity can be considered a possible factor.

The preference for white in China has long been noted in studies beginning with those by Chou and Chen (1935) and Shen (1937). The former study used nine color words and the latter employed six saturated Bradley colors. Chou and Chen (1935) postulated two possible explanations for this preference: association influence and tradition influence. As the most frequently used color word in Chinese literature was the character for white, they thought that the subjects' preference was based on familiarity, that is, frequency of association. The second possibility, namely, tradition influence, was that their preference derived from the color of their national flag. They explained that, although the dominant color in the flag was red, the symbol of the 12 stars in the flag was white and blue (the colors of the flag in 1935 were not the same as the present ones). Shen, however, strongly questioned Chou and Chen's explanations, especially the one concerning tradition influence, because white and blue each occupied no more than one-tenth of the total area of the flag. He, therefore, offered an explanation that combined Chou and Chen's concept of association frequency with language influence. For example, he noted that the Chinese character for white is not only associated with pureness but also everything open, clear, and unselfish, while grayness (gray was the least preferred color in their study) is a symbol for everything negative, disappointing, discouraging, or pessimistic.

In these past two decades, white has been found consistently to be quite popular among the Japanese people in light of the report by the Japan Color Research Institute (1992) and related studies by Saito (1981, 1992). In the present study, the authors further extended the study of preference for white by researching the subjects' stated reasons for their choices. In Tokyo, for instance, white was mostly preferred because of its associative image of being clean, pure, harmonious, refreshing, beautiful, clear, gentle, and natural. The most frequently stated reason in Taipei was association with chastity or purity. Beauty and tidiness were second. Another interesting image mentioned was infinity. In Tianjin, the reasons for the choice were mainly association with chastity or purity. They also liked white because it is elegant, clean, beautiful, and "pure white." It is also a symbol of sacredness for them. A few subjects stated that white was the source of every color so it was substantial and unique. These associative images stated above are assumed to be responsible for the strong preference for white.[1]

However, it should be noted that in Tianjin, and to a slight degree in Taipei, white was sometimes disliked, especially by male subjects (Tianjin: like, 29.7%; dislike, 8.9%; Taipei: like, 19.2%; dislike, 3.8%; Tokyo: like, 17.5%; dislike, 0.6%).

---

[1]Another possible explanation of the preference for white in Japan is as follows: Literature on ancient Japanese religion and mythology states that ancient people believed in the power of the Sun. This belief still can be seen in Japanese Shintoism. The Sun Goddess is called *Amaterasu-omikarmi*. As white represented the color of the Sun or sunshine, people accepted it as a sacred color. This is shown by Shinto priests wearing holy white costumes and also holding a sacred wand called a *gohei* with white strips of paper, a purifying implement, while they pray. Such evidence, which implies white is a sacred color, can still be seen throughout the country. There are examples especially among folk beliefs of white objects or animals becoming objects of worship at times. In this way, white had special meaning for people who revered the sun. For those people, the color quite naturally came to be favored and admired.

The color was disapproved of in Tianjin because of its lifelessness, emptiness, loneliness, and reminiscence of death. People in Taipei disliked it also because it was lifeless. In short, white was sometimes rejected in Chinese cultural spheres because of its association with the image of death. This tendency was not found among the subjects in Tokyo in the present study.

In conclusion, the following findings were obtained. First, cross-cultural differences were found in preferences for both hues and tones. Each area had an unique tendency in color preference. While the subjects in Tokyo were characterized by their preference for BG and silver, the people in Taipei had a remarkable preference for PB and P hues, and the people in Tianjin most strikingly preferred the G hue and vivid tone colors. Second, a strong preference for white was observed not only in Tokyo but also, to an even greater degree, in all Asian areas where the survey has been carried out so far. One of the reasons for this common preference in these neighboring areas was believed to be geographical and cultural proximity. Moreover, some cross-cultural agreements were found to some single colors as well. Vivid blue, vivid green, light violet, and light blue were preferred highly in common. Meanwhile, olive, dark gray, gold, and dark yellowish brown were commonly unpopular.

Last, the associative images of colors were assumed related to color preference. These associative images mostly reflected the subjects' mental images toward colors themselves. This also suggested that the psychological affective value of colors should be investigated by presenting colors without any association with or influence from particular objects, although contrasting views may propose that the affective value of color cannot be studied separately from other variables such as form and texture. The fact that subjects can associate and evaluate color images more freely by observing single color chips is probably the main reason why many past studies have carried out the investigation of the affective value of color not in combination with other variables, but free from their influence.

Moreover, in the present study, the associative images of colors sometimes reflected cultural and environmental aspects in each area. Therefore, one could consider that geographical, environmental, and cultural aspects are important factors that regulate color preference. Further psychological and anthropological studies are necessary to clarify the components of the preference for color more extensively.

NOTES

Part of this study was presented at the 22nd International Congress of Psychology, Brussels, Belgium, 1992, and the 54th Annual Convention of the Japanese Psychological Association, Kyoto, 1992.

This chapter is a brief summary of the following paper:

Saito, M. (1994). A cross-cultural study on color preference in three Asian cities: Comparison between Tokyo, Taipei and Tianjin. *Japanese psychological Research*, 36(4) 219–232. If anyone would like to obtain more details about the study, please contact M. Saito at Waseda University, School of Human Sciences, Mikajima Saitama 359, Japan, or by Internet (E-mail: miho@human.waseda.ac.jp).

REFERENCES

Adams, F. M., & Osgood, C. E. (1973). A cross-cultural study of the affective meaning of color. *Journal of Cross-Cultural Psychology, 7,* 135–157.

Birren, F. (1950). *Color psychology and color therapy.* New York: McGraw-Hill.

Chou, S. K., & Chen, H. P. (1935). General versus specific color preferences of Chinese students. *Journal of Social Psychology, 6,* 290–314.

Choungourian, A. (1968). Color preference and cultural variation. *Perceptual and Motor Skills, 26,* 1203–1206.

Eysenck, H. J. (1941). A critical and experimental study of color-preferences. *American Journal of Psychology, 54,* 385–394.

Japan Color Research Institute. (1992). *12th annual report on consumers' color preference.* Japan Color Research Institute.

Nishisato, S. (1980). Dual scaling of successive categories data. *Japanese Psychological Research, 22,* 134–143.

Saito, M. (1981). A cross cultural research on color preference. *Bulletin of the Graduate Division of Literature of Waseda University, 27,* 211–216.

Saito, M. (1992). A cross-cultural survey on color preference in Asian countries: Comparison between Japanese and Koreans with emphasis on preference for white. *Journal of the Color Science Association of Japan, 16*(1) 1–10.

Saito, M. (1990). A cross-cultural survey on color preference. *Bulletin of the Musashino Art University, 21,* 81-88.

Saito, M., & Lai, A. C. (1992). A cross-cultural survey on color preference in Asian countries: (2) Comparison between Japanese and Taiwanese with emphasis on preference for white. *Journal of the Color Science Association of Japan, 16*(2), 84–96.

Saito, M., Tomita, M., & Kogo, C. (1991). Color preference at four different districts in Japan: Factor analytical study. *Journal of the Color Science Association of Japan, 15*(1), 1–12.

Saito, M., Tomita, M., & Yamashita, K. (1991). Color preference at four different districts in Japan: Classification of characteristics of life style by cluster analysis. *Journal of the Color Science Association of Japan, 15* (2) 99–108.

Shen, N. C. (1937). The color preference of 1368 Chinese students, with special reference to the most preferred color. *Journal of Social Psychology, 8,* 185–204.

Chapter 9

MODE OF BEING IN PLACES
A Case Study in Urban Public Space

Takeshi Suzuki

INTRODUCTION

The purpose of this chapter is to advance a language to describe and analyze "a person in a place or in a scene." Two problems are combined here.

The first problem is the placelessness of urban public space in Japan in recent years. Words such as amenity, landscape, and genius-loci have become popular in urban design and the facade of streets has become "beautiful." However, the number of appropriate spaces for being, where we can spend time comfortably as long as we like, has not increased.

The second problem is the lack of a method for analyzing a scene. When I see an attractive scene from fieldwork or an impressive photo taken, for example, by Cartier-Bresson (Figure 9.1) or Kertèsz, (1971), I often notice that I cannot explain the meanings of them. Of course, we have many methods for the study of people environment relations: proxemics, (Hall, 1966), behavior setting (Barker, 1968), townscape (Cullen, 1961), etc. However, our experience of a place is a total one. An important field that falls between proxemics and townscape is open for study. Canter (1984) is correct when he says:

> The experience of a place is always some combination of plan and elevation! Somehow or other we have to find ways of exploring and representing and describing these different perspectives, of which the whole experience of a place and its various components consist. (p. 45)

Takeshi Suzuki • Department of Architecture, Faculty of Engineering, University of Tokyo, Tokyo 113, Japan.

Handbook of Japan–United States Environment–Behavior Research: Toward a Transactional Approach, edited by Seymour Wapner, Jack Demick, Takiji Yamamoto, and Takashi Takahashi. Plenum Press, New York, 1997.

FIGURE 9.1. "Square du Vert Galant et Pont des Arts" (Paris) (Cartier-Bresson, 1953)

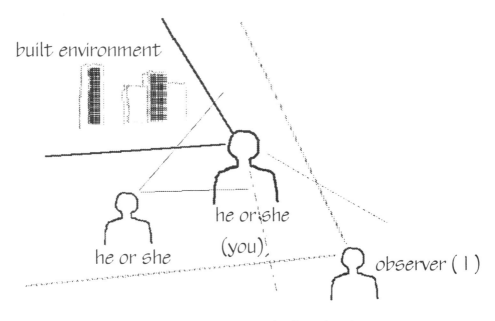

FIGURE 9.2. Relations in the mode of being in a place.

FIGURE 9.3. "Mode of being in places" belongs between proxemics and townscape.

When a person takes a position in a public space, whether he or she desires it or not, many relations and modes are generated around him or her. The built environment has the power to generate and visualize people environment relations (Figure 9.2). In contrast, in cyberspace almost all relations are programmed in advance.

For these modes and relations, I define the term *i-kata* in Japanese (Suzuki 1993, 1994, 1995). The word *i* means "be" and *kata* means "way or style or situation." Provisionally, I will translate the Japanese term to mean "mode of being in places" in English (Figure 9.3).

SEVERAL TYPES OF "MODES OF BEING IN PLACES"

Through field observations and examination of my photo and video data of urban public spaces in Tokyo, Taipei, Paris, and Tianjin, several types of "modes of being in places" were obtained. These models are not independent of each other however.

Placing Oneself (Personal World) in Public

In many spaces in Paris, single-person iron chairs, which you can move and set in any direction you like, are a common sight. The Palais Royal (Figure 9.4), the Tuileries garden, and the Luxembourg garden (Figure 9.5) are some of the attractive places where you can see people sitting in these chairs, ome reading newspapers under a tree or seated side by side in their chairs talking.

The same kinds of chairs can be seen at Bryant Park in Manhattan and Sinlong-park in Taipei. Sinlong-park is a neighborhood park in a residential area, where residents bring their private chairs from home and place them in the park—to be used not only by them and their friends but also by complete strangers (Figure 9.6) (Lee, Suzuki, & Takahashi, 1994).

FIGURE 9.4. Palais Royal (Paris).

FIGURE 9.5. Luxembourg garden (Paris).

FIGURE 9.6. Sinlong-park (Taipei).

I think the mode afforded by these chairs is quite different from the mode afforded by a bench. I call this mode of sitting in a chair in a public space "placing oneself (personal world) in public." This mode of being has two features.

The first is the appearance of privacy: a chair for a single person looks like one's own private seat because of its size and because it is the person who selects the chair's position. Sitting on a chair is a positive behavior. In a sense, it forms and emphasizes one's own personal world in an open space.

The second feature is the occupied chair's appearance to the public. When somebody is sitting in a chair, other people are able to (and permitted to) see him or her from behind. They clearly can see his or her whole body and what he or she is doing. Then they can guess what the person is watching or thinking. It is also easy to take a photograph of the person.

Design guidelines for an open space often say "Don't place benches in the center of the plaza. You should place them at the edge of the space so as to defend the backs of users" (e.g., Gehl, 1987). However, if one sits on a bench placed at the edge of a plaza, we hardly see the person because the relation between the person and us is a face-to-face relation.

Therefore, the phrase "placing oneself in public" makes clear both one's place and one's access to other people. Of course, a chair is not the only form of sitting that affords this mode (Figure 9.7). Steps or some benches have the same effect, and a chair does not always provide it.

Figure 9.7. Court of Palais Royal (Paris) (artwork by Daniel Buren).

Happen to Be Present (*Iawaseru* in Japanese)

Even a traveler can easily find and spend a comfortable time in a cafe in Paris. Much writing has been done on the cafe as a public space. The most important point to note is that one can enjoy time and space in common with others there (Figure 9.8). I am going to call this situation and mode of being "happen to be present."

This mode of being a common experience in urban public space. According to my observations, it has the following special features:

- It can be had in spaces such as a restaurant, cafe, gallery, small park, station (Figure 9.9), and the like (but not in a dwelling).
- People are permitted to see others naturally.
- Everyone has access to the space.
- Many of those present are alone or in small groups.
- Some of them come frequently, but not all.
- Many are strangers to each other.
- There is no direct communication such as conversation between strangers, but all are conscious of others. (We can hardly say "I happened to be with him without being noticed by him.")
- Through keeping a close watch on another person, one can get information about him or her and imagine what he or she is doing.
- Size of the space or area is not so large that people are unable to recognize others. Space layout is also important.

FIGURE 9.8. Bar (Paris).

FIGURE 9.9. Cafe in Lyon Station (Paris).

Figure 9.10. "Paris vu de Notre-Dame" (Cartier-Bresson, (1955).

Figure 9.11. Tuileries garden (Paris).

• There are several common norms and rules on how to behave there, and a newcomer can pick them up easily.

Though "happen to be present" is a basic and ideal mode of being in urban public space (the city itself), it is difficult to find it in the modern Japanese city. We have merely crowds, and people are not conscious of others there.

Being in Urban Structure

Relations within "modes of being in places" are not limited to what is called the human scale. In some settings, such as a high place (Figure 9.10), urban axis (Figure 9.11), or along a (Figure 9.12), river, a person (and the observer) can clearly see large urban structures. The young woman in Figure 9. 9 is sitting on the axis of Paris, which runs from the Louvre museum to the Triumphal Arch (and to the Grande Arche in Defense). Her mode of being allows not only her own awareness of Paris, but also ours (the observers'). In a sense, the people in these photographs guide the observer's awareness of the cities.

Each in One's Own Way (*Omoi-Omoi* in Japanese)

In Sinlong-park in Taipei, especially in the early morning, we can see many kinds of activities: training on *chiikon* and *taijiquan*, disco dancing, ballroom dancing, walking, chatting, and "placing oneself (personal world)." Old people,

FIGURE 9.12. Frozen river (Tianjin).

Figure 9.13. Sinlong-park (Taipei).

women, and young men stay in the park, each in his or her own way (Figure 9.13). Some of them are alone and others are in small or large groups; many of these groups are open to newcomers.

This mode of being is not so popular in public spaces in Japan. In most of the planned parks and open spaces in Tokyo, users and activities are homogeneous. For instance, mothers and their babies play in the new town, office workers break in the center of the city.

5. Come and Go (*yuki-Kau* in Japanese)

The movie *Nuovo Cinema Paradiso* of Giuseppe Tornatore showed us impressive scenes of a plaza in a small town in Siciliy. This plaza is a important space for daily life and the symbolic space of the town or stage. At the same time, it serves as a crossing (Figure 9.14). People come from all quarters and go in every direction. One person is waiting for a bus, another is passing by on a bicycle.

I was surprised to find a similar scene in Tianjin (Figure 9.15). These scenes visualize people's lives (history) and reveal a city where all manners of people live and pass by. This is also one of the typical types of modes of being in places in cities. Perhaps Grand Central Terminal in New York is the most magnificent example.

FIGURE 9.14. Plaza in *Nuovo Cinema Paradiso* (Giuseppe Tornatore, 1990).

FIGURE 9.15. Plaza (Tianjin).

FIGURE 9.16.   On the street (a town near Taichung).

FIGURE 9.17.   Bridge (Tokyo).

Stand Still (*Tatazumu* in Japanese)

We occasionally meet a person standing alone doing nothing (Figure 9.16-9.17). In Japanese, we call this situation *tatazumu*. According to a Japanese dictionary *tatazumu* is simply defined as "stop for a while," but this is not a sufficient definition. For instance, we do not use the term *tatazumu* when a person is watching something at close range. This mode of being generates an interesting awareness of the environment around an individual.

DISCUSSION

Methodology

In this chapter, modes of being in places were examined through observations of scenes. For a more systematic analysis, new methods for describing scenes and modes of being have to be developed. A tool (e.g., a multiwindows perspective system or hypertext media on a personal computer) that can represent how people perceive others describe scenes, and give information to each other in a space would be helpful (Figures 9.18 and 9.19).

Two kinds of field studies are possible. One is focused on the space: What kinds of modes of being does the space afford?. The other is focused on the person: What kinds of modes of being does a person have in his or her neighborhood and city and where are they?

Moreover, it is important as well to examine the necessary and sufficient conditions for each mode of being. This will require some experimental methods.

Cultural Factors

The problem of cultural factors was omitted in this chapter. We can say with fair certainty that every mode of being in a place is greatly influenced by one's race and culture. Though architects and planners have designed or imported many foreign-style urban spaces in the cities of Japan, the behavioral pattern and lifestyle of the Japanese people are quite different from the behavioral patterns of the people in Taipei or Paris. However, a person's culture is not the only factor responsible for his or her behavior. I think several types of modes of being are common in urban areas in many countries. Furthermore, Japanese people who visit foreign cities sometimes adopt a new (unaccustomed) mode of being, and foreign people who visit Japan will have to abandon some of their modes of being. These cases are caused by the built and social environment.

We must also draw attention to a historical viewpoint. For example, we can find many attractive modes of being in old street photographs taken Tokyo by Ihee Kimura (1901–1974) (Figures 9.20 and 9.21), a famous Japanese photographer. Some of these modes cannot be seen nowadays. Old Tokyo, before the 1960s, seemed to have more kinds of urban modes of being than the Tokyo of today.

FIGURE 9.18. Prototype scystem based on 3D-LOGO, a computer programming language by LINKed & BYNAS DIVISION, UNY CO.LTD, Japan.

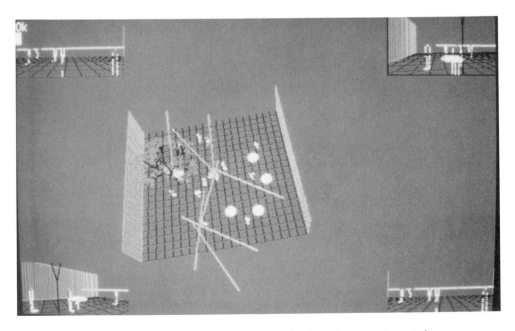

FIGURE 9.19. Each person ("turtle" on LOGO) has his or her own view window.

FIGURE 9.20. *Morikawa-cho, Hongo* (Tokyo) (Ihee Kimura, 1953).

Toward a Design Language

Architects have a variety of design vocabularies for an urban space, but have few for modes of being. They have only terms such as "gathering," "territorial behavior," and the actors and spectators model. If we develop a proper typology of modes of being in places, (see Alexander, et al., 1977), it will be possible to bridge the gap between design and environment behavior studies in the field of public space design.

NOTES

This chapter is based on Suzuki (1993, 1994, 1995), Suzuki and Takahashi (1992), and Lee, Suzuki and Takahashi (1994, 1995). All photographs (except Photos 1, 8, 12, 16, 17) are by Takeshi Suzuki. Photos 1 and 8 from H. Cartier-Bresson (1984). Photo 12 is from the film *Nuovo Cinema Paradiso* (Giuseppe Tornatore, 1990). Photos 16 and 17 from Tokyo Metropolitan Museum of Photography (Eds.) (1990).

FIGURE 9.21.   *Ginza* (Tokyo) (Ihee Kimura, 1949).

REFERENCES

(J): in Japanese

Alexander, C., Ishikawa, S., Silverstein, M., Jacobsen, M., Fiksdahl-King, I., & Angel, S. (1977). *A pattern language*. New York: Oxford University Press.

Barker, R. (1968). Ecological psychology: Concepts and methods for studying human behavior. Stanford, CA: Stanford University Press.

Canter, D. (1984). Beyond building utilization. In J. A. Powell, I. Cooper, & S. Lera (Eds.), *Designing for building utilization* (pp. 41–47). E. & F. N. Spon.

Cartier-Bresson, H. (1984). *Paris vue d'il*. Munich: Schirmer/Mosel Verlag Gmbh.

Cullen, G. (1961). *The concise townscape*. London: Architectural Press.

Hall, E. T. (1966). *The hidden dimension*. New York: Doubleday.

Gehl, J. (1987). *Life between buildings: Using public space*. New York: Van Nostrand Reinhold.

Kertèsz, A. (1971). *On reading*. New York: Grossman.

Lee, W., Suzuki, T., & Takahashi, T. (1994). A study of the user's behavior pattern in Taipei's Sinlong-park. Summaries of Technical Papers of the annual meeting of the Architectural Institute of Japan (Architectural Planning and Design, Rural Planning), 1073–1074. (J)

Lee, W., Suzuki, T., & Takahashi, T. (1995). A case study on the users' mode of being in places, quality of communication of Lonsan-Temple and its surrounding area in Taipei: Study on an affordable place for being in the city, Part 1. *Journal of Architectural Planning and Environmental Engineering 0 (Transactions of AIJ), 468*, 133–141. (J)

Suzuki, T. (1993). Environmental design through the mode of being in places: part 1. Mode of being in urban public space. *The Kenchiku Gijutsu, 517*, 204–207. (J)

Suzuki, T. (1994). Environment from the view point of 'Ikata: Mode of being places'. *Gendai-Shiso (revue de la pense d'aujourd' hui), 22*(13), 188–197. (J)

Suzuki, T. (1995). Reading a quality of environment in a scene. *Journal of Architecture and Building Science, 1367*, 48–49. (J)

Suzuki, T., & Takahashi, T. (1992). Study on human positioning in public urban space. Summaries of Technical Papers of the annual meeting of the Architectural Institute of Japan (Architectural Planning and Design, Rural Planning), 719–720. (J)

Tokyo Metropolitan Museum of Photography (Eds.) (1990). *TOKYO/A city perspective*. Tokyo Metropolitan Museum of Photography. (J)

Part III
SOCIOCULTURAL ASPECTS OF THE PERSON

Chapter **10**

URBAN RENEWAL AND THE ELDERLY
An Ethnographic Approach

Hirofumi Minami

INTRODUCTION

The present study investigated people's understanding of their own community in the context of urban renewal. In particular, elderly residents were taken as focal informants who provided narrative accounts on the insiders' view of the urban renewal process and their relationship with the community in the course of life-long development. In addition, elderly residents were assumed to constitute the population that is most vulnerable to radical changes induced by the renewal process and, therefore, in need of particular understanding and care.

The chosen site for the study was a downtown community in the city of Hiroshima. This district (D district, hereafter) exhibits characteristics of a close-knit urban community equivalent to Boston's West End, as described in Gans' (1962) seminal work, *The Urban Villagers*. Unlike the West End community, however, the residents of D district came back to the rearranged site and engaged in renewed community life there. In this respect, the longitudinal observation of community life of D district seemed to provide unusual opportunities to study a macroscale critical person-in-environment transition (Wapner, 1981) before and after urban renewal.

Previous research on the impact of urban renewal and relocation has revealed a variety of grief responses experienced by residents as well as the feeling of loss of "spatial identity" (Fried, 1963). These reactions can be understood in the context

Hirofumi Minami • Faculty of Education, Kyushu University, Fukuoka 812, Japan.

Handbook of Japan–United States Environment–Behavior Research: Toward a Transactional Approach, edited by Seymour Wapner, Jack Demick, Takiji Yamamoto, and Takashi Takahashi. Plenum Press, New York, 1997.

of the residents' strong emotional ties to the lost community and differential subculture developed in such a close-knit community (Gans, 1962; Ryan, 1963).

These studies, however, did not focus on the recovery processes after renewal. From the perspective of community development, a longitudinal follow-up in the postrelocation period will be more important in understanding how the residents manage radical changes and reconstruct the community. The integration of the old or lost community and the renewed one will constitute a significant task for the residents. This might be more difficult or challenging to elderly residents who had formed a strong cognitive and emotional commitment to the previous town image.

To investigate the transactional processes of this radical person-in-environment transition (Wapner, 1981, 1991), the following assumptions were adopted to guide the present investigation.

1. *Ethnographic approach*: The present study employed an ethnographic approach, which emphasizes the in-depth understanding of insiders' view and "thick description" of contextual background of the target phenomenon (Geertz, 1973). It was expected that the transactional and contextual quality of community life (Altman & Rogoff, 1984) could be best described through a longitudinal and in-depth participation in the target phenomenon. Such descriptive understanding will enable us to generate more meaningful hypotheses with higher ecological validity (Stokols, 1987).

2. *Urban environment as life-world*: In line with the ethnographic approach, the lived experiences of the urban environment from the residents' points of views are conceptualized by using Schutz and Luckmann's (1973) notion of the "life-world." The life-world as "the taken-for-granted basis of everyday life" becomes notable or "problematic" when a significant portion of the assumptions are questioned under critical changes in one's life circumstances (Minami, 1987; Parkes, 1971). It was expected that by studying the residents' experiences of urban renewal we could touch upon the strata of their life-world that are indispensable to maintaining a sense of community.

3. *Social ecological dimension underlying urban dynamics*: To investigate the urban renewal processes by using both individual and collective units of analyses, a social ecological perspective developed by Binder (1972) and Stokols' (1987) notion of group × place transaction were incorporated in the present study. The environmental changes induced by the renewal projects affect not only the individual's life-world but also group dynamic processes such as intergenerational conflicts and mobilities in the community. A modified social ecological approach, which conceptualizes the transaction between group dynamic processes and the physical environment (Minami & Tanaka, 1995) was applied in understanding the complex nature of the community changes. In particular, sociopetal versus sociofugal components (Osmond, 1957) in the pre- and postrenewal environments that facilitate or inhibit social interactions between residents were investigated through interviewing and behavioral observations.

4. *Multilevel analyses of the residents' transactions with the urban renewal processes*: As a methodological guiding principle to implement a transactional perspective in structuring the research design, Wapner's (1991) in-

terpretation of "person-in-environment transactions" as "experiencing and acting" of the agent was employed. The experiential aspect of the transactions was tapped through in-depth interviewing that induced "self-talk" or narrative accounts (Bruner, 1987) of the renewal process by the residents. The action layer of the transaction was investigated through participant observations of community activities. Other observational techniques focusing on salient changes in the physical environment and their behavioral implications (Zeisel, 1984) were used to substantiate the environmental conditions accounting for significant changes in the everyday life patterns of the residents. Such multilevel analyses were expected to generate basic data and hypotheses for "ecologically valid environmental planning" (Minami & Tanaka, 1995).

Based on the above assumptions, this study aimed at understanding (1) how the residents perceived and construed the renewal process and made sense out of discontinued life circumstances, (2) how the community as a whole managed to maintain its sense of continuity in a prolonged transitional period, and (3) how there were differential impacts of urban renewal on the life of elderly residents.

METHOD

Research Site and Its Social-Ecological Background

D district in the city of Hiroshima was chosen as the research site. This district, located 2 km from the center of the city, had eight subdivisions with a population of 13,000 residents prior to the renewal project.

This district has typical old-fashioned, downtown characteristics with lower wooden houses, narrow mazelike streets, and densely crowded housing. Such physical characteristics together with its tightly knit social networks remained virtually unchanged during the rapid postwar redevelopment period in Hiroshima, because a neighboring mountain served as a barrier to the area's destruction by A-bombing. Such historical conditions surrounding the D district resulted in the conspicuous outlook and subculture of this area.

Large-scale renewal planning started in 1973 and will be completed within the next 5 years. During the renewal process residents are provided with temporary housing arrangements and are to return to the reconstructed cite. There is a marked change in the land use by this renewal. The area for housing, which constituted 90.2 % of the whole land, is decreasing to 66.1 %. The area for roads and parks is increasing from 5.5 % to 29.3 % and from 0.7 % to 3.0%, respectively. Two main large avenues with widths of 36 m and 25 m cut across the district and divide the existing town structure into several different subareas. Temporal relocation and reconstruction proceeds in a gradual and successive manner based on the five divisions of the town. The construction of the first area was completed in 1987 and the last area is to be finished toward the end of the 20th century.

FIGURE 10.1. A scene from S town: Old-fashioned narrow streets with wooden houses.

Participants

Two subareas were chosen as the focal field for this study. One of them, M area (called M town, hereafter), was the division that was completed in 1991, the earliest in the whole project. Another area, S (S town, hereafter), will be the last division to be completed. The selection of these two areas was planned so as to observe communities at different stages of the renewal process. On the one hand, by observing M town, we could see what the community is like after renewal and how the new community is taking shape. By observing S town, on the other hand, we could see community life prior to renewal.

Residents living in the two areas were approached through community organizations (*Chonai-kai*), which were run by mostly elderly members. Elderly leaders of the two areas functioned as focal informants as well as conveners for arranging participant-observations of various community activities and individual interviews. In-depth interviews with 11 elderly residents in the two areas provided narrative data on insiders' views of community life and the renewal processes.

FIGURE 10.2. A scene from M town: Renewed residential area with enlarged streets and modern housing.

Procedure

Initial contact with D district occurred in 1987. Occasional visits and participant observations of seasonal festivals have continued since then. A more intensive set of observations and interviews started in 1991 and has continued.

Participant observations of community activities such as town meetings and festivals are conducted to collect ethnographic data on the role of elderly people and the interpersonal and social organizational properties of community life in general.

In-depth interviews with 11 elderly residents were conducted by three interviewers who had acquired rapport with the residents and the community through participant observations. The interview was conducted in the interviewee's home, tape recorded, and transcribed. To design the interview schedule, Plimmer's (1983) methodological perspective and guidelines for studying individual life history were used. The main questions included (1) the person's relationship with the area in his or her life history; (2) the person's view on the characteristics of the community, interpersonal relations and social life; and (3) perceived changes in the area by the renewal project, its impact on the community, evaluations of the change, and expectations for the future of community life.

To increase intersubjective validity in the administration of interviews and interpretation of the interview data, regular case reporting and discussions among the three interviewers were arranged. Other information gained by photographs and videotaping were also used to supplement our understanding of the community life and daily activities of the elderly residents.

RESULTS

Qualitative analyses of the ethnographic data (field notes, interview transcriptions, photographic records, and videotaped recordings) revealed several topical issues related to the aforementioned research questions. Since the fieldwork and data interpretation are still in process, the following remarks are temporary in nature and mostly illustrative accounts, which require further cross-validations. These analyses were directed toward formulating meaningful hypotheses on the life of elderly residents in the process of large-scale urban renewal.

Environmental Disengagement of the Elderly Residents

The comparison of community activities between the two areas that were under different stages of the renewal process revealed several interesting differences. One of the main findings of the participant observation study during the autumn festival season concerned the gradual disengagement of the elderly residents from communal activities.

In S town, which remained in a prerenewal state, a significant portion of the area's social activity, such as managing town meetings and festivals, was achieved by elderly residents (those over 65). All the leading figures in the community organization are over 65 years old and are "known faces" in the area who had been settled in the town since the prewar period. For communal activities such as autumn festivals, only elderly community members are familiar with unwritten codes and know-hows to prepare ceremonial settings and to perform minutely prescribed procedures.

Through participant observation of the autumn festival, the following characteristics were observed: (1) the project gets done in a piecemeal manner: no single person knows or controls the total project and the works are divided and voluntarily conducted by everyone on hand; (2) work is done on the spot through trial-and-error and continual negotiations among members.

Against our expectations, the management and administration of large-scale community activities are conducted without formal plans. Decisions are made "in the process" of doing. The necessary knowledge was induced through occasional inquiries to the elderly members. The central role of the elderly residents during communal festivals can be understood by considering such oral communication as a principal means of achieving the festival works.

During the festival performance, elderly members and young children occupied the center of the stage. The only middle-aged adults we observed were mothers of children, who took back-stage roles such as cooking and preparing clothes. The presence of elderly residents who came out to look at the ceremonial activities was also noticeable. There were voluntary conversations among these elderly onlookers in the street.

The management of and actual participation in the community activities in M town, which was in a postrenewal state, showed marked differences from those in town. Following the renewal, the main role was taken by the middle-aged population. Although an elderly member remained a leader of the community organiza-

FIGURE 10.3. A festival group of S town: A high proportion of the elderly members.

tion, the actual management and administration of the festival were now carried out by middle-aged male members of M town. They were mostly fathers of the children who took leading roles in the ceremonial act. The middle-aged generation was also predominant among the onlookers.

In accordance with generational changes in the constitution of administrators and performers, there were significant modifications in the manner of performing the ceremony. A set of precise rules and prescribed ways in preparing the setting were reduced. More ready-made commercial goods and foods were used instead of handmade instruments handed down from previous generations and communal and collaborative cooking. The main reason for such simplification and the use of commercial services was that "everyone is too busy to bother with time-consuming preparation." The lack of knowledgeable figures among the participants also produced awkward trial-and-error situations and negligence of ceremonial routines.

Through observation of the physical environment of the two towns and the interviews of residents, such intergenerational changes were related to the changes in the physical environment after renewal. The renewed town—with high-rise

FIGURE 10.4. A festival group of M town: Increased participation of the middle-aged population.

buildings, sophisticated security systems, such as auto-locked gates and the use of interphones, widened roads with busier car traffic, and the replacement of individually owned local stores with "convenience stores" run by a large corporation—inhibited or put extra barriers to the social activities of the elderly residents. In contrast, the younger generation took advantage of the new facilities and traffic systems.

The social disengagement processes to which the elderly population is normally prone, along with the decrease in physical abilities and social roles, seem to be reinforced through the "environmental disengagement" caused by the city's renewal constructions. Sociopetal components for the younger residents such as widened roads enabling easy access by automobiles would function as sociofugal components for the elderly residents, to whom walking is a more important means of locomotion.

Despite the discontent of "being locked up in the top of a tall building," most elderly residents seem to be accepting the general circumstances. According to them, it is "inevitable in this time of technological development and probably desirable considering the generations to come who need a renovated and more active town that will match the needs of younger people." Such consideration of the living conditions for younger generations rather than their own and of the future rather than the present and the past seems to circumscribe the elder residents' passive acceptance of environmental deprivations on their part caused by the renewal construction.

FIGURE 10.5. A social encounter in S town: Less physical barriers between the inside and outside of the house.

"The Lost Landscape" and People's Recovering Work through Festive Activities

It is a known saying in D district that elderly people talk about this renewal project as "another *pikadon* (colloquial expression of the A-bombing)." The metaphoric use of A-bombing in the elder residents' description of the renewal construction and destruction tells the intensity of experienced impacts of this process, as well as the deep-rooted underpinnings of the A-bombing in the collective representation of the residents.

In fact, the experience of A-bombing in Hiroshima City casts a shadow on various aspects of the life of the residents of D district. This is partially due to the historical background of this district. Since this area escaped from the total destruction by virtue of the guard of the neighboring hill, it was used as a refuge by sufferers from other areas of the city. Many of them remained in this area through the reconstruction period of the city, which resulted in a high ratio of *hibakusha* (A-bombing sufferers) in D district.

FIGURE 10.6. A social encounter in M town: Extended entrance approach with security system.

There are several heroic stories associated with unusual hospitality by the residents of D district in helping with the immediate and subsequent problems of the A-bombing sufferers. The first medical investigation of the aftereffects of radio-active exposures was done by an individual physician who ran a family hospital in D district. Dr. M, who is now over 70 years old, kept the hospital open and provided an inexpensive residential ward for long-term sufferers until the hospital was re-moved by the renewal project. When the renewal project was set and the time for temporary closure of the hospital came close, Dr. M started to videotape the area near the hospital site and keep records of radical changes in the familiar landscape of the area caused by the renewal operation. Although he does not talk explicitly, his maniacal insistence on keeping regular records of the progress of the destruc-tion of the old town seems to suggest a deeply felt loss and some kind of mourning period for the lost community and its landscape.

Through interviews with the residents, we heard of a number of episodes that dramatically illustrate the effects of the destruction of their own homes, especially for the senior residents. One episode described the intense refusal of an old lady

FIGURE 10.7. The last festival at S town prior to renewal construction.

to move from her home, and her severe reaction to the removal of a pine tree in her garden. After the enforced relocation to temporary housing, she exhibited signs of dementia and disorientation and soon passed away.

These and other similar episodes seem to confirm Fried's (1963) account of posttransitional syndromes during and after relocation. Such severe reactions to loss, however, were only one effect of the process. From our participant observations and interviews, most residents, including elder ones, showed a steady adjustment to the new situation and engaged in active reconstructive work for their own home and the community.

Communal seasonal festivals among other more ordinal community activities, such as town meetings, seem to function as "integrative collaborative actions." These festivals played an important role in both the pre- and postrenewal stage.

In S town, the elderly residents insisted on producing the autumn festival, in which children play central roles, when the majority of residents were already removed and the remaining ones were busy preparing for the temporary relocation of their homes. During this seasonal festival, children together with adults walked around every street and corner, stopped in front of every home of the residents, played chants there, and were cheerfully welcomed by the house owners. In participating in these ceremonial activities, the children, it seemed, would acquire tacit memory traces of the physical makeup of the community. In fact, it was the last opportunity for them to see a detailed landscape of the community, which would soon be demolished and remodeled.

Figure 10.8. The first festival at M town after the completion of renewal construction.

In M town, the elderly leaders insisted on resuming this autumn festival immediately after completion of the renewal project. As described above, the actual
administration of the festival in this renewed town was rather awkward and only
partially conducted compared to the traditional methods in S town. Nevertheless,
here, again, children walked around every street and corner and stopped by every
house in the area. It was as if the activities were designed to teach children the
intricate physical makeup of this renewed community.

In both the old and new towns in D district, avoidance of dissolution and
disintegration of the community was present in the very act of administering the
ceremony. In particular, the tacit learning of the landscape of the community
provided to new generations seemed to underlie the cheerful festive event
where children play a central role. Accumulation of annual experiences of the
ceremonial routine together with everyday transactions may result in the formation of an internalized image of the community—"the original landscape"—which provides shared meaning and the context of community life
across individual members and generations. This hypothesis is further clarified
through the narrative analyses of the interview data with the elderly residents
on their accounts of community life.

Continuity of the "Internalized Community" through Narrative Acts

One of the most notable characteristics of the narrative accounts of 11 elderly
residents on their perception of community life and their experiences of the re-

newal construction was the similarity in their expressions. This town, in their accounts, has a distinctive color or identity, "D townness," which is not found in other areas in Hiroshima City. Unlike the "cool" transactions of people in other urbanized area, the people in D district have closer relationships, which are almost "rural" or old-fashioned in their naive nature.

The most used expression was *mukosangen ryodonari* (literal translation: three houses on the opposite side of the street and two neighboring houses), meaning very intimate transactions among neighboring households. Families within this neighboring unit would visit freely ("knows kitchen outlet"), pass dishes on occasion, and pay careful but not intrusive attention to the welfare of their neighbors. Such intimate relationships are similar to those described in Gans' (1962) ethnographic accounts of the community life of Boston's West End. Unlike the West Enders, for whom kinship is the strengthening base of bond, the intimate relationships in D district are based mainly on physical proximity rather than kinship. As one informant described, such relationships develop "in decade's of affiliation in the area, chatting on the street, and having children go to the same local school."

Despite repeated expressions of intimate transactions of neighbors, based on our observations we could not confirm actual occurrences of sharing household goods and dishes in daily life. We have formed a hypothesis that the omnipresence in the narrative accounts of close-knit neighborhood relationships is not a reflection of actual community life in this area but rather an act of narrative creation, or a kind of a "myth" shared by the residents. It is a community as it exists in the collective representation of the residents. Although the expressions described above may have occurred in the real transactions in this community in the old days, they are not observable events in the present life of the residents. In this respect, they are stereotyped narratives, which correspond to the "original landscape" of the long-term inhabitants of this area.

Even if the community life as expressed in these narratives is not alive in this area any longer, the very act of narrating in this way by the local elder residents may constitute continuity in the collective representation and, thus, creates a context in which the distinctive "D townness" is passed to future generations. One of the rationales to resume the autumn festival described above was that, without it, people especially, young children, would forget the good part of D district. These stereotyped narratives, in this sense, are functional in guiding the social ecology of the community under the radical changes in the physical environment induced by the renewal project.

CONCLUSION

Humanization of Urban Environments through the Mediation by Elderly Residents

The ethnographic investigation into the long-term impact of city renewal on the life of community residents revealed both regressive responses and recovering actions. Since the renewal project in the observed town is still in progress, our estimation of the long-term aftereffects is a tentative one at best. In addition, some quantitative evaluation with larger samples of community residents on the impact

FIGURE 10.9. Environmental maintenance activities by the elderly: An elderly sweeping a street.

of the renewal project will be needed to supplement our understanding obtained by an ethnographic approach.

With these restrictions, it can be said that the progression of the renewal project, with the accompanying changes in the physical environments of the habitat, induced or facilitated generational changes in the social organizational dynamics of the community. There was a shift in the constituency of principal actors in community activities from the elderly members to the middle-aged members. The renewed environmental infrastructure, which was designed to match the lifestyles of younger generations, was experienced as a barrier for locomotion by the elder residents, which might be related to the gradual disengagement of elderly residents from daily encounters on the street. The necessity to build a new house in a renewed site also was taken as a chance for the middle-aged members to take more initiatives in managing the household. The ownership of local stores was also handed over from the older generations to the middle-aged family members. Since the owners of local stores take central roles in community organizations, such changes result in displacement of community leadership from the elder residents to the younger generation.

These intergenerational transformations of the social organization in the renewed town are exactly what this project planned. As one community leader, who is one of the eldest in the town, commented, "the whole project is in the service of generations to come, not for old ones like ourselves." In this respect, the project can be said to be successful.

It does not mean, however, that the elderly residents should retire and withdraw from communal activities. Quite the opposite might be true. From our obser-

FIGURE 10.10. Interpersonal care for the community: An elderly man watching a small child in a neighboring park.

vations of daily life as well as festival activities of the town, the elderly residents were found to play tacit but significant roles in maintaintaining a pleasant and safe atmosphere in the street life of the community. Regular cyclical patterns of the morning sweeping, routine walking around, store visiting, and gardening activities by the elder residents provide a steady presence of familiar figures on the street and the dramaturgical playing out of community scenes. They are active participants in the "transactional" features (Altman & Rogoff, 1987) of the street life. These roles may not be "productive" in terms of the economic system, but they are significant in terms of their "reproductive" function in maintaining the life-world of the community.

The renewed town with its adjustment to new requirements for the commercial and industrial social system produced a number of benefits at the cost of some aspects of communal street life that were once the distinctive character of this district. The elderly residents, with their stored memory of this area and daily activity patterns freed from the pressure of economic production, may be the most capable agents to implement "humane" components in the transactional quality of community life.

This hypothesis can be examined through further follow-up observations of this social experiment.

REFERENCES

Altman, I. & Rogoff, B. (1987). World views in psychology: Trait, interactional, organismic and transactional perspectives. In D. Stokols & I. Altman (Eds.), *Handbook of environmental psychology* (pp.1433–1465). New York: Wiley Interscience.

Binder, A. (1972). Psychology in action: A new concept for psychology: Social ecology. *American Psychologist,* September, 1972, 903–908.

Bruner, J. (1987). Life as narrative. *Social Research, 54*(1), 11–32.

Fried, M. (1963). Grieving for a lost home. In L. J. Duhl (Ed.), *The urban condition* (pp.151–171). New York: Basic Books.

Gans, H. J. (1962). *The urban villagers: Group and class in the life of Italian-Americans.* New York: Free Press.

Geertz, C. (1973). *The interpretation of cultures.* New York: Basic Books.

Minami, H. (1987). A conceptual model of critical life transition: Disruption and reconstruction of life-world. *Hiroshima Forum for Psychology, 12,* 33–56.

Minami, H., & Tanaka, K. (1995). Social and environmental psychology: Transaction between physical space and group-dynamic processes. *Environment and Behavior, 27*(1), 43–55.

Osmond, H. (1957). Function as a basis of psychiatric ward design. *Mental Hospital, 8,* 23–29.

Parkes, C. M. (1971). Psycho-social transitions: A field for study. Vol 5: Social Science and Medicine (pp. 101–115). London: Pergamon Press.

Plimmer, K. (1983). *Document of life.* London: Gerge Allen & Unwin.

Ryan, E. J. (1963). Personal identity in an urban slum. In L. J. Duhl (Ed.), *The urban condition* (pp. 135–150). New York: Basic Books.

Stokols, D. (1981). Group x place transactions: Some neglected issues in psychological research on settings. In D. Magnusson (Ed.), *Toward a psychology of situations: An international perspective* (pp. 393–415). Hillsdale, NJ: Wiley-Interscience.

Stokols, D. (1987). Conceptual strategies of environmental psychology. In D. Stokols & I. Altman (Eds.), *Handbook of environmental psychology* (pp. 41–70). New York: Wiley-Interscience.

Schutz, A., & Luckmann, T. (1973). *The structure of the life-world.* Evanston: Northwestern University Press.

Wapner, S. (1981). Transactions of person-in-environments: Some critical transitions. *Journal of Environmental Psychology, 1,* 223–239.

Wapner, S. (1991). An organismic-developmental systems oriented approach. In S. Wapner & T. Yamamoto (Eds.), *Developmental psychology of life transitions.* Kyoto: Kitaohji Shuppan.

Zeisel, J. (1984). *Inquiry by design: Tools for environment-behavior research.* Cambridge: Cambridge University Press.

Chapter **11**

INTEGRATING ENVIRONMENTAL FACTORS INTO MULTIDIMENSIONAL ANALYSIS

William Michelson

INTRODUCTION

The problems generating environmental research by social scientists vary greatly over time. How many researchers who started with an interest in housing, school, or hospital design imagined that they might one day apply their studies to mass migration, waste disposal, earthquakes, or the safety of government buildings? The absence of firm predictability in applications suggests attention to methods that cover a wide range of considerations.

There are alternative ways to approach such demands on methodology. One way is to devise and maintain for use as appropriate a wide range of research techniques. Insofar as information must respond to research demands, this approach resembles the creation of a massive tool box, whose individual items are drawn upon as necessary. Sommer and Wicker (1991) suggest that if this box becomes too large for a single practitioner to carry or if its contents are too diverse for expert use by the same person, then specialization is needed. Such an approach surely reflects the growing diversity of applications and of tools over time (cf. Bechtel, Marans, & Michelson, 1987).

William Michelson • Department of Sociology and Centre for Urban and Community Studies, University of Toronto, Toronto, Ontario M5S 1A1, Canada.

Handbook of Japan–United States Environment–Behavior Research: Toward a Transactional Approach, edited by Seymour Wapner, Jack Demick, Takiji Yamamoto, and Takashi Takahashi. Plenum Press, New York, 1997.

Nonetheless, the opposite approach should be considered for its potential value. This would be the creation of research techniques which *integrate* certain kinds of information that are commonly relevant to varying applications. This is one of the appealing features of Barker's behavior settings scheme in his ecological psychology perspective (Barker, 1968). This second approach has the positive attribute of utilizing dimensions with inherent interrelationships, instead of great numbers of simple measures without necessary interlinkage. In addition, if integrative measures are shown to be fruitful to diverse applications, there is less need for constant creation of unique measures.

These opposing approaches are surely complementary. Clearly, no one technique answers all demands, and, it is unlikely that all applications demand absolutely unique research techniques.

This chapter will detail the extension to emerging research applications of a specific integrative technique—time use analysis. I shall emphasize how to optimize the integrative aspects of this technique as well as the relative ease to which it accommodates to emerging research applications.

TIME USE AS AN INTEGRATIVE RESEARCH METHOD

Time use protocols can integrate environmental factors into multidimensional analysis because they take the form of a matrix. Each row in the matrix represents a discrete episode of behavior, in a serial-order listing of activities for a specific period of time (typically the current or immediately past day). For each such episode, there is a series of data in columns, representing generally useful information in explanation. These almost always include specification of the nature of the behavior/activity, its temporal dimensions (starting time, termination time, and, hence, duration), persons present (hence, aspects of social interaction), and place (hence, environmental considerations). Some time use studies expand the matrix through columns on subjective dimensions such as degree of enjoyment, tension, or choice in doing the activity (cf. Michelson, forthcoming).

This matrix is integrated in at least three ways. First, it is a compact, neutral data gathering format, in which great amounts of data are gathered quickly and economically because of the standardized, repetitive format; these data are integrated into the matrix format and structure. Second, all the episodes within the time period are also integrated as part of a system of sequences and trade-offs, representing the ordering and functioning of everyday life. Third, what integrates the diverse dimensions of behavior is that they are all rooted in episodes that have actually occurred and been part of respondents' experiences.

The first type of integration simply comes as a data-gathering advantage of the research method. It is a given regardless of subsequent analytic strategy and the kinds of inference made possible.

Most analytic strategy has centered on the second form of integration. The pattern of activity over the day (or week)—what activities are integrated, in what order and amount, and in the absence of what other activities—is a common concern. When taking this approach, one can compare the major patterns and trends of the different columnar dimensions. For example, are people with extensive time devoted to paid employment more likely to be absent from home and/or higher in perceived tension? However well integrated these data are within the structure of the day, information on the separate dimensions, when aggregated over the time period involved, is not fully integrated. In the current example, it is not given that the components of tension in the time period represent or reflect the extent of employment. It is just as logical that a workaholic loves his or her work but has a terrible pattern of activity outside work that heavily influences aggregate measures of tension.

Where environmental and other dimensions are integrated most fully is at the level of the episode—in the third type of integration. There, each form of behavior is attached to a place, to interpersonal relationships, and to subjective feelings and outcomes. The multidimensional characteristics of an episode are specific to that episode and only lost in meaning if extracted beyond the context of the episode. Where aggregation occurs positively in this approach is across examples of the same kind of episode.

Harvey et al. (1984) have suggested an approach called hypercoding, which creates unique, multidigit codes reflecting the joining together of individual codes for the respective dimensions. But since such disaggregation leads to very many, nonrepeating contexts, it is best used to isolate modal contexts. For multidimensional analyses that integrate environments, a more practical approach is to analyze two or three aspects of episodes conjointly. For example, in previous research on the extent that child care became easier in housing areas with specific experimental features, time use data on where adults spent time with their children proved useful in explanation (Michelson, 1990). This involved multidimensional analysis aggregating from episodes the time in a day spent in the company of one's children in particular kinds of places.

To do this kind of analysis, which optimizes integration, the structure of the data file has to remain in its nearly raw format. This is necessary to retrieve any combination of variables at the episode level at any time. But it creates problems in use of standardized software that demand an equal number of variables or records per case. People simply do not have the same number of episodes of behavior during the day. While aggregating the episodes over the time period to get a total number of minutes devoted to each category of activity, person present, place, or subjective evaluation, one gains a uniform set of variables but loses optimal integration. For this latter objective, more flexible software or programming is necessary. The data presented in this chapter were analyzed with self-written programs in GW-Basic.

WOMEN AND ENVIRONMENTS

The subject of women and environments has presented opportunities for examining new applications of environment behavior research. Not only has the range of environments for study expanded once the needs of women have been considered more fully (cf. Franck & Ahrentzen, 1989) but it has been shown as well that the same environments are often experienced differently by women than by men (cf. Michelson, 1985a).

I argued in a recent discussion (Michelson, 1994) that the impact of the great trend in many societies for women to participate in paid outside employment, even during the child-rearing years, is understood more fully through the perspective of time use. Justified attention has been placed on questions of equity, availability, and affordability for women in such specific environments as workplaces, child care, transportation, and housing. What I suggested, based on a conceptual analysis of the main components of time use, is that the experience of daily life is more than the sum of what happens within the specific environments encountered. What has changed even more is the *pattern* of everyday life that employed women now encounter. The dynamic and qualitative aspects of daily life have changed in terms of the balance and combinations of the various dimensions of everyday time use: what, when, with whom, where, and with what personal outcomes. Thus, employed women not only encounter a workplace environment, but while there spend much more time with workplace colleagues and much less time with relatives and neighbors. Women not only travel somewhat more, but they do so to different places and at different times. Even the discrete environments themselves take on different meanings and objectives. Housing, for example, serves different purposes in the life of a person who is there nearly half as often as under other conditions, and its location is subject to different criteria when different trip destinations and time pressures are considered (Wekerle, 1984).

Among the considerations that gain in clarity under this conceptual spotlight is safety in public places. It is generally recognized that crimes against women have increased during recent decades. What the time use perspective adds is some understanding of changes in the contexts under which women are at risk. Women are now less confined to familiar and often carefully designed and chosen residential areas as the contexts for much of everyday life. Suburban settings had once been intended as safe havens for women and children. Regardless of whether this intention (largely by male planners and decision makers) had ever been justified, the logistical fact is that women are increasingly spending time in different places, at different times of day, and in different company than previously. The daily pattern of time use of the employed woman is more likely to involve travel and presence in a greater variety of local areas, usually different and possibly distant from her residential area, often during periods of darkness, and without accompaniment. It may involve travel (alone) on public transportation and/or the need to access an automobile in a parking lot or garage. This situation has implications for public safety practices.

While conceptual analysis can help identify ways to explain and understand such emerging applications, empirical substantiation must nonetheless follow. Everything that appears logical does not necessarily have a basis in fact. As Hanson and Hanson (1993, p. 263) note, "Surprisingly little is currently known about how activity participation (e.g., the timing and location of episodes) is related to concerns about personal safety." In the present case, not only could time use data be employed to assess these conceptual interpretations, but existing data could be drawn upon for an entirely different application. The respective dimensions of time use data contain such basic information integrating environments and other crucial variables at the episode level that an examination of this question was possible through secondary analysis of time use data gathered originally for different purposes.

PERSONAL SAFETY AND TRAVEL CONTEXTS

Several hypotheses flow directly from the preceding conceptual analysis:

1. Employed women take more trips on weekdays than do those without external jobs. (This is well entrenched in the empirical literature, but is mentioned as part of the current chain of reasoning [cf. Michelson, 1985].)
2. Trips made by employed women are spread out around the day more than are those taken by homemakers, including during times more likely to be dark during the colder months of the year.
3. More of the trips that employed women make are taken alone.
4. Tension levels are higher among women travelling alone, particularly so in the evening and among employed women (whose jobs take them out of their local residential areas more frequently).

Two time use files were analyzed regarding these hypotheses. One was a sample of 593 women, which was the third phase of a longitudinal sample of families in Metropolitan Toronto that were making residential moves to either single-family homes or high-rise apartments in central and suburban locations (Michelson, 1977).[1] This sample consisted of intact, middle-class families. The interviews took place in 1970 and 1971. The other was a stratified random sample of households with children taken in 1980, also in Metropolitan Toronto. This was the sample taken for the purpose of better understanding the everyday lives of employed women and their children (Michelson, 1985b).[2] The sample size contained 557 adult women for present purposes.

The time use measures were not identical in these two studies. Only the latter had subjective dimensions. However, the structure of the former enabled more detail on certain points. Because the results of the two are compatible where common analyses were possible, results from both samples are drawn upon as appropriate.[3]

Trips Per Day

In both samples, the employed women, as expected, made more trips per day than the full-time homemakers. In the 1970 sample, the employed women made

[1]These data were gathered with grants from The Canada Council and from The Canada Mortgage and Housing Corporation.

[2]A contribution from the Ministry of National Health and Welfare (Canada) made this study possible.

[3]This analysis was conducted during an academic year as a visiting researcher in the Institute for Building Functions Analysis, University of Lund, Sweden. I am grateful for the stimulating working conditions in Lund. This analysis was conducted within the framework of a grant from the Social Science and Humanities Research Council of Canada.

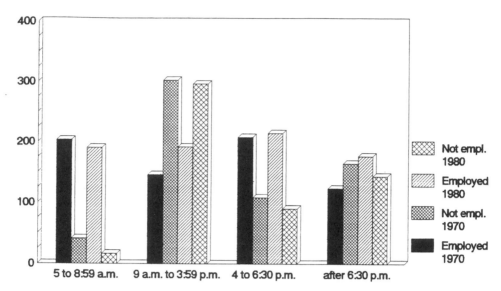

FIGURE 11.1. Number of trips by time of day for employed vs. nonemployed women.

2.83 trips, compared to 1.74 for those without paid employment. In the latter sample, these figures were a nearly identical 3.12 and 1.75. If one can assume safely that two trips among the employed women were dedicated to commuting, compared to none for the housewives, then the latter made more noncommuting trips. Nonetheless, the total number of trips is still greater for those with formal jobs. Previous studies have also shown that the employed women spend more weekday time travelling (cf. Michelson, 1985b).

Time of Day

The time of day in which trips were made could be calculated by integrating time parameters with place information specified as in transit. The meaning of time of day is slightly different in the two samples, inasmuch as the travel data in 1970 reflect when the trip started, while in 1980, when it terminated. But even with this difference, the patterns are quite similar.

Figure 11.1 shows, as expected, that the women without jobs travelled largely between the hours of 9 a.m. and 3:45 p.m.; about half (49.1% and 54.3%) of all daily trips were made within this time period alone. In contrast, trips made by the employed women were spread out throughout the day; nearly half were before 9 a.m. and after 6:30 p.m., while only 21.4% and 24.7% were during the midday time slot used so much by the housewives. The differences between the employed and nonemployed women in terms of time of day for travel are highly significant statistically (chi square, $p < .0005$).

What was not expected was the finding in the 1970 data that the nonemployed women took more trips in the evening. However, when standardized according to the number of employed and nonemployed women in their respective subsamples, this finding assumes its expected direction (though by a narrow margin). Employed

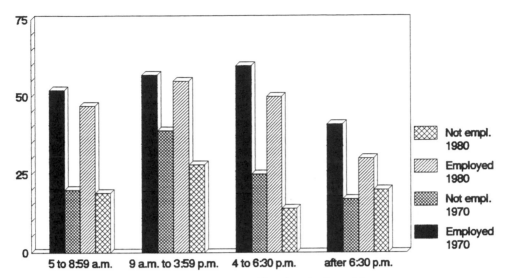

FIGURE 11.2. Percentage travelling alone by time of day.

women took a mean of .51 trips in the evening, compared to .46 for those without paid employment. Nonetheless, the qualitative nature of the trips was not distributed the same way between the two categories. Nearly half of the trips were for recreational or civic activity, regardless of employment status. Among the other half of trips, the employed women took them in connection with work primarily, but also for shopping and courses. The nonemployed women took these other trips largely for shopping, but also took a number of trips for children's activities. Work and courses are likely to be found in more distant locations, with less discretionary choice than shopping and children's activities. Thus, even though the number of trips in the evening is relatively equal according to employment status, their qualitative nature has implications for personal safety.

Qualitative differences in the context of travel become stronger yet when integrating the persons present dimension of time use to the dimensions of time of day and travel behavior.

Accompaniment during Travel

Figure 11.2 substantiates the hypothesis that employed women are more likely to travel alone. There are again marked (and statistically significant) differences between the employed and the nonemployed. In every comparison, the percentage of trips in which a woman travelled alone (or collectively with strangers on public transport) is greater for the employed women, often by a factor of 2 or more. This is true in both the 1970 and the 1980 samples. Only during the evening period do employed women take substantially less than 50% of their trips alone. But even then, this percentage is clearly greater than among the nonemployed women.

The 1970 data suggest that during the evening, women are most likely to be accompanied by their spouses, regardless of their employment status. When not

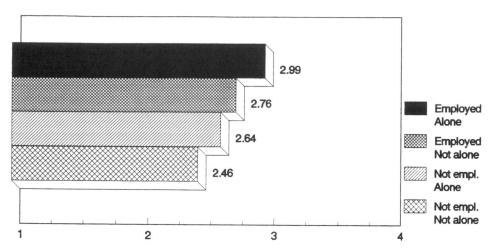

FIGURE 11.3. Mean travel tension by employment and whether travelling alone.

with their spouses, the homemakers are more likely to have children with them, consistent with the number of trips made for children's activities.

During the midday period of peak travel for the homemakers, they are most likely to have children with them. On 42% of such trips, women are accompanied only by children. Moreover, during the two time periods when nonemployed women are least likely to travel, the persons present data suggest that they do so largely because of their children. For example, 20 of the 40 trips that nonemployed women make before 9 a.m. are with their children only. Between 4 and 6:30 p.m., 50 of the 107 trips are with children only. Taking children to and from school, activities, and services is less likely to take women far afield or into uncharted territories. And when it is most likely to be dark outside, their spouses are with them in the great majority of cases. In any case, the homemakers are less likely to be alone while travelling, regardless of the time of day.

If these contexts for travel are relevant to personal safety, then women's subjective perceptions of their trips should reflect the differing degrees of risk.

Subjective Aspects of Travel

The 1980 sample filled out time diaries with two subjective dimensions. Each was a scale between 1 and 7. One was on the degree to which respondents perceived themselves as tense or relaxed while doing the activity. As presented here, the scale is inverted, with a higher score showing greater tension. By and large, tension levels were perceived as low for many activities. But some differences reflected subcategories consistently (cf. Michelson, 1985b). Figure 11.3 portrays mean tension levels for trips, broken down by employment status and whether or not the women were traveling alone.

The hypotheses are clearly supported by these data. Employed women perceive more tension when making weekday trips, as do those who travel alone. The

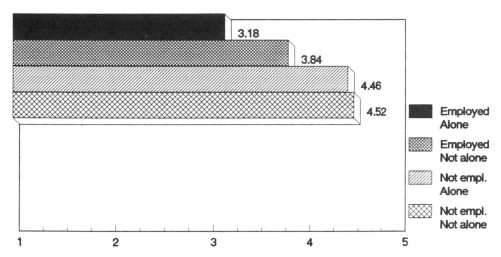

3.18

3.84

4.46

4.52

■ Employed
Alone

▦ Employed
Not alone

▨ Not empl.
Alone

▨ Not empl.
Not alone

1   2   3   4   5

FIGURE 11.4. Mean choice in travel behavior by employment and whether travelling alone.

combination of employment and unaccompanied travel produce the greatest mean level of tension. This additive relationship reflects well the combination of factors involved: time, place, and persons present.

These relationships hold across all the time periods, with but two anomalies. In the early morning time period, the employed women who travel with somebody else are likely to be more tense than those who travel alone. But this is not a time of day with high crime rates. And our previous analyses suggest that the trip to work is a sensitive activity, lying as it does between getting children off to school or daycare and arriving on time to work. It may be that having to coordinate with other persons increases whatever tension is found at that time of day. The women who fit this subcategory have the highest trip tension levels of any single subcategory in the analysis (i.e., throughout all time periods and for all combinations of employment and accompaniment). The rest of the day, the employed women who travel alone report the greatest tension.

Among the homemakers, it is in the evening that the anomaly occurs. Then, it is those who travel alone who show the lowest tension among all subcategories in the analysis. This probably is best understood with respect to the purpose of travel and what has happened earlier in the day. After driving children around throughout the middle of the day, a trip in the evening by oneself for recreational or shopping purposes, most likely close to home, is more likely to be a relief than a strain. This is very different than the locations and the circumstances of the evening trips taken by the employed women connected with work and study.

The time use data on a second subjective dimension, the degree of choice involved in trips, helps clarify how tension is attached to the context of different trips according to employment. In this case, as shown in Figure 11.4, there are also differences in mean scale values by employment and by whether or not a woman travels alone. But the factor of accompaniment is major only among the employed women. The employed women who travel alone experience far less choice in the circumstances of their travel than do employed women who travel with somebody else. But the nonemployed women perceive more choice in their travel conditions regardless of whether or not they travel alone. The difference in this latter case in accompaniment is in the expected direction, but it is not great. This helps support the tension data insofar as the employed women, especially if alone, are more likely to be forced to take trips outside their usual standards of control.

Bringing in considerations of time of day helps to refine certain points, once again. Among the employed women, those travelling alone perceive less choice during every period of the day. The least difference, however, is in the early morning time slot, in which the tension level of those travelling with others is greater. The greatest differences are during the late afternoon and evening periods, when women are more likely to travel in darkness from randomly located workplace locations. In contrast, the greatest degree of choice is reported by nonemployed women making trips alone in the evening. Again, this is consistent and likely to have a meaningful relationship with the low level of tension perceived by these women in the kinds of trips taken then.

CONCLUDING REMARKS

Doubtless more complete and sensitive information on the question of women's increasing exposure to risk that accompanies vigorous participation in the paid labor force can be gained from research dedicated to the subject and employing well-chosen research techniques. Such research would be a useful contribution to emerging research needs.

Nonetheless, the common dimensions of the time use approach have been able to take the implications of conceptual analysis closer to empirical confirmation. Data from two studies on different applications were able to provide support for all four hypotheses arising from the conceptual analysis. This provides greater credence to the consideration of personal safety issues arising as an unexpected and unwanted byproduct of the great change in women's everyday activities. The brunt on society is to focus on how to control personal safety at different times and in different places, not to suggest that people remain home. For this, research efforts have to suggest where such increased attention might be focused. If the contributions of this one integrative research approach are useful, it will be in part to influence the use of broader-scale efforts into this emerging question.

REFERENCES

Barker, R. (1968). *Ecological psychology*. Stanford: Stanford University Press.

Bechtel, R., Marans, R., & Michelson, W. (Eds.). (1987). *Methods in environmental and behavioral research*. New York: Van Nostrand Reinhold.

Franck, K., & Ahrentzen, S. (Eds.). (1989). *New households, new housing*. New York: Van Nostrand Reinhold.

Hanson, S., & Hanson, P. (1993). The geography of everyday life. In T. Gaerling & R. G. Golledge (Eds.), *Behavior and environment: Psychological and geographical approaches* (pp. 249–269). New York: North-Holland.

Harvey, A., Szalai, A., Elliott, D., Stone, P., & Clark, S. (1984). *Time budget research.* Frankfurt: Campus Verlag.

Michelson, W. (1977). Environmental choice, human behavior, and residential satisfaction. New York: Oxford University Press.

Michelson, W. (1985a). Divergent convergence: The daily routine of employed spouses as a public affairs agenda. *Public Affairs Report, 26*(4) [entire issue].

Michelson, W. (1985b). From sun to sun: Daily obligations and community structure in the lives of employed women and their families. Totowa, NJ: Rowman & Allanheld.

Michelson, W. (1990). Measuring behavioral quality in experimental housing. In Y. Yoshitake, R. Bechtel, T. Takahashi, & M. Asai (Eds.), *Current issues in environment-behavior research* (pp. 173–182). Tokyo: University of Tokyo.

Michelson, W. (1994). Everyday life in contextual perspective. In A. Altman & A. Churchman (Eds.), *Women and the environment* (pp. 17–42). New York: Plenum.

Michelson, W. (forthcoming). Analysis and exploration of meaning and outcomes in connection with time use data. In W. E. Pentland, A. Harvey, M. P. Lawton, & M. A. McColl (Eds.), *Application of time use methodology in the social sciences.* New York: Plenum.

Sommer, R., & Wicker, A. (1991). Gas station psychology. *Environment & behavior, 23*(2), 131–149.

Wekerle, G. (1984). A woman's place is in the city. *Antipode 16*(5), 11–19.

Part **IV**
PHYSICAL ASPECTS OF THE ENVIRONMENT

Chapter **12**

EXPERIENCING JAPANESE GARDENS

Sensory Information and Behavior

Ryuzo Ohno, Tomohiro Hata, and Miki Kondo

INTRODUCTION

Japanese circuit-style gardens have long been appreciated for their sequential scenes of beautiful landscapes. It has often been mentioned that Japanese gardens have been designed so as to control visitors' experience, particularly the vistas, as they move along the garden paths (e.g., Hall, 1970). If we can learn from these sophisticated skills of landscape design, we could naturally direct people's attention to something we want to be viewed (e.g., signs in urban streets) without using harsh colors and brutal forms. With some exceptions (Miyagishi & Zaino, 1992), the relation between the physical arrangements and visitors' behavior in the garden, however, has never been analyzed based on objective data.

The present study attempts to explain visitors' behavior in the garden through sensory information in the environment. The following hypotheses were examined through a case study of a Japanese circuit-style garden in Kobe city:

1. People's behavior in terms of (a) change of moving pace including complete stop, and (b) change of viewing direction, which commonly occurs at certain places in the garden path.
2. These behaviors can be explained by the following two aspects of sensory information: ambient and focal visual information perceived from sur-

Ryuzo Ohno, Tomohiro Hata, and **Miki Kondo** • Department of Built Environment, Tokyo Institute of Technology, Yokohama-shi 226, Japan.

Handbook of Japan–United States Environment–Behavior Research: Toward a Transactional Approach, edited by Seymour Wapner, Jack Demick, Takiji Yamamoto, and Takashi Takahashi. Plenum Press, New York, 1997.

FIGURE 12.1. General view of the garden.

rounding scenes, and nonvisual information such as tactile and kinesthetic senses perceived when walking the garden paths.

The roles of ambient and focal visual information in environmental perception have been distinguished (Ohno, 1991). Ambient visual information plays a role in orienting people in space and guiding their larger movements, while focal visual information is used for the detailed examination and identification of objects. Ambient vision can be regarded as a preattentive visual system, which cannot process complex forms, yet can almost instantaneously detect differences in a few local features regardless of where they occur. Julesz and Bergen (1983) noted that "the preattentive process appears to work in parallel and extends over a wide area of the visual field, while scrutiny by local or foreal attention is a serial process, which at any given time is restricted to a small patch" (p. 1638).

As for nonvisual information, the senses of touch and pressure experienced by the foot when we walk along garden paths of different surfaces, and the motion or kinesthetic sense experienced when following a change of direction and height of garden paths are also significant variables of the visitor's behavior. Hall (1970) has noted:

> The designer makes the garden visitor stop here and there, perhaps to find his footing on a stone in the middle of a pool so that he looks up at precisely the right moment to catch a glimpse of a unsuspected vista. The study of Japanese spaces illustrates their habit of leading the individual to a spot where he can discover something for himself. (p. 154)

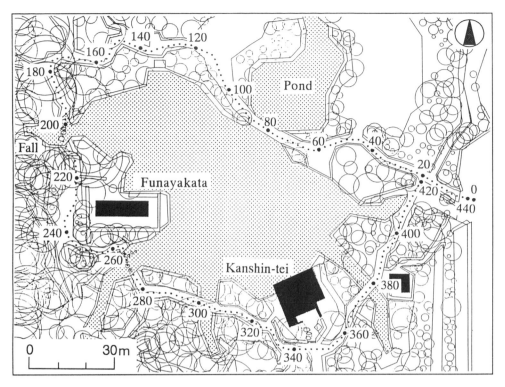

FIGURE 12.2. Site plan and main circuit path of the garden.

BEHAVIOR IN THE GARDEN

Study Site and Participants

A typical Japanese circuit-style garden, Soraku-en in Kobe city, was chosen as the study site (see Figures 12.1 and 12.2). A total of 21 participants, ten male and eleven female students who have never visited the garden, were employed in this experiment.

Experimental Design

Each of the participants was asked to stroll at will the main circuit path. The route to follow was shown to them beforehand on a map of the garden. Participants were allowed to spend as much time they liked. Participants' behaviors were recorded on videotape by a TV camera from a position of about 5 meters behind them.

Procedure

The experiment was conducted on both fair and cloudy days in October and November 1994. From the videotape each participant's motion (viewing directions estimated by head and body rotations) and walking pace were observed and re-

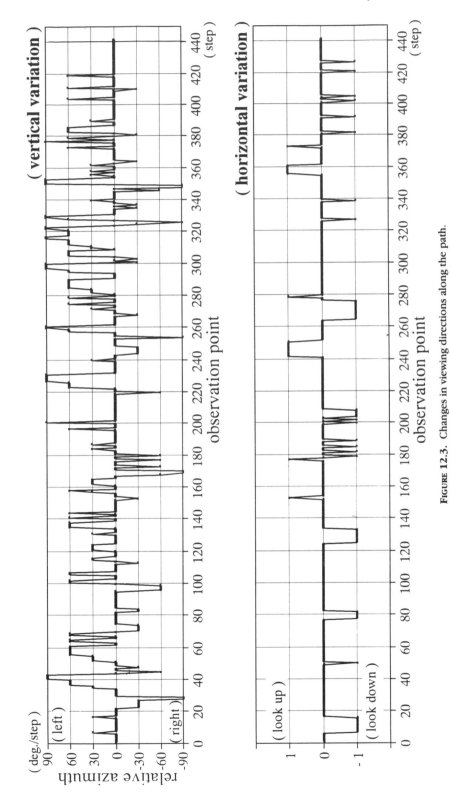

FIGURE **12.3.** Changes in viewing directions along the path.

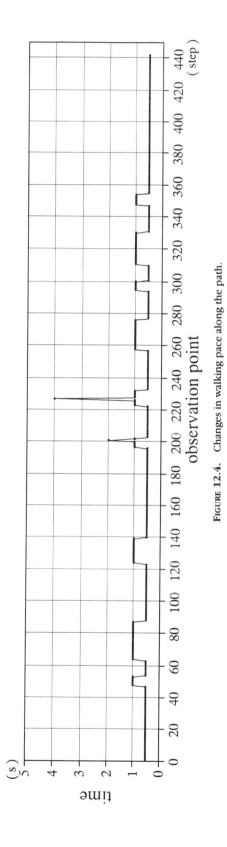

FIGURE 12.4. Changes in walking pace along the path.

corded at 0.5-meter consecutive points (one pedestrian step) along the garden path.

Results

The viewing directions estimated by head and body rotations were described by the relative azimuth from the direction of movement (left: +, right: −) in three grades (angles of 30, 60, and 90 degrees), while the vertical changes were described by upward (+1) and downward (−1). Figure 12.3 is an example of the results for one participant, in which the scale of the horizontal axis is the serial number of observation points along the path.

Figure 12.4 is an example of the results for a participant showing the estimated time spent in walking from previous viewpoints 0.5 m apart. In this figure, a slow pace (0.5 m/s) and a fast pace (1 m/s) were distinguished, and observation points where the participant came to a stop were indicated by a sharp rise in time spent.

To examine the general tendency of the results for all the participants, Figures 12.5 and 12.6 show the points where each participant looked to the left (+90) and to the right (−90), respectively. Similarly, Figure 12.7 shows the observation points where each participant looked downward. Figure 12.8 shows the zones where they slowed down, and the points where they made a stop.

These figures, in which dots form several vertical lines or clusters, indicate that the places where these actions tended to occur are fairly common among the participants. Furthermore, the tendency to choose similar places to stop and look in a certain direction is readily noted in Figure 12.9, in which the viewing directions are shown by arrows on the map. The length of the arrows indicates the number of participants who viewed the same direction from an observation point.

MEASUREMENT AND DESCRIPTION OF SENSORY INFORMATION

To analyze visitors' behavior in the garden with objective data, an attempt was made to describe the changes of measurable sensory input as people moved along the path. A set of personal computer programs was developed to measure some aspects of sensory information, and it was applied to the environmental data extracted from a survey map of the garden.

Environmental Data Creation

Based on a survey map of the garden, the following data were created in the memory of a personal computer by using an image scanner and CAD software.

1. *Land configuration data*: Graphic data of land elevation changes every 0.5-m were identified by coded colors.
2. *Site plan data*: Graphic data of building height and land-covering materials such as path, grass, and water, identified by coded colors.
3. *Tree data*: Numerical data recording location of trees on the site, their type (14 types : 7 different shapes, 2 different kinds) and size of crown.

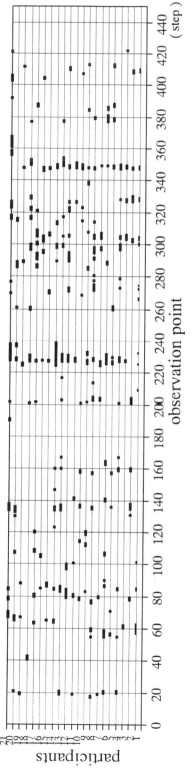

Figure 12.5.   The observation points where each participant viewed to the left side (+90 deg.)

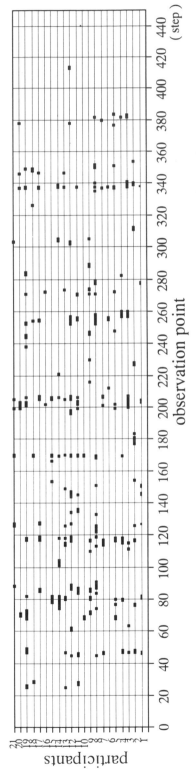

FIGURE 12.6.   The observation points where each participant viewked to the right side (−90 deg.)

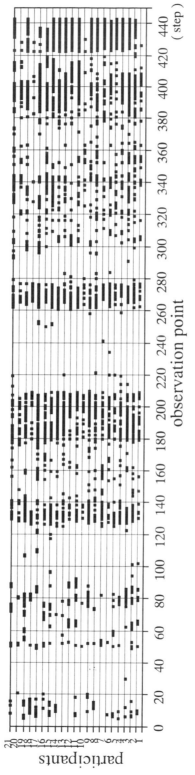

FIGURE 12.7.   The observation points where each participant looked downward.

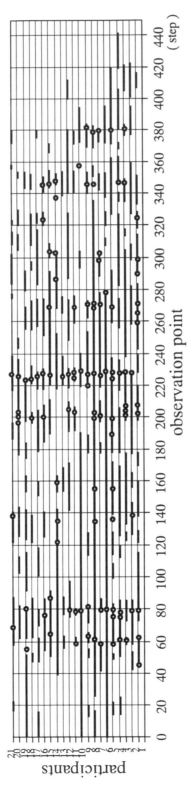

Figure 12.8. The slow walking zones and the points where the participants stopped.

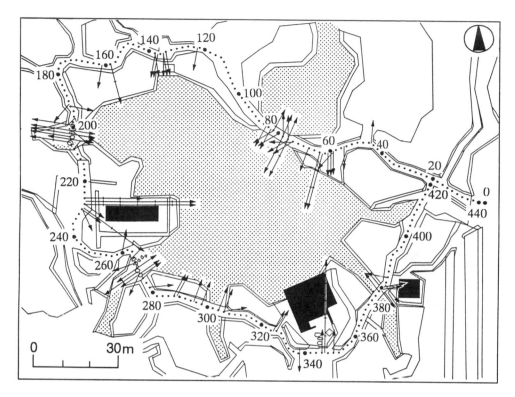

FIGURE 12.9. The stopped points and viewing directions.

4. *Observation points data*: Numerical data recording location and direction of movement created along the garden path every 0.5-m.
5. *Garden path texture data*: Graphic data for the ground textures of the path were classified into 6 types: soil, gravel, paving stones, bridge, stepping-stones, and steps and were identified by coded colors.

Description of Visual Information

In this study, continuous environmental surfaces were considered to be the source of ambient information. The basic units that convey ambient visual information were postulated to be areas of visible surfaces distinguished by differences in their meaning for basic human behavior, or their "affordance" (in Gibson's [1979] term). The components are, in this study, path surface, grass, trees, building, water, and sky. Path, for instance, affords walking but water doesn't.

In a previous study, a personal computer program was developed to assess an array of visual surfaces that surround an observer (Ohno & Kondo, 1994). The program, using the data (1), (2), and (3), assesses surrounding scenes by numerous scanning lines radiated from a station point in all directions with equal density, and records the array of visible surfaces of various components and the distance be-

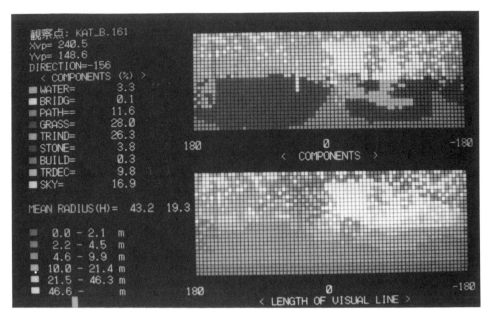

FIGURE 12.10. An example of the assessment results.

tween the surfaces and the station point. The assessment eventually generates two charts as shown in Figure 12.10. Two numerical measures were extracted from the charts that were relevant to ambient information: a ratio (%) of total area of solid angle for each visible component and a measure of spatial volume represented by mean distance (m) from the surrounding surfaces. The program was applied to a sequence of observation points along the path using the data (4). The results, which were described in a similar manner to Thiel's (1970, 1976) "notation," showed changing profiles of the solid angle for each visible component and the spatial volume as one moves along the garden path.

Description of Nonvisual Information

A program was developed to identify and record changes in texture and altitude of garden paths. Ground textures were measured by using the observation point data (4) and the garden path texture data (5).

The changes of relative altitude as one moves along the path were measured by using the land configuration data (1) and the observation points data (4).

DISCUSSION

Most obvious correspondences were noted between the behavior of looking downward and the changes of altitude of garden paths and changes in ground texture (compare Figures 12.7, 12.11, and 12.12). At those places where the path descends (around observation points 52, 128, 181–197, 266, 340, 391–403, and 427–442), most participants look down continuously, and where it goes up (around 1–16, 142–166, 208, 274–277, 310–320, and 353–362), they also frequently look down. In level places participants look down while walking over stepping-stones (around 128–141, 198–207, 266–273, 326–339, and 343–352) and bridges (around 72–86, 302–308, 377–380, and 404–416). These clear relations were obtained because such behavior is required for safe navigation in space.

As to the basis for the horizontal changes in viewing direction, the results suggested that there were at least two different possibilities: one related to ambient or preattentive visual information, and the other related to focal or attentive visual information.

In the former case, ambient vision seems to detect the sudden changes in surrounding scenes (1) when a participant is moving from enclosed space to open space, and (2) when a participant is passing over water on a bridge or stepping-stones. The peaks of the profiles showing the spatial volume in Figure 12.13 (around observation points of 20, 48, 62, 82, 115, 228, 348, and 380) and the visible area of the water in Figure 12.14 (around 80, 203, 270, 305, 379, and 410) correspond with the observation points where the participants looked aside (see Figures 12.5 and 12.6). These changes in a wide area of the visual field seem to activate focal vision and lead it to those directions that should be attended. However, since the above two measures only indicate the average state of the surrounding scenes, they cannot be used to predict the direction to be looked at. Figure 12.15 compares the participants' viewing directions and an asymmetrical distribution of the mean distance from the surrounding surfaces measured by the program. This figure may suggest that people tend to extend their attention to open areas.

In the latter case, focal vision seems to detect such dominant objects as a tea pavilion within the visual field, and the participants tend to stop in order to acquire detailed information. Compare Figure 12.8 and 12.16, in which the visible area of buildings increases at around 135, 228, 326, 348, and 382.

Although most places where the participants looked aside can be explained either by sudden change in the state of surrounding scenes or the visibility of prominent objects, some exceptional cases were observed. At around observation point 205, where the path approaches a small waterfall, their attention was attracted by the sound, which is nonvisual ambient information. At around 255 and 338, where a branch path joined the main path, the participants tended to look in that direction.

In summary, one's viewing direction is first used to acquire information for safe movement in space, and then ambient information shapes the frame of visual field; if one detects something within the visual field, focal vision operates to get detailed information from it. With this interplay of two aspects of vision, we can acquire desired information from a wide area of the environment with limited attentional effort.

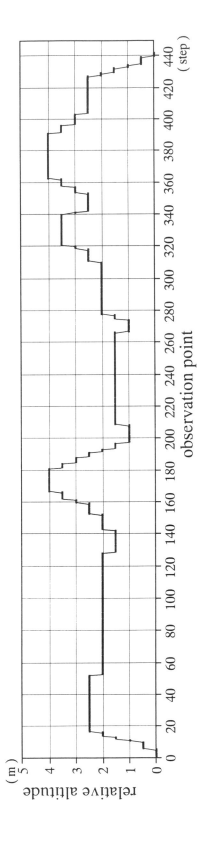

FIGURE 12.11.   Changes in altitude of the path.

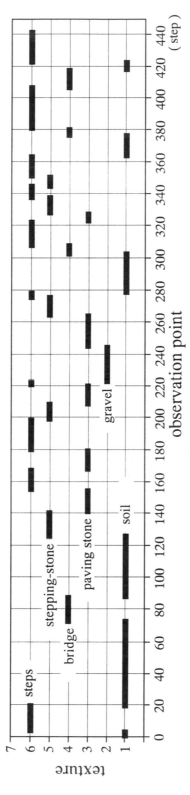

FIGURE 12.12.   Changes of ground texture along the path.

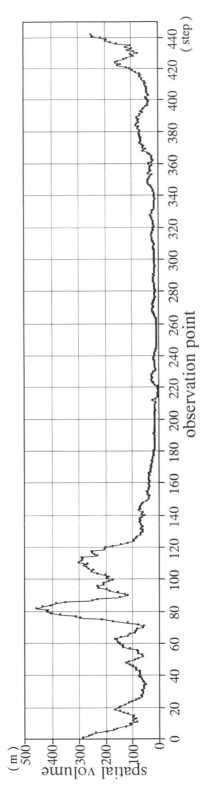

FIGURE **12.13.** Changes in spatial volume along the path.

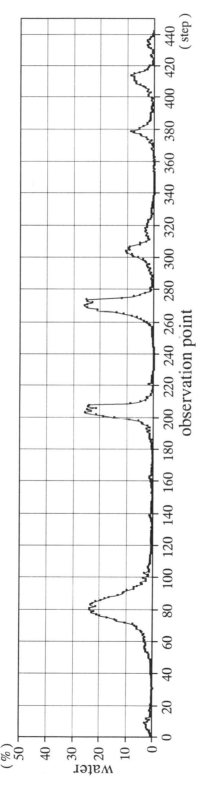

FIGURE 12.14.   Changes in visible area of the water along the path.

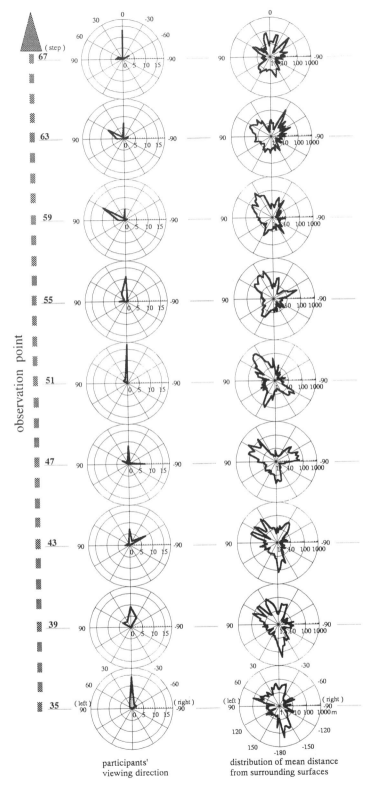

FIGURE 12.15. The participants' viewing directions and the mean distance from the surrounding surfaces (shown only every 4 observation points from 35 to 67).

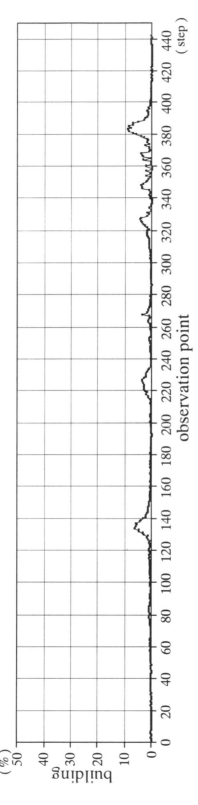

FIGURE 12.16.   Changes in visible area of buildings along the path.

CONCLUSION

The data from the present empirical study generally support the hypotheses that people's behavior commonly changes at certain places in the garden path and can be explained by the sensory information in the environment. Although more comprehensive analysis including other senses such as hearing is necessary to construct an explanatory system, the measurement of sensory information in the environment has been found to be useful for the prediction of human behavior.

REFERENCES

Gibson, J. J. (1979). *The ecological approach to visual perception*. Boston: Houghton-Mifflin.

Hall, E. T. (1966). *The hidden dimension*. NY: Doubleday.

Julesz, B., & Bergen, J. R. (1983). Textons, the fundamental elements in preattentive vision and perception of textures. *The Bell System Technical Journal, 6*, 1619–1645.

Miyagishi, Y., & Zaino, H. (1992). The relation on sequence of landscape between open-close, impact in space, and walker behavior in landscape: A study on visual sequence of landscapes from the view point of walker behavior in landscape. *Journal of Archit. Plann. Environ. Engng (Transactions of AIJ), 440*, 119–125. (In Japanese)

Ohno, R. (1991). Ambient vision of the environmental perception: Describing ambient visual information. *Proceedings of EDRA22*, 237–252.

Ohno, R., & Kondo, M. (1994). Measurement of the Multi-sensory Information for Describing Sequential Experience in the environment: An application to the Japanese circuit-style garden. *Proceedings of IAPS13*, 425–437.

Thiel, P. (1970). Notes on the description, scaling notation and scoring of some perceptual and cognitive attributes of the physical environment. In H. M. Proshansky, W. H. Ittelson, & L. G. Rivhn (Eds.), *Environmental psychology* (pp. 593–619). New York: Holt, Rinehart & Winston.

Thiel, P. (1996). *People, paths, and purposes*. Seattle: University of Washington Press.

Chapter 13

PREFERENCE FOR TREES ON URBAN RIVERFRONTS

Yoichi Kubota

INTRODUCTION

Waterfronts, especially riverfronts, are places of value as open spaces in urbanized areas, providing citizens with sites for mental and physical refreshment. To create and enhance such riverside environments, the planting of specific kinds of trees has been applied since ancient times in Japan. As a result, such places came to be noted in those days as places of interest with landscapes of importance.

Landscape designers of rivers should be aware of the effects of combining trees and riverfronts. Since the introduction of the functionalist methodology of river engineering, however, the planting of trees tends to be done with lesser consideration to landscaping. The planting of trees on riverfronts has been strictly prohibited for security reasons (e.g., possible damage by rivers by floodwaters).

Because of recent progress in the improvement of riverside environments (e.g., reconsideration of the necessity of permanent rivetment by concrete channels, which degrade the riverscape) the guidelines that prevented the planting of trees on riverfronts have been altered (see Figure 13.1). Traditional methods using natural materials (e.g., plants such as bamboos for preventing damage to riverfronts by flood) are now being reevaluated from the standpoint of environmental recovery and natural protection. Such usage of natural materials has already become prevalent in Europe, being categorized as the ecological design method of riverside environments.

The meaning of trees on the waterfront has been analyzed by theorists on landscape from various perspectives. Litton, Tetlow, Sorensen, and Beatty (1980)

Yoichi Kubota • Department of Construction Engineering, Faculty of Engineering, Saitama University, Saitama 338, Japan.

Handbook of Japan–United States Environment–Behavior Research: Toward a Transactional Approach, edited by Seymour Wapner, Jack Demick, Takiji Yamamoto, and Takashi Takahashi. Plenum Press, New York, 1997.

FIGURE 13.1. Functionlly altered river channel with concrete walls.

were among the first generation to conduct studies on landscape with water; the studies, however, did not examine the visual and compositional relationship between trees and river, but rather, categorized inventories of landscapes with water.

One of the most important discussions was presented by Appleton (1975), who developed a symbolistic interpretation of landscape through classical paintings. Appleton (1975) and Nakamura (1982) also discussed that the viewer of a landscape with trees on the waterfront finds symbolic connotations in how the trees are related to the water. The viewer interprets a tree standing on a waterfront as an imaginal or alternative self (i.e., the substitution of virtual or imaginal accessibility for real access to the waterside).

This is observable not only in Western landscape paintings but also in Eastern paintings, that is, monochrome ink drawings or woodcuts from China and Japan. This kind of mental projection as an interactive relationship between the human being and his or her surrounding environment is sometimes regarded as sentimental devotion.

In the modern civilized urban environment, trees tend to be planted geometrically (Arnold, 1980), rather than randomly as they occur in the natural environment. How trees should be arranged along rivers provokes contemporary issues, that is, how human society should perceive rivers with trees in the marginal space lying between artificial urban settings and the natural environment. The river is a continuum, running with an evocative image of nature.

In a similar context, the preference for environmental elements including trees in public places can be regarded as a key to the value and attitude attached to them. Trees are important environmental elements in the ecological and sociocultural sense. A preliminary study on this issue was carried out by Hirada and Kubota (1995).

METHOD

Field Investigation

Saitama Prefecture, on the north of Tokyo, which has wide, flat land with many rivers, canals, and irrigation channels developed since the 16th century, was regarded as a suitable area for field study. A field investigation was conducted in this area to collect examples of trees found on riverfronts.

Documentary Survey

As the physical environment of rivers in Japan has been altered drastically by modernization, especially in this century, it is almost impossible to find well-preserved examples of historic riverfronts with the traditional pattern of tree planting. For this reason, a documentary survey was conducted by reviewing old woodcut prints published especially in the Edo era from the 17th to 19th centuries. To obtain an overview of the traditional tendency of planting trees on riverfronts, prints containing trees on the waterfront were collected and species of trees and patterns of planting were classified.

Experiment A

A simple experiment was conducted using a computer graphic simulation system, which enables users to superimpose the most appropriate image of landscape elements into a given background scene according to their preference. Fourteen subjects were asked to place a preferable visual image of the tree they chose from an inventory consisting of 15 species of trees into the scene of a river without trees, that is to "plant" favorite trees on personally preferable locations along the riverfront, a full-color scene of which was presented on a monitor screen. Subjects could change the position and size of the tree image freely, and they could create more than one planting pattern. Background images of two rivers, Shingashi River and Shoubu River, were selected as representative rivers in the area of southern Saitama that was investigated.

Inquiry Survey

A general inquiry survey by interview was carried out in connection with Experiment A to hear opinions on the basis for selection of tree species to be planted on the riverfront.

Experiment B

As it is inevitable to plant trees in a linear way along rivers already improved from the technical standpoint of flood control, criteria for such planting patterns should be supported from a fairly specific viewpoint. The relationship of theory on spatial scale and planting patterns of trees, especially concerning the interval of trees arranged in a row, has not been analyzed well. As it is not possible experimentally to create new landscapes with different patterns of tree plantings on actual sites, experiments using a technical method was applied. By producing a set of visual images of rivers with different arrangements of trees on riverfronts by means of a computer graphic simulation system, the preference for spatial configuration of trees, namely, the interval of trees in a row, was investigated by asking subjects to compare and choose one of the two paired images; that is, a paired comparison method was applied and several series of visual images with different sets of spatial parameters were created for comparison.

RESULTS

Actual State of Riverside Planting

Rivers in the southern part of Saitama Prefecture, within a circular area around Saitama University of 20 km in diameter, were investigated (see Figure 13.2).

Along rivers stretching a total of 171.2 km in this area, 38,105 m of riverside are planted with tree rows. More than three-quarters, or 85%, of the riverfronts in this urbanized area of southern Saitama were planted with cherry trees (see Figure 13.3).

This preference for cherry trees derives from the 10-century-old cultural tradition of contemplating seasonal change indicated by the appearance of the cherry tree: the blooming of its flowers or the changing colors of its leaves.

The willow tree was second to the cherry tree but found only 6% of the investigated waterfront. The willow, which has branches and twigs that hang over the waters surface to create a sense of depth through shadow and reflection, seems to have fallen out of favor today, probably because of its ambiguous form that does not match with surrounding modern environments.

Traditional Images of Trees on Riverfronts

Seventy landscape images of rivers with trees on riverfronts were collected from reprints of pictorials and graphics of the Edo era. The dominant kind of tree was the pine tree, found in 19 of 70 pictures; the willow was second, followed by the cherry tree. This finding differs from the above result of the field investigation (see Figures 13.4, 13.5, and 13.6).

Figure 13.2. River network in the investigated area in Saitama Prefecture.

The pine tree connotes the good fortune of long life because it takes a long time to mature. Following the pine tree and the willow, the cherry tree was usually planted with other species. The combination of species enriches the meaning of the place where they are planted.

The number of them was also an important key to the grade of place. In 22 cases, a single tree was drawn without other trees in the neighborhood. This isolated image of the tree attracts keen attention by the viewer, generating a sense of personification of the tree, based on the viewer's mental projection of the self. The second category consisted of trees planted in a row and the third contained trees in groups distributed randomly. Dominant locations of trees are bridgehead plazas and watersides. The tree standing at bridgehead plaza is usually a single

FIGURE 13.3. Cherry trees along Kirishiki River.

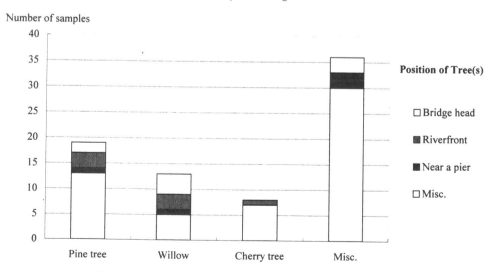

FIGURE 13.4. Species and positions of trees in traditional pictures.

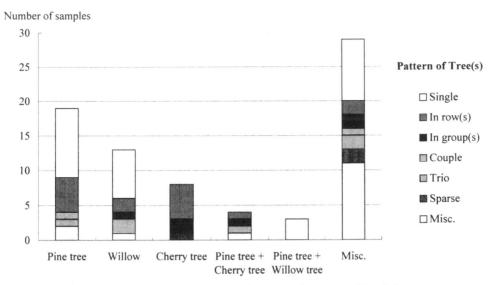

FIGURE 13.5. Combinations of species and patterns of trees in traditional pictures.

willow, and trees along the waterside are generally cherry trees or pine trees arranged in a row or rows or at random (see Figures 13.7, 13.8, and 13.9).

Preferable Visual Images of Trees on Riverfronts

Evaluation by the contemporary generation on how trees should be planted on riverfronts was obtained by experimental procedures. Thirty-one different images created by the subjects' simulations were obtained (see Figure 13.10 and Table 13.1).

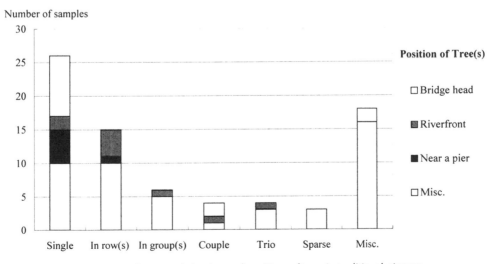

FIGURE 13.6. Patterns of planting and positions of trees in traditional pictures.

FIGURE 13.7. Traditional landscape of a river with a single pine tree on riverfront. "Onagigawa Gohonmatsu" from "Meisho Edo Hyakkei"; wood-cut print by Hiroshige Ando (1856).

FIGURE 13.8. Traditional landscape of a river with a single willow at bridgehead. "Yatsumi-no-hashi from "Meisho Edo Hyakkei"; woodcut print by Hiroshige Ando (1856).

FIGURE 13.9. Traditional landscape of a river with plural cherry trees planted at random. "Sendagi Dangozaka Hanayashiki" from "Meisho Edo Hyakkei"; woodcut print by Hiroshige Ando (1857).

FIGURE 13.10. Simulated image with a cherry tree on the riverfront of the Shingashi River.

Cherry trees in full blossom were the favorite choice, followed by willows and other kinds of trees without leaves in winter. The most conspicuous tendency observed was that the order of the species selected in the simulation was different from that of the documentary survey on traditional graphical materials. A linear arrangement of cherry trees was a favorite in the simulations; this arrangement was also found in old prints.

(4) Preferred Species of Trees for Riverfronts

Preference for species of trees by contemporary generations, declared orally without seeing visual images as in Experiment A, was also noted. Twenty-eight samples were obtained.

Visual associations were identified from verbal replies (e.g., cherry trees were always accompanied by the image of full bloom). Pine trees were not mentioned. Instead, trees usually planted along streets (e.g., zelcova, ginko) appeared, probably because of the frequency of daily encounters with them (see Figure 13.11).

(5) Spatial Balance of Tree Arrangement

In Experiment A, it was shown that a linear pattern of arranging trees was preferred to the traditional random pattern. The major reason for this is that the compositional balance between trees and riverscape is strongly affected by the spatial form of the river, which has been transformed from a natural form into a geometric one by modern public works.

TABLE 13.1. Percentages of Preferred Trees in the Visual Simulation

Species	Riverfront image	
	Shingashi River	Shoubu R ver
Cherry tree (in full bloom)	24	10
Poplar	9	24
Broadleafed tree (cone shape/defoliate)	16	14
Willow	12	14
Ginkgo tree (cone shape)	3	14
Broadleafed tree	3	14
Dogwood	6	5
Pine tree	9	0
Zelkova	3	0
Plum tree	3	0
Mixed	6	0
Cherry tree	0	5
Total	100	100

The most obvious result was obtained concerning the interval of trees planted in a row. Independent of river width, there is a preferable interval between trees, which seems to be about 10 m, slightly wider than the value observed along streets, which is usually 8 m. This interval may correspond to the preferable density of trees on riverfronts for providing pleasant views of the water (see Figures 13.12 and 13.13).

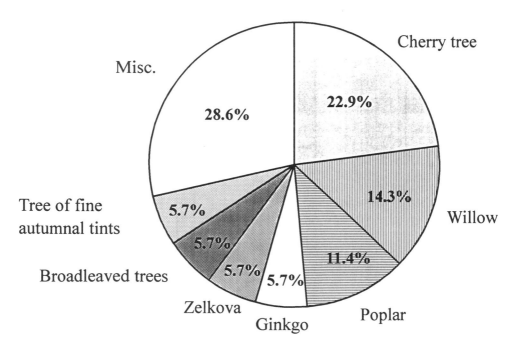

FIGURE 13.11. Verbal answers on preferable trees for riverfronts.

FIGURE 13.12. Simulated images of trees along a river: (above) Interval of trees = 10 m, viewed from riverside; (below) Interval of trees = 10 m, viewed from a bridge.

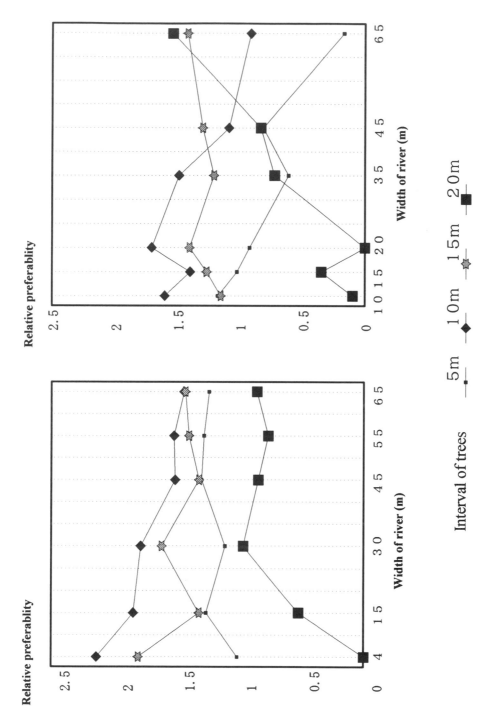

FIGURE 13.13.   Preferred interval between trees in a row along a riverfront; (left) riverside viewpoint; (right) viewpoint from a bridge.

The preferences for tree plantings viewed from a bridge are rather uneven, probably because of the difficulties in judging differences among intervals of trees from an angle; viewing from the riverside provides a better view of each tree.

DISCUSSION

Continuity in the preference for trees consists basically in the life span of trees, which is usually longer than that of human beings. The life of the cherry tree, however, is not as long as other species of trees; its life span is almost the same as that of human beings. This fact seems to be the major reason why people prefer the cherry tree; its seasonal change directly denotes the transient phenomena in nature and corresponding activities in human society. The ephmerality of transient phenomena, such as the blooming of flowers or the falling of leaves, mirrors the life of the human being. The similarity between two lives is the basis for resonance as the background of preference (see Figure 13.14).

The river has similar characteristics, and its flow of water represents the irresistible duration of time. The combination of symbolic trees and the connotative stream is the basis for percciving the aliveness and transiency of life in general. In addition, the river is a space dividing land into two sides, and its waterside or riverfront is marginal to the inland area. This marginality is quite characteristic in riparian environments where land meets water, or human society encounters nature.

As long as the natural elements of landscape, whether artificially introduced or not, are sustained, the preference for them may not change abruptly because of the commonness of daily experience in the society. Even if the natural elements are replaced with new ones, continuity of memory will last for a certain length of time. This can be verified by examples of people who protest against the cutting of trees necessitated by the construction of a public works project to alter the environment of the river for the sake of security against floods.

CONCLUSIONS

After a rapid growth in economy, some Japanese landscape designers and researchers have been seeking ways to reconsider cultural backgrounds in the relationship between citizens and surrounding environments and have tried to recompose them as a new method for creating new landscapes. Trees on riverfronts have been, as mentioned above, out of their coverage owing to technical reasons for security against floods.

In recent years, growing environmental concerns have changed the situation and environmental rehabilitation, as well as preservation and protection, has become first priority in improving riverside places, which can hopefully lead to the amelioration of the artificial environment. The socialized preference for natural elements in the urban context is expected to prevail in parallel with environmental concerns.

A regressive attitude toward species of trees, contrary to the pattern of tree arrangement on riverfronts, is one of the aspects of environmental reconciliation. Many topics are left to be investigated concerning whether the Japanese environmental-cultural background is continuous in its preferences for trees on urban

FIGURE 13.14. Cherry trees in full bloom (Tetsugaku-no-michi, or Alley of Philosophy in Kyoto).

riverfronts after the long influence of modernization and whether there is any new inclination in preference pertinent to the future.

REFERENCES (* in Japanese)

Appleton, J. (1975). *The experience of landscape*. London: John Wiley & Sons.

Arnold, H. F. (1980). *Trees in urban design*. Van Nostrand Reinhold.

*Hirada, S., & Kubota, Y. (1995). *Visual effect of riverside planting*. Proceedings of Infrastructure Planning, Japan Society of Civil Engineering.

Litton, R. B., Jr., Tetlow, R. J., Sorensen, J., & Beatty, R. A. (1980). *Water and landscape: An aesthetic overview of the role of water in the landscape*. Port Washington: Water Information Center.

Mann, R. (1978). *Rivers in the city*. Van Nostrand Reinhold.

*Ministry of Construction (Ed.). (1985). *Guidelines for planting trees on riverside*. Tokyo: M. C.

*Nakamura, Y. (1982). *Introduction to scenics*. Chuko ShinshoL Chuou-Kouronsha.

*Study Group on Landscape/JSCE (Ed.). (1990). *Landscape design of riverfront*. Tokyo: Gihodo Publishers.

Chapter 14

LANDSCAPE VALUES
Congruence and Conflict in Desert Riparian Areas

Ervin H. Zube

INTRODUCTION

Common descriptors of deserts include "desolate," "hot," "arid, uninhabitable," and, occasionally, "wasteland." This chapter reports on research that spans approximately 15 years. The primary focus is on congruencies and conflicts among the perceived values attributed to desert landscapes by diverse interest groups. The specific desert landscapes that were the focus of study are riparian areas places that defy the common descriptors of deserts. They are the places where runoff water from infrequent rains collects and supports linear, green oases that exist in the midst of very extensive desert landscapes that are frequently characterized by a muted spectrum of browns, tans, and grays. More significant characteristics of desert riparian areas, however, are their scarcity and their importance for sustaining life: plant, animal, and human. They have been the sites of human occupantion in Southwestern American deserts for thousands of years (Lister & Lister, 1983). Nevertheless, Ffolliott and Thorud noted in 1974 that significant amounts of riparian areas had been lost within the state of Arizona and that they then accounted for only 0.4% of the surface area of the state. It is very likely that they occupy even less area than that today.

Ervin H. Zube • School of Renewable Natural Resources, University of Arizona, Tucson, Arizona, 85721.

Handbook of Japan–United States Environment–Behavior Research: Toward a Transactional Approach, edited by Seymour Wapner, Jack Demick, Takiji Yamamoto, and Takashi Takahashi. Plenum Press, New York, 1997.

BACKGROUND

The decision to focus on riparian landscapes was based on data from 87 open-ended exploratory interviews and a subsequent statewide mail survey. The interviewees included: elected governmental officials, urban and regional planners, conservationists, recreationists, farmers and ranchers, seasonal visitors (also known locally as snow birds), newly arrived residents, resource managers, and real estate sales persons. The interviews explored:

1. Individual's descriptions of their proximate environments, what was important to them.
2. Their perceptions of changes in the landscape in the vicinity of their place of residence and the relationships of those changes with the perceived quality of those landscapes.
3. Their preferred outdoor activities.
4. Their attitudes about which resource management activities were most important for receiving limited available public funds for management, activities such as recreation, habitat protection, grazing, forestry, and mining.
5. Their favorite places for outdoor activities.

Of primary interest for this chapter are the responses to the second and the fifth questions. The second question is of interest because of the implications that land use changes can have for water demands and the sustainability of riparian areas. The fifth question is of interest because it was the question for which there was the greatest agreement among interviewees, for whom the three most favored places for outdoor activities were mountains, canyons, and water features—lakes, rivers, and streams (Zube & Law, 1981). Within the desert, these elements of the landscape frequently exist in close proximity.

The Statewide Survey

The mail survey included a comprehensive list of land use and resource management issues and values that were perceived to be important by the panel of interviewees. The survey was distributed to a stratified random sample of 1,500 Arizona households. Of the 1,500 questionnaires mailed, 1,303 were deliverable and 847 were returned, for a response rate of 65%. Data from that survey reinforced interpretations of the interview data (Zube, Law, & Carpenter, 1984). Fifty percent of the respondents placed high priority on spending available resource management funds for protecting wildlife and managing recreation resources, activities that are strongly related to riparian areas. The three land use activities that were most preferred around where respondents lived were conservation areas (65.6%), recreation areas (57%), and wildlife habitat (51%). There were, however, notable exceptions found in areas of the state that are dependent upon natural resource based industries such as farming, forestry, and mining. For respondents living in those areas, the land use activities most preferred by 60 to 68% were farming, forestry, and mining. Of equal interest are the land use activities that were least preferred: commercial developments, subdivisions, and second-home or retirement communities. These are the uses that are most frequently associated with growth and change. In a state that, at the time of the survey, had one of the fastest population growth rates in the country, respondents in general were surprisingly antigrowth, at least in reference to their proximate environments.

The survey also addressed questions of favorite outdoor experiences and preferred places for recreation. Paralleling the interview data, favorite experiences were sightseeing, camping, and picnicking. Also paralleling the interview data, preferred places for such activities were mountains, canyons, and water features (Law, 1984).

While the decision to focus on riparian areas was a direct outgrowth of the findings from the interviews and mail survey, the decision was reinforced by the fact that all mountain ranges in the state are publicly owned and have some degree of protection. In contrast, most riparian areas are privately owned and, hence, more susceptible to development and significant change.

STUDIES OF RIPARIAN AREA VALUES

Figure 14.1 illustrates the evolution of this research program, starting with the interviews and ending with the riparian area mail surveys. The mail surveys addressed, among other topics, respondents' description of the specific area, attitudes about appropriate land uses in and adjacent to the area, and perceptions of environmental change in the area. The identification of study areas was guided by a set of criteria that called for emphasis on the southern desert area of the state; a diverse set of land use conditions that ranged from areas of rapid change due to urbanization or suburbanization to protected natural areas; and variation in stream flow from ephemeral (stream flow only following rainfall) to perennial, continuous, year-long, flow. The five areas selected for study and listed in Figure 14.1 include (1) the upper San Pedro River, which has been designated as a National Riparian Conservation Area, the first such designation in the United States; (2) a section of the upper Gila River, a major agricultural area; (3) a rapidly urbanizing section of the Rillito River adjacent to the city of Tucson; (4) a suburbanizing area downstream (to the north) from the U.S./Mexico border city of Nogales; and (5) a section of Sonoita Creek which is adjacent to the town of Sonoita and also to a Nature Conservancy Reserve (Simcox & Zube, 1990; Zube, Friedman, & Simcox, 1989).

This chapter focuses on the upper San Pedro and the upper Gila rivers (Figure 14.2) because they provide contrasts in (1) kinds and intensities of land uses, (2) degrees of land use change, and (3) resident-expressed landscape values. In addition, both study areas center on communities that exert strong influences on the riparian areas and adjacent lands. Table 14.1 presents demographic data that illustrate major differences in the economic bases of the two communities Sierra Vista/San Pedro River and Safford/Gila River, and in their population dynamics.

Notable in these data are the rates of population growth and the differences in employment sectors. While the town of Safford grew at that rate of 10.75% during

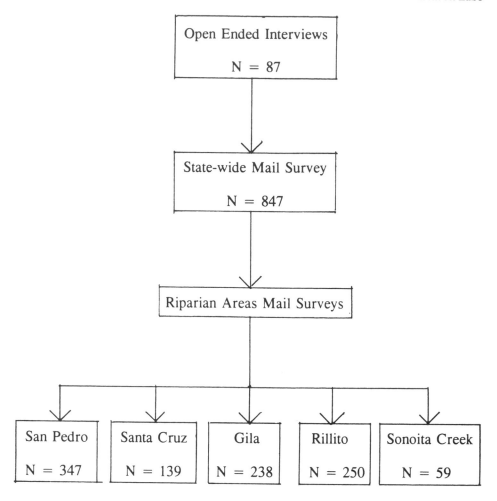

Program N = 1967

FIGURE 14.1. Research program.

the decade of the 1980s, Sierra Vista grew at a rate of 37.6%, more than three times the rate of Safford. Employment statistics provide another useful indicator of important differences between the two communities. The emphasis on agriculture indicated for Safford is contrasted by larger employment in the administrative, service, and retail sectors in Sierra Vista. This latter phenomenon, as well as the differences in growth rates, is directly attributable to the existence of a major military facility in Sierra Vista, Fort Huachucha. This U.S. Army base has been increasing in size as other bases are being closed and selected components are transferred from some of those bases to Fort Huachucha.

Potential for values conflicts can be found in these basic population and employment statistics, particularly as they influence the demand for available water

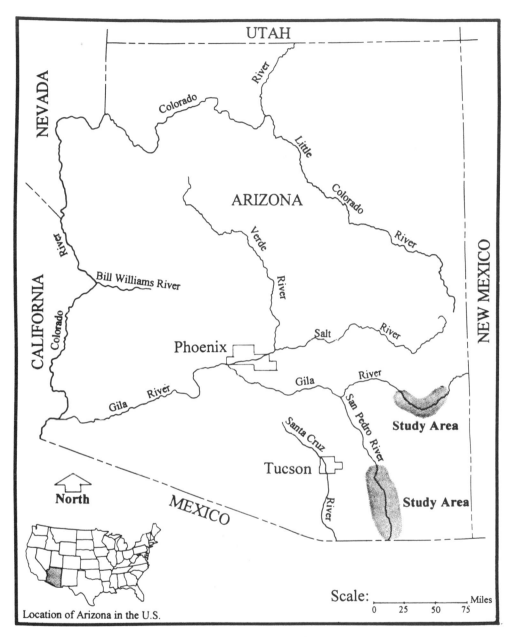

FIGURE 14.2. Riparian study area.

resources. Currently, both rivers are classified as perennial streams in the areas under study. This means that they have flowing waters in their channels year-round, even though at certain periods that flow can be minimal. The demand for water in the Gila River area is related primarily to farming operations. The major crops produced on these farms are cotton, barley, sorghum, and alfalfa, all crops with high

TABLE 14.1. Comparisons of Sierra Vista and Safford

	Safford	Sierra Vista
Population		
1980	7,010	24,937
1989	7,760	34,300
% Change	10.7	37.6
Employment, %		
Admin, service, & retail	65.8	79.2
Ag & Mining	13.2	1.1
Other	21.0	19.7

water demands. A factor that contributes to the potential for value conflicts in the San Pedro River area is the previously noted designation of a section of the San Pedro River as a National Riparian Conservation Area. That section extends for approximately 40 miles (64 km) from the Mexican border to a boundary north of Fort Huachucha and Sierra Vista. In 1978 the area was identified as nationally significant because "its mammalian diversity . . . is the richest assemblage of mammal species in the entire United States and the second known richest in the world" (USDI Fish and Wildlife Service, p. 24). This biological diversity was the driving force to have the area designated as a National Riparian Conservation Area in 1986. Demands for housing, related services, and commercial development have accompanied the growth of the Fort as have increased demands for water. Nevertheless, this 40-mile sector of the San Pedro River is still regarded as a significant wildlife habitat. In 1995 the American Bird Conservancy named it as their first "Globally Important Bird Area in the United States" (Foust, 1995). The Conservancy noted that it is home for more than 100 species of birds and of seasonal importance to another 250 species.

Subjects for the mail surveys included random samples of area households plus the populations of local decision makers (elected and appointed officials), farmers, and realtors. In addition, there were groups of conservationists (members of the Audubon Society) and resource managers included in the San Pedro study. Comparable groups were not available in the upper Gila area.

RELATED LITERATURE

In a review of literature in the then young field of environment and behavior studies, Craik (1970) focused on "the environmental dispositions of environmental decision makers." Within this group he included architects, urban planners, and natural-resources managers. On the basis of that review he advanced the hypothesis that "environmental decision makers differ from their clients in their perception, interpretation, and evaluation of the every day physical environment" (p. 89).

Culhane (1981) published a study of two broad categories of interest groups and a group of professional resource managers employed by the U.S. Forest Service and the Bureau of Land Management. He explored the congruence and conflicts of values among these groups. The first interest group category represented commodity interests, such as timber harvest, grazing, and mining. The second category represented noncommodity interests such as wildlife, wilderness, and recreation. Culhane found that the managers were perceived by each interest group to be biased toward the other. When he reviewed their positions on an environmental-utility scale, however, he found that the managers were in the middle, between the commodity and the noncommodity interests groups. He concluded that those managers, who were responsible for managing resources for multiple values and uses, were doing the job they were expected to do. They represented the multiple use mandates under which their agencies were charged to operate.

Pitt and Zube (1987), in a review of literature relating to the management of natural environments, concluded in 1987 that increasing demands on noncommodity resource values "will sustain the need for research in the use of conflict management among competing users and managers" (p. 1035), thus suggesting a stronger display of conflict than congruence between users and managers.

RESEARCH QUESTIONS

Based on the early literature review by Craik (1970), the study of the perceptions of resource managers and commodity and noncommodity users of natural resources by Culhane (1981), and the later literature review by Pitt and Zube (1987), the questions for this paper are the following:

1. Do professional resource managers of riparian areas reflect the traditional multiple use orientations of the agencies they represent or do they reflect resource preservation attitudes?
2. Do the public and special interest groups perceive growth and concomitant change adjacent to riparian areas to be a positive factor?
3. Will the public and local interest groups support commodity uses of lands adjacent to riparian areas, or will they support uses that are more supportive of sustaining the viability of the riparian areas?

FINDINGS

What Management Position Do Resource Managers Support for the San Pedro Riparian Area and Does It Differ from Other Survey Respondents?

Data from resource managers were available only for the upper San Pedro area. No comparable group with similar management responsibilities exists for the upper Gila area.

Mean scores from discriminant analysis of both 7-point semantic landscape description scales and 5-point (strongly agree-strongly disagree) planning and management attitudinal statements indicate that the resource managers occupied extreme positions on both landscape description and planning and management attitudes. They clearly represented a resource protection value orientation that was not shared or, perhaps, not understood by the general public, real estate agents, local decision makers, or farmers. A notable exception was the conservation group from Sierra Vista whose responses were closely related to those of the resource managers. Also notable is the extreme opposite positions of farmers and real estate agents in reference to both landscape descriptions and attitudes about the adequacy of planning and management (Zube, 1987). The resource managers clearly reflected a protectionist position for the upper San Pedro area (Zube & Simcox, 1987).

What Are the Public and Special Interest Group Perceptions of and Attitudes about Growth and Change?

Analysis of the awareness of change scales indicated no significant differences among respondent groups from either Sierra Vista or Safford, including the resource managers. There were two scales, however, that resulted in means that merit mention. On a scale of 1 to 7, the "more surface water—less surface water" scale had a relatively high mean of 5.0, suggesting perceptions of decreasing surface water flows, a perception supported by annual stream flow data for the 10 previous years. The second scale related to increased off-road vehicle use in the riparian area, a use that tends to be destructive of riparian plants. Perhaps more revealing of attitudes about growth and change are those related to land use and resource planning. The following statements were identified as the best discriminating variables for the upper Gila and the upper San Pedro.

Upper Gila River

- This area can accommodate a lot more growth.
- Farming/orchards should be maintained here.

Upper San Pedro River

- Growth and change can be accommodated without significant change in the quality of living here.
- Protection of water resources and water related lands is very important.
- There is adequate planning in this county to manage growth and development.

Upper San Pedro and Upper Gila Rivers

- Trees and shrubs along rivers, creeks, and washes should be protected for wildlife and birds.

What Land Uses Do the Public and Special Interest Groups Perceive to Be Appropriate for Lands Adjacent to Riparian Areas?

Survey participants were asked to select the most and second-most appropriate and the least and second-least appropriate land uses from the following list: wildlife area, farming, hunting, grazing, protection of water flow, preserved natural area, mining, off-road vehicle driving, flood protection, subdivisions, and recreation areas. No significant differences were found among groups. Frequencies for the total sample revealed the following strong choices:

- Most appropriate: farming, protection of water flow, wildlife area, and flood protection.
- Least appropriate: subdivisions, off-road vehicle driving, and mining.

DISCUSSION

What do the answers to these questions suggest about the long-term viability of these riparian areas?

Perhaps, the single most important insight to be extracted from these data is the apparent low probability of sustaining the viability of the two riparian areas without significant changes in values. The focus of this chapter has been on the congruence and conflict among landscape values. There are at least two dimensions to this concept: the first is congruence and conflict among the diverse user groups and managers: the second is congruence and conflict between those users' values and the sustainability of the riparian areas. Serious points of conflict exist in both dimensions.

Resource managers appear to have relatively little local support for the kinds of management practices that would contribute to the sustainability of the two riparian areas. National designation of both rivers as among the 20 most threatened rivers in the United States sends a strong warning about their futures (American Rivers, 1995). Notably absent in the survey data are indications that respondents are aware of the relationships between land use activities adjacent to the riparian areas and the health of those areas. Equally notable is the apparent lack of knowledge about relationships between surface water and groundwater; of the fact that overpumping of ground water leads to the demise of surface water. Also of significance is the failure of the U.S. Army to acknowledge the probable impacts of the continued growth of the base and the probable concomitant demise in the sustainability of the riparian habitat (Ibarra, 1995).

In both the upper Gila and the upper San Pedro rivers, the continued growth that is supported by much of the population equals continued demands upon groundwater and, eventually, the disappearance of surface water flows. In the San Pedro area, that will also equate with the loss of biodiversity for which the area was designated the nation's first National Riparian Conservation Area. In the Gila area that will probably mean ever-increasing costs of agricultural production as water tables fall and the costs of pumping from ever-deeper aquifers increase.

NOTE

There was a 5-year interval between the data collection for the upper San Pedro and upper Gila rivers studies. During that time, and continuing today, annual meetings with the planning staff in Sierra Vista provide assessments of current conditions relating to development, growth, and change. In addition, news media have been and continue to be monitored for current information. All indicators suggest that values and attitudes related to growth, change, planning, and management have not changed significantly in either area.

REFERENCES

American Rivers (1995). *America's ten most threatened rivers.* Washington, DC: Author.

Craik, K. H. (1970). The environmental dispositions of Environmental Decision Makers. *Annals of the American Academy of Political and Social Science 389*, 87–94.

Culhane, P. J. (1981). *Public lands politics.* Baltimore, MD: Johns Hopkins Press.

Ffolliott, P. F., & Thorud, D. B. (1974). *Vegetation management for increased water yield in Arizona.* Agricultural Experiment Station Technical Bulletin 215. Tucson: University of Arizona.

Foust, T. (1995, August 31). Bird area picked among world's best. *Arizona Daily Star,* p. C6.

Friedman S. K., & Zube, E. H. (1992). Assessing landscape dynamics in a protected area. *Environmental Management, 16*(3), 363–370.

Ibarra, I. (1995, September 1). Army found at fault on river area growth. *Arizona Daily Star*, p. B2.

Law, C. S. (1984). *An experiential assessment of the Arizona landscape.* Ph.D. dissertation,University of Arizona, School of Renewable Natural Resources, Tucson.

Lister, R. H., & Lister, F. C. (1983). *Those who came before.* Tucson: University of Arizona Press.

Pitt, D. G., & Zube, E. H. (1987). Management of natural environments. In D. Stokols, & I. Altman (Eds.), *Handbook of environmental psychology: Vol 2* (pp. 1009–1042). New York: John Wiley.

Simcox, D. E., & Zube, E. H. (1990). Public value orientations toward urban riparian landscapes. *Society and Natural Resources 2*, 229–239.

USDI Fish and Wildlife Service. (1978). *Concept plan: Unique wildlife ecosystems, Arizona.* Washington, DC.

Zube, E.H. (1987). Perceived land use patterns and landscape values. *Landscape Ecology, 1*(1), 37–45.

Zube, E. H., & Law, C. S. (1981). *Natural resource values and landscape planning.* Paper presented at American Society of Landscape Architects Annual Meeting, Washington, DC.

Zube, E. H., & Simcox, D. E. (1986). *Public perceptions and attitudes towards land use and land use control in the Upper Santa Cruz and Sonoita Creek Areas.* Technical Report. University of Arizona, School of Renewable Natural Resources, Tucson.

Zube, E. H., & Simcox, D. E. (1987). Arid lands, riparian landscapes and management conflicts. *Environmental Management, 11*(4), 529–535.

Zube, E. H., Friedman, S., & Simcox, D.E. (1989). Landscape change: perceptions and physical measures. *Environmental Management 13*(5), 639–644.

Zube, E. H., Law, C. S., & Carpenter, E. H. (1984). Arizona survey reveals anti-development attitude. *Landscape Architecture, 74*(6), 97–100.

Chapter **15**

ENVIRONMENTAL TRANSITION AND NATURAL DISASTER
Restoration Housing for the Mt. Unzen Volcanic Eruption

Masami Kobayashi, Ken Miura, and Norio Maki

INTRODUCTION

Recently, Japan has had three serious natural disasters. the eruption of Mt. Unzen-Fugendake, starting in 1990, the tsunami caused by Hokkaido Southwest, Earthquake in 1993; and the Great Hanshin-Awaji Earthquake in 1995. Natural disasters such as earthquakes and volcanic eruptions sometimes cause a large number of people, many of whom have lived all their lives in one place, to be uprooted from homes and familiar surroundings (e.g., Miura, 1995).

This study is concerned with the environmental transition of relocation after a natural disaster. Relocation is a life event that requires the individual involved to change his or her perspective about the relationship between person and environment. When one is forcibly moved from one place to another, one's sense of identity may be threatened and, sometimes, even lost. Close relationships with other people or a memento of the past may help restore some continuity (Ohara & Suzuki, 1992).

Some studies done on relocation such as the institutionalization of older adults, have established its negative physical and psychological effects. However, little is known about the effects of relocation related to a natural disaster and how

Masami Kobayashi, Ken Miura, and **Norio Maki** • Division of Global Environmental Engineering, Graduate school of Engineering, Kyoto University, Kyoto, 606, Japan.

Handbook of Japan–United States Environment–Behavior Research: Toward a Transactional Approach, edited by Seymour Wapner, Jack Demick, Takiji Yamamoto, and Takashi Takahashi. Plenum Press, New York, 1997.

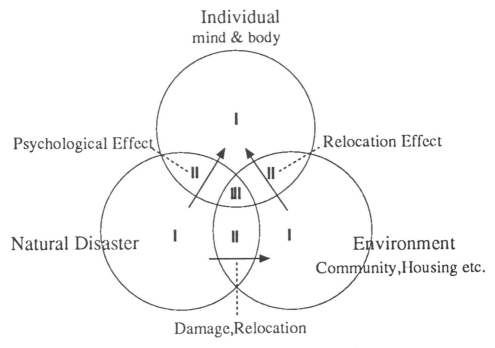

FIGURE 15.1. Environmental transition after natural disaster.

the resulting events affect people environment relations. It was pointed out in one study that relocation is largely determined by the perceived predictability and controllability of the events surrounding a move and differences in controllability between pre- and postrelocation environments (Brenner, 1977). In this sense, relocations due to a natural disaster may have more profound effects than relocations related to other reasons because disaster often deprives people of one or all of the bases for settlement, namely, human relationships, homes and community, and occupation.

We will now provide some background on this study on relocation after natural disaster from a holistic viewpoint. As shown in Figure 15.1, the environmental transition related to the relocation caused by a natural disaster would be conceptualized as consisting of three major elements essential for understanding environment behavior interactions. The environmental transition related to natural disaster has two impacts on individuals. The first concerns built-environment changes, which involve both physical damage and relocations; the second is the acute psychological reactions resulting from the traumatic events of disaster, which are often studied in the realm of psychiatry and psychology. Those who have been exposed to the traumatic events of a natural disaster are in grief and may continue suffering from the pathological condition known psychologically as posttraumatic stress disorder. Through its process, refugees may be required to reorganize their identity in the new environment, which might never be exactly as it was before the disaster (e.G., Raphael, 1988).

FIGURE 15.2. Mt. Unzen-Fugendake.

Environmental transitions after natural disasters are more drastic than any other voluntary relocation. In the transition, one's ego identity may be put to the test because of two simultaneous difficulties: radical environmental change and trauma. Analyzing the relocation offers a good chance to understand the interaction between people and environment, as well as the nature of human settlement.

THE ERUPTION OF MT. UNZEN-FUGENDAKE

Mt.Unzen-Fugendake in Nagasaki Prefecture erupted on November 17, 1990, for the first time after 200 years of silence. The volcanic activity, which has lasted over 5 years has caused two kinds of intermittent natural hazards; debris and pyroclastic material flows into the quiet rural district of this prefecture. Debris flow has occurred frequently in the Mizunashi River with a small amount of rainfall, sweeping away vast former communities and fields. Huge amounts of debris and gravel have drifted with the water and formed tall piles in the Mizunashi basin. In addition to such severe damage, several pyroclastic flows have also reached about 6 km downstream from the top of Mt.Unzen-Fugendake and burned to ashes the community and extensive fields at the foot of the mountain (Figures 15.2 and 15.3). A total of 44 persons, including 19 journalists, have been killed in the pyroclastic flows and 1,399 houses have been destroyed in this disaster. The volcanic activity has done a great deal of damage to this area (Sediment Control Division, 1992).

FIGURE 15.3. House destroyed by debris flow.

RELOCATION TO THE RESTORATION HOUSES

Those who were dislocated by the disaster first got "sheltering" for a few weeks and were then relocated to temporary housing until housing repairs and reconstruction were accomplished. In this disaster, some areas in Shimabara City and Fukae Town were specified as the "Caution District," and an evacuation order was issued for the residents living there. More than 1,455 households had to take refuge in emergency shelters and temporary housing for 2 years or more, and about 800 households are now living their lives in public restoration apartment houses about 2 or 3 km from their former dwellings because entry to some parts of the disaster area has been prohibited.

In an industrial country, shelter and housing issues are less problematic than in other countries. In Japan, the shelter and housing systems after natural disaster are prescribed in disaster relief law. The Disaster Countermeasure Headquarters was established by the mayors to take action to protect residents living in the districts. The government applied the Disaster Relief Act for this disaster area, and in 1991 established the Mt. Unzen Eruption Emergency Countermeasure Headquarters headed by the Director General of the National Land Agency. Since 1991, the local government of Nagasaki Prefecture has supplied 36 housing developments and 1,455 temporary housing units, and has constructed about 800 restoration apartment houses with modern equipment for the refugees' continued use.

FIGURE 15.4. Temporary housing units.

The temporary housing units (Figure 15.4) supplied were low-cost prefabricated units, which were usually used as temporary offices or for storage because of their transportability and easy-to-assemble features. There are three types of units: one-bedroom units (19.8 m^2) for one to two persons; two-bedroom units (29.7 m^2) for three to six persons; and three-bedroom units (39.6 m^2) for seven persons or more. Owing to the limited space available, single-story buildings that consist of two units were erected side by side with hardly any space in between. Some physical problems in temporary housing were identified by our Post Occupancy Evaluation investigation: shortage of storage space, inconvenient bathtub, cramped floor space, absence of a *genkan* (small entrance hall), thin walls, and so on. Neighbors' noise from the thin wall deprived inhabitants of their family privacy, the small floor space irritated them, and the poor variations of the unit type separated families (Maki & Kobayashi, 1994). After some waiting period in the temporary units, refugees relocated to restoration apartment houses.

Restoration houses have been constructed with beautiful finishes and modern conveniences. They mainly consist of two kinds of ferroconcrete and wooden buildings of 50-81m^2; the victims leased these at a low price. There were some differences between modern restoration houses in the housing developments (Figure 15.5) and the large former private dwellings in rural areas. The physical problems in the prefabricated housing units were eliminated in the restoration apartment houses.

FIGURE 15.5. Public restoration apartment house.

AIMS

This study aims to answer the following three questions from the viewpoint of interaction between human behavior and the environment.

1. How is the environment, especially private space, restructured by each individual during the period of adjustment? For example, how does one reorganize one's identity into the new space when one is forcibly moved from one's familiar dwelling?
2. How does the individual feel about the new environment at different?
3. How does a forced relocation related to natural disaster compare with voluntary migration?

METHOD

Interviews were conducted twice, in August and in December of 1994, and took place in both the restoration apartment houses and some rebuilt houses located in Shimabara City and Fukae Town. Out of the 79 households, 24 cooperated in our investigation and agreed to be interviewed. This group is composed of 18 households living in public restoration apartment houses; 2 households living in an apartment house managed by a private owner; and 4 households living in their own houses. All participants who were renters in restoration apartment houses had been homeowners until the disaster.

FIGURE 15.6. A painting put on the floor.

FIGURE 15. 7. Pictures temporarily hung with tape.

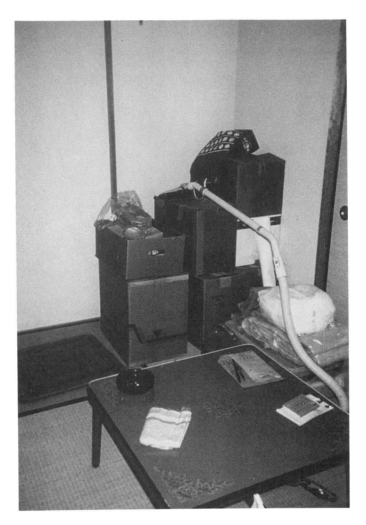

FIGURE 15.8. Boxes unopened in room.

To understand participants' mental condition in detail, we let the individuals talk freely as in normal conversation rather than using a structured interview. The investigation aimed to document the following:

1. Individual's background: age, gender, occupation, family, and so on.
2. Individual's daily life before relocation.
3. Individual's present daily life after relocation.
4. Individual's personal experience and casualties resulting from the disaster.
5. Person environment relations: for example, relationship with other people, approach to the environment, attachment to the environment.
6. Physical traces from sketches and photographs of participants in private spaces.

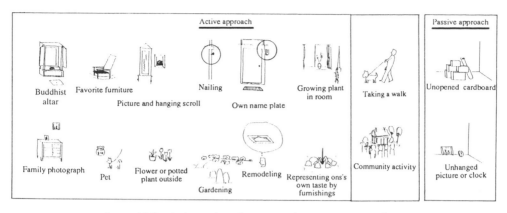

FIGURE 15.9. Active and passive approaches to one's surroundings.

We have examined the damage caused by the disaster, its effect on jobs, changes in people's relationships, and so forth. Natural disasters do not fall on any particular age group or gender, the mental and physical effects of relocation are not always easy to detect. From collecting basic data on differences between the former and latter environments, we have given special attention to two aspects: the approach to one's private space and the attachment to one's home.

When we first started the investigation, we visited some families and felt that their approach to their private space was strongly involved in their mental adaptation to the new environment. As shown in Figures 15.6-15.8, when people are not feeling at home in their environment or do not want stay there permanently, they neglect to hang clocks and paintings on their walls and leave them on the floor or just use tape to hang things up temporarily with boxes unopened and piled on the floor. These would be unusual sights in the homes before the disaster. Figure 15.9 shows 16 (active and passive) behaviors as approaches to one's surroundings.

We examined people's attachment to their environment before and after the disaster. Differences in attachment indicate degrees of mental recovery from the loss. If the victims have converted well to their new life, they can think about which conditions have become better or worse compared to their old life. If, on the other hand, they have not recovered from the shock of losing their homes, they tend to idealize the past.

We also sketched the layout of the physical surroundings of the individual's new dwelling to ascertain how each reorganized his or her identity in the residential space. Effects of each relocation were judged by measures of how the person felt about the relocation and the new environment, as well as how the approach to the environment was restructured.

These points are exemplified in the following case studies:

Case Study 1

- Participants: Mr. A (83)
- Household structure: Mr. A (83), his wife (63)

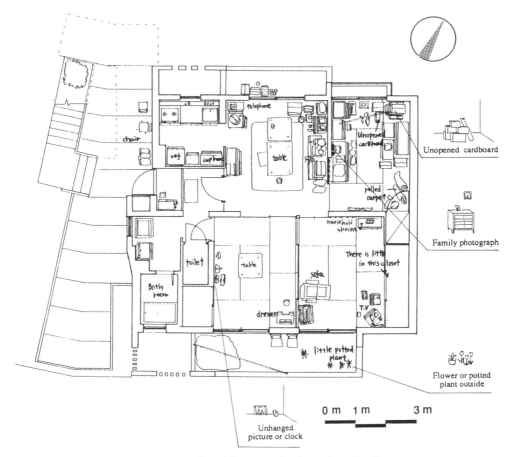

FIGURE 15.10.   Plan of the restoration house (case Mr. A).

- Distance of relocation: 4 km
- Waiting period in the temporary house: 2 years and 10 months
- Former residence: Area of lot 990 m^2, floor space 165 m^2
- Residence: H Restoration apartment house, floor space 63 m^2
- Period after the last relocation: 6 months
- Casualties of the disaster: house, furniture
- Occupation (now/before disaster): jobless/farmer

Mr. A had problems adjusting to the restoration house (Figure 15.10). He is the oldest of the participants. The shock of the disaster was very great for him. His family was not injured but substantial properties were damaged. The house where he lived for over 50 years until the disaster was one that his parents had lost to repay a debt and which he regained with a lot of hardships. It is understandable that he had developed a strong attachment to his house. Unfortunately, he did not take his furniture to his temporary housing but instead kept it on the second floor of the barn, which he believed would be safe from the debris. This fatal mistake increased his loss. The debris flow not only wiped away his old house and field, but also

FIGURE 15.11. Mr. A's furniture is all secondhand.

robbed him of most of his fortune. He even had to be given tableware. Thus, most furniture in his present house is secondhand given as assistance (Figure 15.11). He had been a farmer and he loved to grow all kinds of vegetables and fruits in his field. In addition to these changes, his close relationships did not remain stable. The relationship with his wife has not changed but he has not gone to meet with his former neighbors despite the fact they are only 4 km away. He finds himself with nothing to do in the new environment and spends his day in idleness sitting alone in a chair on the porch (Figure 15.12). He said that he is grateful to the government for such a fine restoration house but he could not accept that he would not die in his own home. His wife said, "He tried to drown his sorrow in alcohol in the daytime." In his room, a clock and a picture hang precariously on the wall. Cardboard boxes left untouched clutter the floor. He does not feel at home in his new surroundings. His sense of incompatibility with his situation remains heavy on his mind.

Case Study 2

- Participants: Mrs. B (80)
- Household structure: Mrs. B (80), daughter-in-law (60), grandson (24)
- Distance of relocation: 9 km
- Waiting period in the apartment house: 5 months
- Former residence: unknown
- Residence: S Restoration apartment, floor space 64 m^2

FIGURE 15.12. Mr. A spends his day in idleness sitting alone on the porch.

- Period after the last relocation: 3 years
- Casualties of the disaster: house, field, furniture, and her son
- Occupation: jobless/farmer

Mrs. B has had some problems adjusting to the restoration house (Figure 15.13). She had lived with her son's family but lost her son in the biggest pyroclastic flow on June 3, 1991. He belonged to the community fire-fighting team and was killed while on duty that day. His family was given preference in the selection of beneficiaries for the restoration housing and was relocated relatively quickly (October, 1991). Mrs. B's close relationships declined not only because of the death of her son but also because of the relocation. Her restoration apartment house is located 9 km northeast of her former residence where she had lived over 50 years and had shared a life with her son. She remains a stranger around her new environment because she does not go out shopping anymore. She loved her former

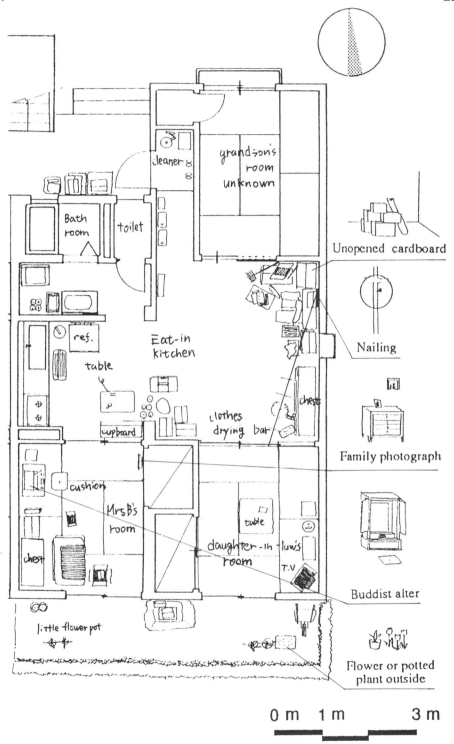

FIGURE 15.13. Plan of the restoration house (case Mrs. B).

Figure 15.14. Piled up cardboard boxes in living room.

community where there was rich vegetation. She could not feel at home in the new restoration house after over 3 years. She said, "Nobody could understand my grief without experiencing the loss of one's own home! No matter how small the house may be, I desire to live in my own home." Her living room was very untidy and cardboard boxes were piled up in confusion (Figure 15.14). No trace of hope or vitality is evident in her house. The absence of the head of the family is undeniable.

Case Study 3

- Participant: Mrs. I (62)
- Household structure: Mrs. I (62), son (40), daughter-in-law (38), grand-daughter (12), granddaughter (11), grandson (9)

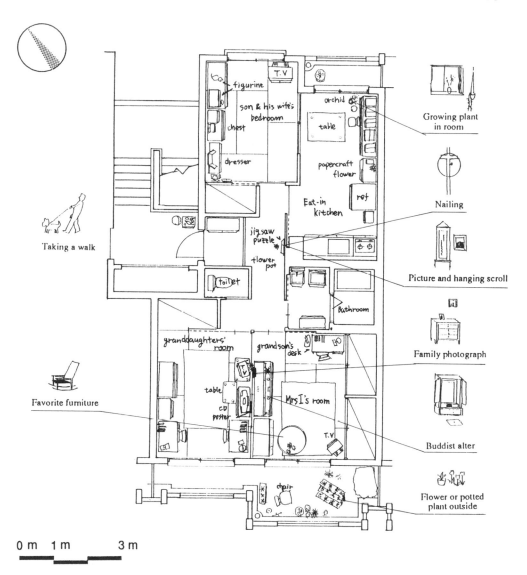

FIGURE 15.15. Plan of the restoration house (case Mrs. I).

- Distance of relocation: 1.3 km
- Waiting period in the temporary house: 3 years
- Former residence: Area of lot 594 m², floor space 198 m²
- Residence: U Restoration apartment house, floor space 69 m²
- Period after the last relocation: 6 months
- Casualties of the disaster: house, some furniture, field
- Participant's occupation: jobless/farmer

Mrs. I has had no major problems adjusting to the restoration house (Figure 15.15). It was consoling for her that her son had carried out some precious mementos before the debris flow swept away her house. Her former dwelling, where she

FIGURE 15.16.   Mrs. I enjoys the view from her balcony.

had been born and where she had spent a long life with her son's family, had been damaged by the debris flow in 1993. For the first 2 or 3 months following the disaster, she could not sleep very well because of the shock of losing both mementos and her favorite activity of growing orchids in her plastic greenhouse. She has, however, good potential for adjusting to her new environment. She has begun to raise potted plants on her balcony. She often sits on her balcony enjoying the cool evening after taking a bath (Figure 15.16). She is also teachaing both her friends and neighbors, who come together at her house every two weeks, how to make paper flowers. In her room she has a small table for this activity (Figure 15.17). It has been helpful for her to maintain her close relationships: her relative's family lives next door to her in the restoration apartment house. By arranging many flowers and mementos (Figure 15.18), she has created surroundings that give her

FIGURE 15.17. Table for Mrs. I's paper flower making.

a feeling of home. She is satisfied with the restoration house, which reflects her identity. She said, "I like this house very much because of its nice view and functional plan. I feel at home in this house." This is because she has taken command of her surroundings and has restructured her identity in the new environment.

Case Study 4

- Participant: Mr. U (70)
- Household structure: Mr. U (70), wife (63)
- Distance of relocation: 3.5 km
- Waiting period in the temporary house: 2 years and 7 months

FIGURE 15.18. Jigsaw puzzle put together by the family is framed and hung on the entrance wall.

- Former residence: Area of lot 330 m^2, floor space 122 m^2
- Residence: own rebuilt house Area of lot 192 m^2, floor space 122 m^2
- Period after the last relocation: 7 months
- Casualties of the disaster: house, furniture, arable land
- Participant's occupation: retired/city officer

Mr. U has had difficulty finding meaning in his new circumstances. A pyroclastic flow burned to ashes his and his wife's former dwelling and community where they had spent over 40 years of their life. He was not injured; however, just before the pyroclastic flow, he had returned his belongings to his former home and they were destroyed in this flow. After the disaster, they had few clothes to wear. They were prohibited by law to enter and ever live in his former community, the Senbongi area. The government purchased his land and house for the construction of super sabo

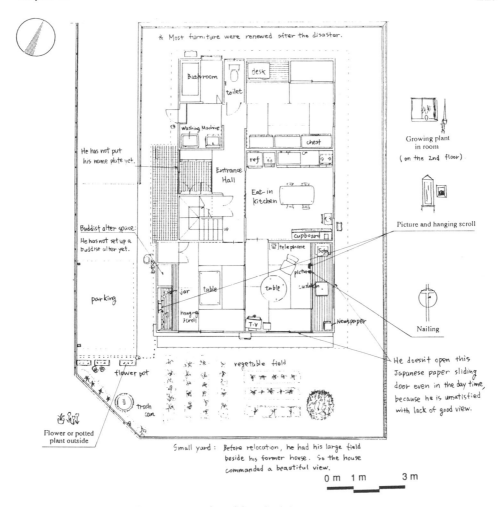

FIGURE 15.19. Plan of the rebuilt house (case Mr. U).

dam. Considering the convenience of living near a supermarket and hospital, he decided to buy 200 m² of land 3.5 km east of his former community, and here he built his new dwelling (Figure 15.19). His close communication with his former neighbors has been cut off by the relocation. He said, "There used to be a lot of contacts, such as seasonal festivals in former days. Today, in spite of not knowing where the neighbors live, I receive only phone calls concerning block meetings of our former community." He continued, "I've become very forgetful these days because not only my house but also most furniture were renewed in this house." The sense of "community loss" forces him to idealize the past. He said, "No house could be equal to my lost house. It has been the best one for us!" Though the square footage of his house has not changed, he is unsatisfied with the lack of the view from windows compared with those of his former dwelling (Figure 15.20) and does not open the Japanese paper sliding door, *Shoji*, even in the daytime. He said, "I

FIGURE 15.20. Mr. U's house is now in a built-up area.

sometimes feel like I am choking in such a house!" It will take more time for him to feel at home in the new place.

Case Study 5

- Participant: Mr. V (66)
- Household structure: Mr. V (66), wife (64), son (40), daughter-in-law (39), granddaughter (15)
- Distance of relocation: 0.6 km
- Waiting period in the temporary house: 2 years and 8 months
- Former residence: Area of lot 2640 m^2, floor space 825 m^2
- Residence: own rebuilt house, area of lot 825 m^2, floor space 264 m^2 (Figure 15.21)
- Period after the last relocation: 9 months
- Casualties of the disaster: house, arable land
- Participant's occupation: retired/farmer

In spite of serious chronic liver trouble (he must be given a blood transfusion at least once a week), Mr. X has obviously succeeded in adapting to the new environment. His former dwelling beside the Mizunashi River was specified as a "Caution District" and was gradually swept away by some debris flows. His family had been relocated to a narrow temporary house in July 1991. It was not only because his days were numbered by his chronic disease but also because he hoped

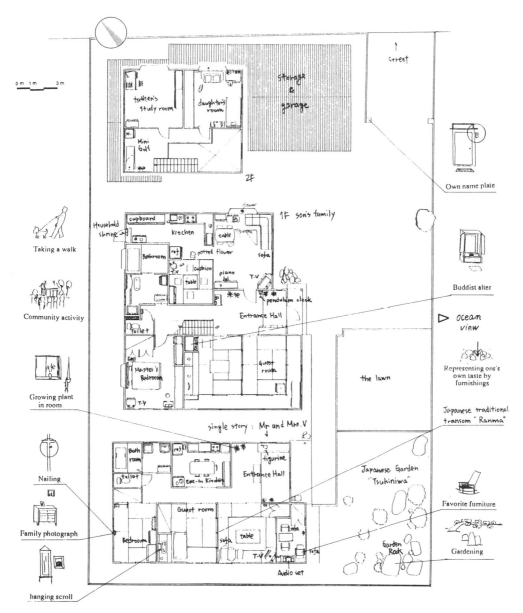

FIGURE 15.21. Plan of the rebuilt house (case Mr. V).

that his brave decision would encourage his neighbors to rebuild their community again. Although some people were concerned with the danger of the debris flow on his site 600 m south of the Mizuanashi River, his family rebuilt two houses on one site; one house for a young household and another for an old household (Figure 15.22). He said it was really a shame since that house was where he had grown up. However, his family was satisfied with the new houses because they were

FIGURE 15.22. Mr. V loves the view from this window.

guaranteed privacy by the adequate distance between the two households. They reused the Japanese transoms, *Ranma*, from the former houses (Figure 15.23). By arranging many flowers and mementos, they created surroundings that give them a feeling of home. Mr. X said, "I prefer this house to the former one with regard to privacy, view, and atmosphere." He has a calm perspective on his relocation.

RESULTS

The 24 households contain diverse cases. Among the participants are families who obviously succeeded in adapting to the new environment as well as families who were unsuccessful.

1. Six out of twenty families who live in restoration housing had a passive approach to their residential space and answered that they did not feel at home and had no attachment to their residential space. This group is composed of five older adults above 60 years and a single person.
2. Eight of twenty families who live in restoration housing answered that they feel at home now.

FIGURE 15.23. *Ranma* taken out of the destroyed house are set up again in zashiki room.

3. Three of twenty families who live in restoration housing answered that they have attachment to their house.
4. All of the four families who rebuilt their own house answered that they feel at home in their new house.
5. Three of four families answered that they have the same attachment to their own rebuilt house as they had to their former one. However, one family idealized the past and has little attachment to the new house.
6. Fourteen of twentyfour families answered that they lost their close relationships with the neighbors of their former community after the relocation.
7. All of the renters in restoration housing desired to become homeowners again.

In accordance with the impact of the disaster and the degree of recovery of the environment, adaptation gaps were identified. Some older adults, who were violently traumatized by the disaster and by the "loss of meaning of life," were vulnerable; they neither feel at home in the modern restoration apartment house nor arrange their residential space well. Further, they do not reorganize their identity to the new environment in a positive manner (Toyama, 1988; Takahashi, 1992).

One reason is that they do not want to accept themselves as socially weak. There would be quite large differences between the houses they had built through their own efforts and the ones that were prepared through help from others. In addition to this, the question of whether they could maintain their identity in the postrelocated environment is an important factor in successful relocation. However, the most substantial factor is social participation   whether the individual has the chance to participate in the revival and society and feel control over his or her surroundings. The negative effects of relocation increase with both the severity and duration of exposure to traumatic events and the lack of control over the environment.

Three of four families who relocated to their rebuilt houses clearly succeeded in adapting to the new environment. Building one's own house and arranging furniture are a representation of one's identity in space. Successful relocation depends not only on the fact that their homes had been reconstructed physically but, more important, on whether they participated in rebuilding their private territory. These factors brought them the mental recovery necessary for mental rehabilitation.

CONCLUSIONS

There are two types of relocation: voluntary and involuntary. These two seem alike, but they are absolutely different. Ending a relationship differs greatly depending on whether you leave your partner or your partner leaves you. The most characteristic feature of such relocation in the case of the Mt.Unzen eruption is the fact the refugees suddenly lost their familiar environment. From psychological viewpoint (Parkes & Weiss, 1987; Underwood, 1995), this could be compared to lost love or severe bereavement.

In the case of bereavement, numerous studies have been done by psychiatrists and psychologists. These studies, have identified that expressing emotion naturally is necessary to integrating the traumatic experience into ongoing life (Noda, 1992).

In this sense, whether each individual lets his or her identity affect the private space by using decorations could become one index of how they feel about their new environment.

How a society deals with disasters shows the true nature of that society. In Japan, impressive public housing is being constructed. However, these buildings are constructed in the same way as any other public housing, with little consideration of the victims' mental recovery. Freud stated that in the process of recovering from the pain of parting by death with a loved one, mourning work, that is, accepting the sorrow, becomes necessary (Okonogi, 1979). What the victims need most of all is the chance to build their identities once again. Readying stately residences and reconstructing the town without giving consideration to the remaining abilities and human dignity of the victims will do more harm than good, putting the victims in an even weaker position in the new environment. In material-rich societies, impressive public housing is built after a disaster. From an outsider's point of view, it seems as though the calamity has ended. When substantial reconstruction is quick, it is difficult to understand that the reconstruction of the psyche requires time.

It is extremely important to reconsider the concept of reconstruction from an environmental psychological point of view. A restoration house should be rebuilt

FIGURE 15.24. Msjit constructed in a restoration housing complex in Flores (Indonesia).

in order to offer these people not only a residence but also a chance to restructure their mental recovery and physical environment.

What can restoration housing do to integrate the traumatic experience into one's ongoing life? As mentioned before, the people relocated to rebuilt homes have adapted to the relocation. What is important is to participate in reorganizing the postdisaster environment.

We also surveyed about restoration housing in Indonesia and the Philippines, where there are many natural disasters such as volcanic eruptions, typhoons, and earthquakes. In these countries, the problem of relocation did not clearly exist, but other problems, such as a shortage of food, unemployment, and the like, are more serious. (Kobayashi, Maki, & Miura, 1994). However, the restoration housing in these countries has potential lessons for Japan. The restoration housing of the Philippines and Indonesia do not have good standards like Japan, but churches and *Msjit* (the praying place for Muslim) are constructed in the restoration housing complexes (Figure 15.24). They play a role in postdisaster trauma as much as the community center does. Core housing is popular in restoration housing in these countries. Core houses have only one room and kitchen; toilet and other improvements are made by the dweller when they have money and time. Through construction, the victims are forced to have a relation to the environment of their new location. Respecting victims' dignity and abilities and allowing them participation in the postdisaster reconstruction of the environment should be taken into account in restoration housing after natural disaster.

REFERENCES

Kobayashi, M., Maki, N., & Miura, K. (1994). *A study on the temporary housing system after natural disasters in Indonesia.* Japan-Indonesia Joint Research on Natural Hazard Prevention and Mitigation, Disaster Prevention Research Institute, Kyoto University (pp. 276–292).

Maki, N., & Kobayashi, M. (1994). *Study on temporary housing: In case of Shimabara-City and Okushiri-Town.* Paper presented at the annual conference of the Institute of Social Safety Science, vol. 4, (pp. 71–79).

Miura, K. (1995). *Study on relocation due to the eruption of Mr. Unzen-Fugendake: Environmental transition in case of natural disaster.* Master's thesis, Division of Global Environmental Engineering, Graduate School of Engineering, Kyoto University.

Noda, M. (1992). *Mono Tojbōnite [Study on mourning of bereaved families.]* . Iwanami Shoten.

Ohara, K., & Suzuki S. (1992). Causes to move into and social activities in the housing for the elderly. *Journal of Architecture, 442,* 65–71.

Okonogi, K. (1979). *Taishō Sōshitsu. [Object loss.]* Chūkō Shinsho.

Parkes, C. M., & Weiss, R. S. (1987). *Recovery from bereavement.* New York: Basic Books.

Raphael, B. (1988). When disaster strikes: How individuals and communities cope with catastrophe. New York: Basic Books.

Schulz, R., & Brenner, G. (1977). Relocation of the aged: A review and theoretical analysis. *Journal of Gerontology, 32*(3), 323–333.

Sediment Control Division, Sediment Control Department, River Bureau, Ministry of Construction. (1992). Eruption of Mt. Fugen-dake of Unzen and Volcanic Sabo Projects.

Takahashi, T. (1992). Housing of the elderly from the viewpoint of environmental transition. *Sumairon, 23,* 18–23.

Toyama, T. (1988). Identity and milieu: Study on relocation focusing on reciprocal changes in elderly people and their environment. Building Function Analysis, The Royal Institute of Technology.

Underwood, P. R. (1995). *Dealing with trauma response syndrome* (pp. 125–129). Misuzu Shobou.

Chapter **16**

ENVIRONMENTAL PSYCHOLOGY AND BIOSPHERE 2

Robert B. Bechtel, Taber MacCallum, and Jane Poynter

INTRODUCTION

This chapter deals with two categories of research in environmental psychology, the ecological psychology of the Biosphere 2 environment and the selection process related to that environment.

ECOLOGICAL PSYCHOLOGY OF BIOSPHERE 2

On September 26, 1991, Biosphere 2 began its first mission when eight persons—four males and four females—were sealed inside a 3.15-acre glass and steel enclosure in which they would have to recycle their water, grow their own food, and generate their atmosphere through plants. The purpose of this 2-year mission was to answer the question of whether a complex, materially closed, energetically and informationally open ecosystem including several biomes could persist and support human life for that period. At the outset it was not known if there was some particular problem inherent to synthetic biospheres in general that would cause them to fail. There were, of course, problems, particularly with maintaining oxygen levels and with producing enough food for eight people. These problems were

Robert B. Bechtel • Department of Psychology, University of Arizona, Tucson, Arizona 85721.   **Taber MacCallum** and **Jane Poynter** • Paragon Space Development Corporation, Tucson, Arizona 85714.

Handbook of Japan–United States Environment–Behavior Research: Toward a Transactional Approach, edited by Seymour Wapner, Jack Demick, Takiji Yamamoto, and Takashi Takahashi. Plenum Press, New York, 1997.

eventually understood and attributable to the specific design of Biosphere 2 and not a generic problem with enclosed ecosystems.

A central question arises as to how to characterize this environment in terms of the human beings who lived in it. In every sense it was intended to be a mini-earth, attempting to duplicate living environments (biomes) such as the desert, marshes, ocean, savannah, and rain forest. Yet this miniworld would be occupied by only eight people. How would they duplicate earth's population? Would they destroy the very environment that was to maintain them?

How can the eight persons who occupied the structure be characterized? Many saw them as the crew of a spaceship. Others saw them as tending a greenhouse. How would they relate to one another? As a crew of a ship in military fashion or as an extended family? What kind of community would they form?

Method

The work of Roger Barker and his colleagues (Barker, 1968; Bechtel, 1977; Schoggen, 1989) permit answering this question in a quantitative form. Barker's principal discovery was the *behavior setting* as the chief unit of human behavior. Behavior settings are the natural units into which people organize themselves to get the business of daily life done. They are part of the community, and at the same time they are recognizable as units of behavior. Each behavior setting is tied to a specific time and place. All communities form behavior settings in order to carry out their existence.

In order to determine whether two observed behavior patterns consist of one or two behavior settings, Barker (1968) created the K-21 scale to quantify differences and then make the determination by counting up overlaps. The K-21 scale consists of seven dimensions: global behavior, population, leaders, physical space, time, objects used, and behavior mechanisms. Technical definitions are given in Barker (1968) and Schoggen (1989). What is important to know is that when any dimension overlaps another by more than 50%, it is counted on the merge side of the decision. If more than four dimensions overlap, the two behavioral events are considered one behavior setting. There is a fixed number at which this decision occurs. Any sum of the seven dimensions adding up to 21 or over means the two entities are seen as two separate settings.

Data were collected from terminal interviews with the eight Biospherians, Mission Control officers, and a set of 50 K-21 scales done by Taber MacCallum and Jane Poynter, two of the Biospherians.

Using the techniques of Barker's methods, which are collectively called *Ecological Psychology*, the question of how to characterize the human environment of Biosphere 2 becomes: How many behavior settings can be discerned and how did these serve the occupants? The Biosphere 2 physical and social structure, presents some interesting questions posed in behavior setting terms. For example, do the five biomes constitute separate behavior settings? One crew member was assigned responsibility for each particular area, but there are two cocaptains among the crew and an associate director of research. How do these leadership roles overlap in each

TABLE 16.1. Fifty Locations Tested by the K-21 Scale

1.	Savannah	26.	Command romm
2.	Thornscrub	27	Roy's room
3.	Desert	28.	Mark's room
4.	Marsh	29.	Lounge
5.	Ocean	30.	Music room
6.	Beach	31.	Taber's room
7.	Tiger pond	32.	Sally's room
8.	Scrubber room	33.	Jane's room
9.	Wilderness basement	34.	Bathrooms
10.	Scrubber area	35.	Gale's room
11.	South lung	36.	Linda's room
12.	Savannah airlock	37.	Spare room
13.	Promenade duct 1	38.	Laser's room
14.	Intensive ag. basement	39.	Exercise room
15.	Intensive agriculture	40.	Recreation room
16.	Orchard	41.	Interview room
17.	Animal area	42.	West lung
18.	Plaza	43.	Terrestrial wilderness
19.	Visitor's window	44.	Ocean biomes
20.	Machine shop	45.	Entire wilderness
21.	Dining room	46.	Entire basement
22.	Kitchens	47.	Habitat
23.	TV room	48.	Int. Ag. & ktichen
24.	Medical facility	49.	Library
25.	Analytical lab	50	Rainforest

biome? In addition, there are overlaps in time, population, objects, and other dimensions of the scale.

In the intensive agriculture section each crew member had agreed to work at least 2 hours per day in order to grow enough food. While this meant 8 hours one day for one person and none for another, how were these "trade-offs" worked out?

In order to test out the reliability of the procedure, the author did a cursory behavior setting survey based on the interview data, and MacCallum and Poynter did an independent survey based on their own notes and diaries. Their survey began with comparing 50 locations inside the structure (see Figure 16.1 with each other by using the K-21 scale. The 50 locations are listed in Table 16.1 and many are shown in the layout of Biosphere 2 in Figure 16.1.

Results

The best initial candidates for separate behavior settings were the apartments of each crew member. Each of the eight had a separate living room and bedroom but shared a bathroom with one another. Technically, all eight were free to crowd into any one of the apartments at any time but seldom did. There was relatively little overlap between apartments and between apartments and the larger areas. The K-21 score on apartments was 28 by Bechtel's rating and an average of 24.9 by MacCallum and Poynter (M = P survey), a fair amount of agreement.

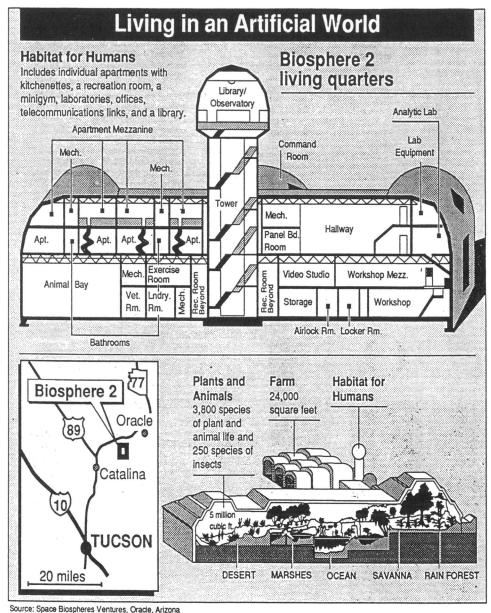

FIGURE 16.1. The layout of Biosphere 2.

The surprises came in the relatively high scores for the seemingly vacant areas such as the spare room (K = 28), the ocean (K = 24) and the scrubber area (K = 25). These were places with low occupancy time but with practically no overlap in population and other dimensions.

The medical laboratory also earned a score as a separate behavior setting on Bechtel's survey (K = 29) but was marginal (K = 20) on the M-P survey. This is largely due to the high score on nonshared behavior objects such as the X-ray machine and by the nonshared leadership of the doctor and his assistant. Most behaviors in the laboratory are complete in themselves and do not affect other settings. Population overlap is extensive. MacCallum and Poynter, however, had better knowledge of the real overlaps in function, so the laboratory merged with the medical facility as a whole.

In another contrast, Bechtel rated the dining room only K = 18 whereas M-P rated it a 23.

Overall, the M-P survey resulted in 28 behavior settings, but 14 of these were at the 21 or 22 level and relatively empty of population. These included the spare room, the exercise room, the scrubber area, and other seldom used places. These were places where the Biospherians could have gone to be alone in addition to their rooms.

The average number of behavior settings available to each Biospherian is 3.5, which is much larger than the slightly over 1 behavior setting per inhabitant of Barker's small towns. Thus, it should have been, in ecological psychology terms, a very *rich* environment, having lots of variety of behavior and varieties of people.

But varieties of people it did not have. There were only eight persons present in the 3.15 acres. *Richness* as Barker defined it meant lots of different kinds of people, people of all ages, occupations, and so on. A richness index is calculated by multiplying the kinds of behaviors observed by the hours of time people spend in the setting and dividing this by 100. A really rich setting would have a richness index of 50 or over.

Discussion

The environment of Biosphere 2 was severely limited by having only eight persons. This puts a cap on the amount of richness in any behavior setting inside the structure by limiting it to about 15. Put in more commonsense terms, the Biospherians were limited by having to see the same people no matter where they went. The option of creating more behavior settings would not alleviate the situation because it would still be the same people all over again. To put it in more simple terms, the Biospherians were stuck with each other.

Or were they? By special electronic arrangements, already built into the structure, each Biospherian had access to private phone lines, FAX, and e-Mail networks. In addition, people could "visit" either by TV at Mission Control, or at the visitor's window next to the airlock. The extensiveness and diversity of these contacts can be illustrated by the fact that MacCallum and Poynter were able to build a house and establish a business corporation while they were locked inside. Three Biospherians stated that they were able to meet more new friends and socialize more than they had been able to do in their lives before they were confined. It appears all eight were very socially active during the 2 years of confinement. It makes the use of the term *confinement* questionable.

The electronic outreach creates a problem in measuring behavior settings in these conditions. If each contact is considered a behavior setting in itself, then is it a behavior setting *inside* or *outside* the Biosphere 2 environment? These overlapping settings could be seen both ways, as adding new behavior settings to the environment or as making the Biosphere 2 environment part of external behavior settings. These are like the extrusive and intrusive behavior settings in small towns (see Bechtel, 1977): some come from outside and enter the town, others begin in the town and go outside.

Although these contacts did not add up to many hours per day, they constituted a major resource for the occupants, and the mere fact that such resources are available significantly reduces the sense of isolation and loneliness (see Bechtel & Ledbetter, 1976). These contacts overcame what Spivack (1984) calls *setting deprivation*, the lack of behavior settings to accomplish what needs to be done in daily life.

The Biospherians could have suffered the same deprivation had they been confined to the actual physical boundaries of the structure. While it was never intended that they endure total isolation, it was also not anticipated they would make such extensive use of the electronic extensions into the outer world. Perhaps it is just this phenomenon that is anticipated for the information highway of the future.

One further aspect of the environment as behavior settings must be explored. The physical demands of the Biosphere 2 routine for living had a very coercive effect on the inhabitants. Each one mentioned the pressure to perform in the agricultural "fields." In addition, the oxygen shortage caused hypoxia resulting in fatigue, difficulty in breathing, limited attention span, and other symptoms. The food shortage meant they got down to a low of about 1,800 calories per day, which was gradually increased to 2,300 calories per person per day after the average Biospherian had sustained a weight loss of 15%. Food became an obsession. One Biospherian reported that watching a movie was difficult if the actors were eating because all he paid attention to was the food.

These experiences made each of them keenly aware of the amount of effort it takes on a daily basis just to survive. One said, "Now when I look at food in a grocery store I am painfully aware of just how much effort it took to put that food on the shelf."

The simple tasks of survival dominated their lives and became a unifying element despite their differences. And differences did arise. There was a clear fractionation between four who supported the "human" view of operations versus those who had a more "mission"-oriented outlook. Nevertheless, all eight felt the need to perform despite whatever differences they might have.

The study of behavior settings in this mission shed light on the usefulness of electronic media in removing a sense of isolation despite the demanding qualities of the environment. The use of behavior setting methodology enabled a quantification of the social fabric of life inside Biosphere 2.

THE SELECTION FACTOR

It was Helmreich's (1974) Tektite studies that provided the extreme example of how professional people are motivated to perform well under the most trying of environmental conditions. Professionals select themselves and are screened to enter into difficult conditions. This *selection* factor is an important aspect in the design and performance of spacecraft, isolated environments, and the testing of new inventions because the motivation of the professionals may be so high as to mask real physical difficulties in the new environment to be tested. The situation was no different for Biosphere 2. The persons who went inside had selected themselves and were selected to be highly motivated to perform.

Method

Crew members were selected by the Biosphere management based on an individual's performance during the design and building of the structure. Scientific methods of crew selection were not employed. However, the crew of the first mission were tested and interviewed just prior to departure in September of 1993 and the next crew were interviewed and tested just prior to entry in January 1994. Tests were administered and scored by Michael Berren of the Arizona Center for Clinical Management. Interviews were conducted by him and the senior author.

The test administered was the Minnesota Multiphasic Personality Inventory—2 (MMPI-2), the revised version of the earlier test. The MMPI is the most widely used of all personality-oriented instruments. It has been used extensively in isolated environments. It is also used as a standard screening instrument for selecting airline pilots (Butcher, 1992).

The MMPI has three validity and ten content scales plus a variety of secondary scales. Scores are reported as total, or "T," scores. The validity scales are the L, or lie scale, which measures a person's willingness to disclose personal information; the K scale, which measures subtle defensiveness and is used to correct the tendency to deny problems; and the F scale which is a measure of symptom exaggeration.

The content scales are: HS, or hypochondriasis, which deals with excessive bodily concerns, selfishness and narcissistic tendencies; D, or depression scale, which measures whether a person feels useless or unable to function (low scores indicate optimism); HY, hysteria, measures reactions to stress, avoiding responsibility, immaturity; Pd, psychopathic, measures antisocial and rebellious behavior, blaming and impulsiveness; MF is a measure of sexual identification, with high-scoring males ($T > 80$) having conflicts and low scores ($T > 30$) indicating a "macho" identification, whereas females who score high ($T > 70$) tend to reject traditional roles and those who score low ($T > 35$) tend to have more stereotypical roles; Pa, paranoia, $T > 75$ indicates pathology while a 65–74 range indicates being "overly sensitive;" Pt, psychasthenia, being anxious, tense, obsessive-compulsive and having self-doubt; Sc, schizophrenia, ($T = 80$–90) being confused, disoriented, and not feeling a part of things; Ma, hypomania, being overreactive, easily bored, creative, and overactive; Si, social inversion, $T > 65$ being introverted and $T > 65$ being extroverted.

Results

Whatever problems clinicians may have with the MMPI, it is generally agreed that it effectively screens for pathology (Helmes & Reddon, 1993). But pathology

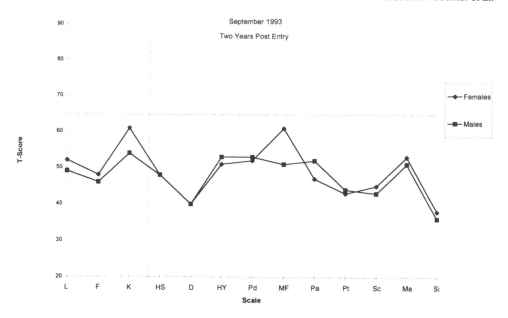

FIGURE 16.2. Biosphere 2 MMPI-2 profiles (males and females).

was not an issue among Biospherians. Their MMPI profiles were well within the normal range. Here MMPI use is more of a matching to task process. Mean scores of the four males and four females from Berren's data are shown in Figure 16.2.

The most outstanding aspect of these data is the close match between male and female scores. Entirely unexpected, it constitutes a major criterion for selection of females. The correlation between male and female scores is $r = .79$, which is statistically significant ($p = .0012$). The second crew had an even higher correlation between males and females of $r = .84$ ($p = .0003$).

Further, when the first crew female scores are compared to those of an all-female Antarctic expedition team (Kahn & Leon, 1994) the correlation is $r = .59$ ($p = .03$). This is some indication that females who volunteer for isolated environments may tend to have similar profiles.

When the biosphere males, however, are compared to males' MMPI scores in an Arctic expedition (Leon et al., 1989), there is a significant *difference* $t = 2.572$ ($p = .0185$). This suggests that males may have a much wider range of scores for selection purposes.

When the first eight Biospherians are compared to astronaut applicants ($N = 106$; Santy, 1994) they correlated, $r = .839$.

Discussion

All of these data, combined with previous literature (Butcher, 1987; Rivolier et al., 1991; Oliver, 1991; Carrere et al., 1991), suggest that selection for isolated environments follows an *adventurer* profile. This is characterized by females having male profiles, and both having high K and high MF scores, and low D, Sc, Pt, Sc, and

Si scores. Translated into English, these scores define people who are somewhat guarded about themselves, highly active, creative, optimistic, sure of themselves, and extroverted.

Other researchers (Gunderson, 1974; Oliver, 1991; Carrere et al., 1991) have mentioned the importance of avoiding depression as a key aspect of endurance in isolated environments. It requires people who are very optimistic. A low D score is a critical element of this profile.

Isolation is a phenomenon that goes beyond isolated environments. Ornish (1991) emphasizes it as a critical element in heart disease and states that professionals of all kinds are especially prone. Thus, a low score on social inversion (Si) is also critical. Extroversion, seeking out other people, is a survival mechanism for isolated and, perhaps, all environments.

Some cautions are necessary in looking at these data. The numbers are small. The first crew had eight persons, the second only six. Much larger numbers are required for a true selection process to be worked out. Nevertheless, these data do suggest that the Biospherians are selecting or being selected for certain traits that are classically useful in stressful situations.

Finally, just as Oliver (1991) reported personnel from her Antarctic experience had problems of reentering U.S. society, so too did the Biospherians. Being on time was at first a severe handicap. Distances inside the Biosphere 2 were small, so it became difficult to estimate traveling time in the larger world. The abundance of food was just too tempting, and more than one reported "ballooning" out from eating too much. Yet, it seems that each Biospherian agrees that the experience was one of the most important if not *the* most important experience of their lives.

CONCLUSIONS

While there are similarities of the Biosphere 2 experience to some small town behavior settings, the small population created a deprived environment when compared to the richness of life outside the structure. Yet this isolation was alleviated to a great extent by the unexpected use of electronic media, which, perhaps, foretells similar uses in the future.

The selection process goes a fair distance in explaining how the Biospherians endured many hardships. They were a group of dedicated professionals who were cautiously adventurous, gregarious, optimistic and persevering. The data especially point to the masculine profiles of women as being the most likely to endure such an experience. They were very similar to astronauts.

NOTE

Since the research reported above, the management of Biosphere 2 has undergone significant restructuring. Along with this reorganization, the scientific goals of the project are being redefined and long-term, human occupation is unlikely in the foreseeable future.

REFERENCES

Barker, R. (1968). *Ecological psychology*. Palo Alto: Stanford University Press.

Bechtel, R. (1977). *Enclosing behavior*. East Stroudsburg: Dowden, Hutchinson & Ross.

Bechtel, R., & Ledbetter, C. (1976). *The temporary environment*. Hanover: Cold Regions Research & Engineering Laboratory.

Butcher, J. (1987). *Use of the MMPI in personnel screening*. Paper presented at the 22nd annual symposium on recent developments of the MMPI, Seattle.

Butcher, J. (1992). Psychological assessment of airline pilot applications: Validity and clinical scores. Unpublished raw data.

Carrere, S., Evans, G., & Stokols, D. (1991). Winter-over stress:Physiological and psychological adaptation to an Antarctic isolated confined environment. In A. Harrison, Y. Clearwater, & C. McKay (Eds.), *From Antarctica to outer space* (pp. 229-238). New York: Springer-Verlag.

Gunderson, E. (Ed.). (1974). *Human adaptability to Antarctic conditions*. New York: American Geophysical Union.

Helmes, E., & Reddon, J. (1993). A perspective on developments in assessing pathology: A critical review of the MMPI and MMPI 2. *Psychological Bulletin, 113*, 453–471.

Helmreich, R. (1974). Evaluation of environments: Behavior observations in an undersea habitat. In J. Lang, C. Burnette, W. Moleski, & D. Vachon (Eds.), *Designing for human behavior* (pp. 274-285). East Stroudsburg: Dowden, Hutchinson & Ross.

Kahn, P., & Leon, G. (1994). Group climate and individual functioning in an all-women's Antarctic expedition team. *Environment & Behavior, 25*, pp. 669 697

Leon, G., McNally, C., & Ben-Porath, Y. (1989). Personality characteristics, mood and coping patterns in a successful North Pole expedition team. *Journal of Research in Personality, 23*, 162–179.

Oliver, D. (1991). Psychological effects of isolation and confinement of a winter-over group at McMurdo Station, Antarctica. In A. Harrison, Y. Clearwater, & C. McKay (Eds.). *From Antarctica to outer space* (pp. 217-228). New York: Springer Verlag.

Ornish, D. (1991). *Program for reversing heart disease*. New York: Ballentine.

Rivolier, J., Bachelard, C., & Cazes, G. (1991). Crew selection for an Antarctic-based space simulator. In A. Harrison, Y. Clearwater, & C. McKay (Eds.) *From Antarctica to outer space* (pp. ...). New York: Springer Verlag.

Santy, P. (1994). *Choosing the right stuff: The psychological selection* New York: Praeger Greenwood.

Schoggen, P. (1989). *Behavior settings*. Palo Alto: Stanford University Press.

Spivack, M. (1984). *Institutional settings: An environmental design approach*. New York: Human Sciences Press.

Part V

INTERPERSONAL ASPECTS OF THE ENVIRONMENT

Chapter **17**

RETHINKING STEREOTYPES OF FAMILY HOUSING IN JAPAN AND THE USA

Sandra C. Howell

INTRODUCTION

This chapter continues the comparative family housing studies offered at our last seminar (Howell, 1990). At that time the focus was mainly on incidents illustrative of family dynamics in the socialization of children in Japanese and U.S. homes. Theories of child development were invoked to explain what was perceived in panel households.

In a recent publication, Goodnow, Miller, and Kessel (1995) attempt to advance developmental theory through contextualization of cultural practices. The authors argue for the need to redefine our approach to analyzing a sequence of activities in a group (including a family) in terms of the state of knowledge and technology within their particular society at a particular point in time. Acknowledged is that the "shared quality" of a practice also means that individuals within a group may "sustain, change, or challenge" the practice. This choice has special relevance when it can be demonstrated that the original significance of the practice is no longer remembered or is no longer salient to the goals of the group. It is noted that breaches in domestic practices are most frequently observed in children's behaviors. Our role as environmental psychologists, thus, becomes to discover the ways

Sandra C. Howell • Department of Architecture, Massachusetts Institute of Technology, Cambridge, Massachusetts 02139.

Handbook of Japan–United States Environment–Behavior Research: Toward a Transactional Approach, edited by Seymour Wapner, Jack Demick, Takiji Yamamoto, and Takashi Takahashi. Plenum Press, New York, 1997.

that the physical context plays into these breaches, leading to changes either in domestic habits or the characteristics of the domestic environment.

The current explorations begin to test hypotheses about social interactions between family members as possibly generated by physical attributes (one aspect of context) of the individual dwelling. We know that archaeologists and anthropologists have frequently inferred systems of activities (behaviors) from the organization of material (including architectural artifacts) on a site (Kent, 1990). In discussions of such reconstructions of past activities, however, these scholars do not, as a rule, infer interpersonal transactions. The question here is: can we go beyond inference in our studies of living and changing social groups by more specifically defining the architectural elements that surround them?

RETHINKING STEREOTYPES

Let us remember the place of "doubt" in science. While theory and intuition often drive our direction in research, sometimes it is necessary to question that direction, to doubt. So it has been with my 17 years of thinking about Japanese and North American domestic life.

The doubt really emerged in 1987 with the remark of the father of one family who agreed to be studied who, on reading my field notes of his Nagoya household, took issue with my use of the word "typical." "We are not a typical Japanese household or house in any way," he said. And yet, my observations during my week-long visit was that many (not all) design decisions they made for their newly constructed urban neighborhood house were, in fact, consistent with so-called traditional rules of entry (*genkan*), some partitioning of rooms (*fusuma*), and special places such as an indoor garden (*Nakaniwa*), and a *tatami* guest room (Morse, 1972). Some of their family practices (habits?) seemed also to have continuities across the society (e.g., mother, father and 12-year-old daughter slept on futons, albeit each in a separate room).

What was it that "typical" represented to him that he had rejected or veered away from it? As it turned out, that he had, as the eldest son, passed his primary inheritance to his younger brother meant that his house displayed no family shrine and few old family memorabilia. Also, his wife was a working professional, and he perceived himself as participating in household tasks more than he thought usual in his society. Kose (1993), in his brief review of contemporary Japanese housing and changing household habits does comment on these trends.

To complicate my doubts, when I returned to the United States I had a similar conversation with my neighbor, an older French-born woman who was the head of a family of five adult children, several of whom intermittently lived with her. Other, younger U.S. families I was studying saw their lifestyle, domestic behavior, and house use as somehow not their conception of the cultural mainstream.

In 1990, on another visit to Japan, a respondent showed me the renovations to their townhouse in a uniform development project built in the late 1970s. Their recent modifications seemed very confrontational to the contextual unity designed for the site, as seen by neighbors. But in the United States, neighbors often express annoyance (and sometimes sue) at the changes householders make in their exterior renovations. In fact, many new suburban developments have explicit rules disallowing certain modifications to residential structures or landscape features.

Given all these experiences, I became aware that I needed to do more than search for behavioral and attitudinal differences within and across households in the two societies. I needed to break from rather unconscious stereotypes imbedded in developmental psychology and architecture, as well as the received stereotypes from sociological studies in both societies.

EXPLORING ALTERNATIVE APPROACHES

The most recent theoretical orientation that I have attempted to apply to my accumulating behavioral material on households is that of Blanton (1994) who utilized some of the principles described by Hillier and Hanson (1984) in the *Social Logic of Space*. In these works, architectural evidence is measured and counted. Blanton, an anthropologist, analyzed the "formal properties" of rural households from plans collected by himself and reported, from very diverse societies, by a wide number of other anthropologists. Blanton interprets the common features of a given community and their houses in terms of what they suggest about contact and communication between occupants or occupants and outsiders.

While Blanton's use of "communications" concepts seems to have intuitive validity, particularly for quite homogeneous rural communities, application to individual houses and households in more urban societies, with considerable behavioral heterogeneity, becomes much more problematic. Further, Blanton acknowledges that he had very little data, from *in situ* observations on daily behavior of occupants of most houses studied, which is a significant problem for an environmental psychologist concerned with individuals in their transactions with one another within bounded domestic space. Nevertheless, certain aspects of Blanton's conceptual framework do have heuristic value.

For example, Blanton describes "canonical" (within house) boundary conditions that suggest the extent to which behavioral interactions between residents may be enhanced or inhibited. Notice I use the term "enhance," rather than the frequently used "reinforced," more typical in such discussions. The reason for this choice is that "enhanced" permits us to allow for behaviors to be perceived that might not be predicted in advance, on the basis of the more usual response to an object or architectural feature. It is in the spirit of Gibson's (1979) concept of "affordances," which predicts human capabilities to conceive of unique or new uses for objects. The act of filling open stair risers with storage drawers, as one panel family did, though not uncommon in Japan, is rarely encountered in the United States. If, for any house we study, we can properly describe boundaries or "partitioning of space" (Kent, 1991) we might, then, be able to test, for a particular domestic group or its individual members, the ways in which boundaries do impact on their behavior.

Where, however, household members abide by culture-based rules of nonverbal communications (e.g., turning one's back to an opening in a partition or averting one's eyes in passing an unbounded area—not uncommon in Japan), the number and location of partitions, alone, will not correspond to behavioral expectations.

In the case of studying children's behaviors within domestic space, as I like to do, the extent to which, at various stages of development, a child consistently defies or tests certain boundaries becomes an interesting aspect of socialization within the family. Additionally, in both U.S. and Japanese households I studied, parents made certain decisions to provide either particular partitions or breaks in partitions in the design of their houses, with specific intentions in mind relative to intrafamily communications. These intentions are not necessarily apparent in analysis of plan, independent of verbal corroboration by or observation of the family in residence.

For example, in one household the parents decided to enclose and extend a second floor balcony area to provide a study for two school-age daughters. Opaque glass doors allow parents to see the figures of the girls at their desks within the space without noises or distractions from elsewhere in the house intruding on their concentration. At the same time, an internal, sliding, clear glass window was punched out of the wall between this study and the parents' bedroom. The explicit intention of the parents and children was said to have been to allow the girls to have ready access to parents for assistance while studying. A window shade was installed over this window in order to indicate when parents did not wish to be disturbed and to control incoming light (I do not know the frequency of its use).

Blanton uses the concept "centrality" to denote the relation, on plan, between the location of a central activity space (e.g., a courtyard or living room) and the location of peripheral spaces such as a kitchen. Among U.S. Urban suburban houses, there is a great deal of variation in such spatial relationships, from that of considerable separation by walls and transitional spaces (halls, dining room) to the actual commonality in space (where kitchen, dining and "family room" are, in effect, all the central activity space both by design and use). In fact, it is not unusual that U.S. households, including those of most of my American-raised students, report almost no regular use of living or dining rooms in their home of origin, but that virtually all family transactions took place in the kitchen, around a table. In my studies of East Boston Italian-Americans in three-level (three-decker) houses, renovations often redirected all occupants and guests, first, into the kitchen itself as the central space (Howell & Tentokali, 1991).

With small exceptions, kitchens in the Japanese houses and apartments I visited reflect a pattern of quite separate location and activities from that of living room, parlor, or guest room, where both family and guests conduct social activities. In one Japanese townhouse development, the planner actually placed the kitchen one-half level below the living room and guest room. The same plan, however, somewhat defies the more traditional off-side location of the kitchen (Morse, 1972), placing it facing the street. The homeowner replaced the clear glass, kitchen window with more opaque, self-screening windows, having commented on his need for greater privacy from the public way. Inside, they added a roll down screen to further visually separate living room from kitchen.

There is value added to our understanding of behaviors within the home by description of key features of the physical environment. Hopefully, we can come to agree on some consistent terms that we may all use for these descriptors. A major weakness, in fact, of much of our past environmental psychology research was the absence or vagueness of a descriptive typology and the primary use of psychosocial theories, only, as generators of hypotheses and as a guide to research protocols. The other side of the coin, however, is that we can become so enamored of counting and measuring architectural elements that a too-ready leap to behavioral conclusions seems logical. The rapidly changing character of urban households, in terms, at least, of gender roles, household composition, consumption patterns (lifestyle), and generational relations makes generalization from plan problematic, but not undoable.

BEYOND PRIVACY

Consider, for example, the concept of "permeability," as used by Rapoport (1990) to refer to one boundary condition. Do the extent and types of permeability within the domestic setting, or between the inside and outside of a dwelling, indicate the probable forms of human interaction, the rules and rituals of the society, or the preferences and values of the contemporary residential designer? Does permeability inhibit or enable behaviors consistent with the preferences and needs of household members for various levels of communication? It is the latter type of question that requires more delicate analysis by behavioral scientists.

Hillier and Hanson (1984) define "permeability" as " how the arrangement of cells (i.e.interior spaces), and entrances controlled access and movement" (p. 14). The graphs used by these authors, with circles representing spaces and linking lines representing entrances, are claimed to have the advantage of simplifying the plan for purposes of better comparison of spatial relations across buildings. Figures 17.1–17.3 represent floor plans and associated "permeability maps" for three houses in my study panel. While the differences in plan are simplified by the maps, some of the limitations in interpreting the potential behavioral implications are exemplified by the absence of (1) linear distance indicators, (2) visual line specifications, (3) change of level descriptors and (4) nonpassage permeabilities (e.g.,. interior windows), as illustrated on plans. Note, also, that in all three houses there are two-function spaces, not partitioned.

TRUE CONTEXT MEETS TRUE BEHAVIOR

Until now it would seem we have been too unspecific in our discussions of intrafamily dynamics. Take the concept of "communication."

If, as a psychologist, I am asked: "How do mother and father communicate?" I have to respond from direct observation and interview. For our purposes, those observations could well take into account the spatial characteristics within which a communication occurs. Often in the United States, but also in more refined Japanese homes, incidents of communication take place across considerable distance

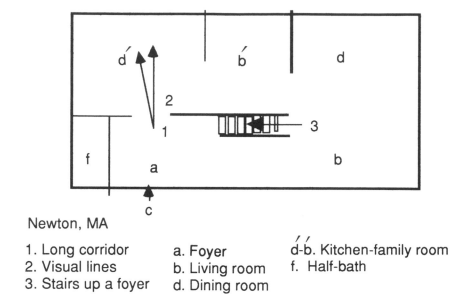

Newton, MA

1. Long corridor   a. Foyer   d-b. Kitchen-family room
2. Visual lines   b. Living room   f. Half-bath
3. Stairs up a foyer   d. Dining room

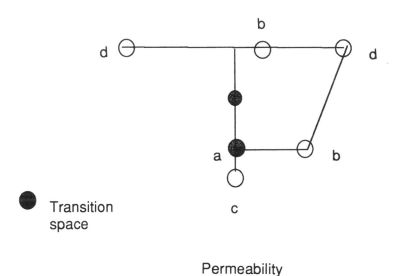

● Transition
    space

Permeability
map

FIGURE 17.1. Newton, Massachusetts, USA. Drawn by Matt Howell.

and through several partitions. The following incidents have been recorded in my field notes:

The father, sitting with an infant on his lap, calls out to his wife, who is three spaces (nodes) plus one partition plus one transition space plus a level change (Blanton's descriptors) away: "The baby is wet!" The mother, busy cooking in the kitchen, calls back: " Well, then change her!" (field notes, Japan, 1980).

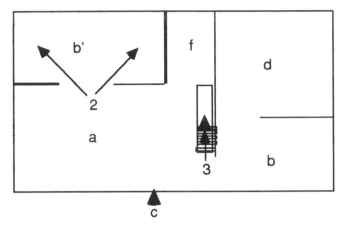

Ashton, MD

a. Living room
b. Den
d. Bedroom
f. Bathroom
b'. Kitchen-dining room

2. Visual lines
3. Stairs up

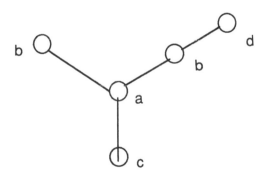

Permeability
map

FIGURE 17.2. Ashton, Maryland, USA. Drawn by Matt Howell.

Similarly, we captured this parent–child transaction in the United States (1993): The mother calls out from the kitchen to her teenage daughter, across five nodes plus three partitions and two transition spaces (a long flight of stairs and hallway): "Shimara, dinner is ready!" The daughter does not respond verbally but descends in her own time, after all other family and guests have been seated (see Figure 17.1).

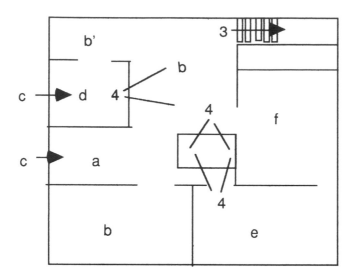

Nagoya, Japan

a. Foyer   b-b'. Living-dining room   3. Stairs up
b. Guest room   d. Kitchen   4. Interior window
e. Study   f. Bathroom

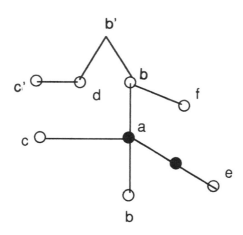

Permeability   ● Transition
map   space

FIGURE 17.3.   Nagoya, Japan. Drawn by Matt Howell.

Both of these examples tell us a great deal about the nature of interpersonal dynamics and the relative lack of inhibition provided by the within-dwelling boundary conditions. Family rules may not explicitly say "do not yell," but do define under what circumstances it is an accepted part of their dynamic to yell, leave a room, shut a door, or knock. Also from field notes, this vignette: A very polite young boy, in a contemporary Japanese village house, was doing his homework at about 8 p.m., in the room next door to the one I had retired to. All *Fusuma* (sliding doors) were closed. He courteously knocked before calling out to me. When I answered, he came through to try to interview me for his next day's class. I suspect from other experiences, that if *Fusuma* had been partially open, he would have quietly expressed apology for interrupting and waited for my acknowledgment and invitation to enter, but it would all be a visually accessible interaction, rather than through a partition (door) as is more typical in a U.S. household.

DEVELOPMENTAL CHANGES INFLUENCING SPACE USE

While we tend to think of developmental issues in terms of children's physical and cognitive growth, the concept may also apply to families, as they evolve and respond, variously, to endogenous and exogenous forces by revising domestic space and its use. Some of the current endogenous forces we know are affecting families both in Japan and the United States include the following:

1. Variations and changes in husband's and/or wife's work schedules, including work at home.
2. Participation, inside or outside the dwelling, in special activities by children (e.g., music lessons, athletics).
3. Varying types of entertaining within the dwelling.
4. Changing "perceptions" by parents of changing needs of children or selves.
5. Parental observations and reactions to sibling behaviors of children.
6. Change in health status of a member of the family in residence.

Some examples of exogenous forces acting on today's urbanized families:

1. Peer influences on children's attitudes and behaviors.
2. Adult perceptions of their peers' opinions and lifestyles.
3. Changing needs of nonresident members of family (health or economic status of members of extended family that may result in additions to household).
4. Availability of external child care.
5. Affects of technology and media.

In virtually all of my studied families there was evidence of more than one of the above forces operating on the ways family members used, modified, or significantly changed, their dwelling. The advantage of longitudinal over single-visit data collection is that you can follow a process rather than relying solely on retrospective recall of a process. This is particularly interesting when families express intentions to make major modifications in the configuration of their dwelling. The most dramatic example of this is a Tokyo couple whose residence was, in 1978, a quite traditional house of the grandparents; in 1980, the couple had completed a separate adjacent house, of rather Swedish-modern style for their two young children; by 1993, with three teenagers and an increasingly frail grandfather, they had demolished the original house and constructed a much larger, very up-scale, house on the same plot. While the new house incorporated small elements of the original, within a *Tatami* room, for the most part it is dominately of Western style in spacial arrangements and furnishings. About this house, Japanese *Home Plans* magazine reports, "Here is the realization of the ideal home, considering it in detail and retaining the family history" (p. 86).

Much more typical of panel families, in both the United States and Japan, were small modifications in the dwelling itself, and a good deal of change, over time, in how existing spaces were used, by whom, for what activities, and with what props. Two families, one in Kyoto and one in Seattle, made comparable structural additions of a basement study to provide work privacy for one or both parents. Additions of rear patios or decks were common among U.S. households in Massachusetts, Maryland, and Michigan, for the stated and observed purpose of providing outdoor family space, a modification that did not occur among Japanese families.

FUTURE DIRECTIONS

As Japanese and U.S. dwellings become increasingly similar in plan, partitioning elements, and other architectural features, it will become possible for psychological researchers to study the spatial dynamics of households, as these transactions change over time. At present, we do not understand well the extent to which families deploy their dwellings to advance or control their developmental goals.

As this chapter suggests, dwelling descriptors proposed by Blanton, Rapoport, and others do have some utility for the advancement of our research but, I believe, require refining through collaborative efforts between architects and social scientists. It is probable that the same aspect of a fixed, semipermeable boundary may have salience for some intrafamily transactions but to be irrelevant or countereffective for others and, further, that a given family, at a given time in their history, chooses its own saliences. One Nagoya mother told me, of a late evening, that she no longer appreciated that her teen daughter's bedroom was only separated from hers by *Fusuma* (as she had carefully planned). Her daughter now insisted on playing rock music while studying at night and refused to use earphones.

The use, by both Japanese and U.S. panel households, of within-dwelling elements that can readily be moved, put up or taken down, opened or closed, raised or lowered, relative to particular household perceptions of need or event, attests to this temporal variability. Let us consider, too, that households and builders do make mistakes. Some fixed architectural features initially thought aesthetically, culturally or functionally appropriate turn out not to work or to be unacceptable by householders (Boudon, on Pessac, 1969). Perhaps creative architects will develop, as have furniture manufacturers, new types of space boundaries to allow families more options. The general rejection by U.S. designers and builders of pocket doors (equivalent to *Fusuma*?) conflicts with even our current understanding of the varying needs of households.

CONCLUSION

Further development of collateral explorations of dwelling features with behavioral evidence will continue because it seems to be a fruitful direction. However, there should be a clear caveat provided with the recommendation to move in this direction. There are, as most social scientists know, many variables that determine how and why individuals and households use their domestic environments in particular ways at particular times, especially among contemporary urbanized, more middle-class households. Consequently, we cannot ever expect congruence to be consistently validated, but then Rapoport (1990) has already asserted this. What we may be able to provide are elaborations of Gibson's (1979) concept of "affordance." Revisiting this concept with more detailed transactional data between environment and behavior could, indeed, illuminate our understanding.

REFERENCES

Blanton, R. E. (1994). *Houses and households: A comparative study*. New York: Plenum.
Boudon, P. (1969). Lived-in architecture: LeCorbusier's Pessac revisited. Cambridge, MA: MIT Press.
Gibson, J. J. (1979). *An ecological approach to visual perception*. Boston: Houghton Mifflin.
Goodnow, J., Miller, P. J., & Kessel, F. (Eds.). (1995). *Cultural practices as contexts for development*. San Francisco: Jossey-Bass.
Hillier, W., & Hanson, J. (1984). *The social logic of space*. Cambridge: Cambridge University Press.
Howell, S. C. (1990). Family life: habit and habitability. In Y. Yoshitake, R. B. Bechtel, T. Takahashi, & M. Asai (Eds.), *Current issues in environment-behavior research* (pp. 163–172). Tokyo: Univ. of Tokyo.
Howell, S. C., & Tentokali, V. (1991). Domestic privacy, gender, culture and development issues. In S. Low & E. Chambers (Eds.), *Housing, culture and design: A comparative perspective* (pp. 281–300). Philadelphia: Univ. of Pennsylvania Press.
Kent, S. (1990). *Domestic architecture and the use of space*. Cambridge: Cambridge Univ. Press.
Kent, S. (1991). Partitioning space: Cross cultural factors influencing domestic spatial segmentation. *Environment & Behavior, 23*(4).
Kose, S. (1993). Changing lifestyles, changing housing form: Japanese housing in transition. In M. Bulos & N. Teymur (Eds.), *Housing: Design, research, education* (pp. 193–205). Brookfield, VT: Ashgate.
Morse, E. S. (1972). *Japanese homes and their surroundings*. Rutland, VT: Charles E. Tuttle.
Rapoport, A. (1990). Systems of activities and systems of settings. In S. Kent (Ed.), *Domestic architecture and the use of space* (pp. 9–20). Cambridge: Cambridge University Press.

Part VI
SOCIOCULTURAL ASPECTS OF THE ENVIRONMENT

Chapter **18**

SOCIOPSYCHOLOGICAL ENVIRONMENTS OF JAPANESE SCHOOLS AS PERCEIVED BY SCHOOL STUDENTS

Masaaki Asai and Sonomi Hirata

INTRODUCTION

Based on Barker's ecological psychology, Gump (1980) reviewed research relevant to perception of the school setting in the framework of "the school as social situation" and emphasized the important role of the social aspect of school settings.

Hirata (1994) developed an original Japanese scale to assess the classroom environment, based on the three dimensions, or domains, proposed by Moos (1974). Using this scale, Hirata compared junior high school students with students who refused to go to school (nonattendant students) and found significant differences in mean scale scores on the three dimensions. The nonattendant group showed lower scale scores on Relationship and Personal Growth. A bimodal distribution of scale scores was seen in the System Maintenance and Change domain of the nonattendant group. When compared with U.S. data, a detailed analysis showed that in both Japan and the United States, nonattendant groups who lack support in the domain of relationship exhibit excessive competition in the domain of personal growth, and lack of clarity in the system maintenance and change domain as

Masaaki Asai • Department of Psychology, College of Humanities and Sciences, Nihon University, Tokyo 156, Japan. **Sonomi Hirata** • Department of Human Sciences, Graduate School, Waseda University, Saitama 359, Japan.

Handbook of Japan–United States Environment–Behavior Research: Toward a Transactional Approach, edited by Seymour Wapner, Jack Demick, Takiji Yamamoto, and Takashi Takahashi. Plenum Press, New York, 1997.

predicted by Moos (1991). On the basis of these data, Hirata concluded that similar perceptions of the classroom environment are responsible for the standing behavior, the refusal of school attendance among the students in Japan and the United States, despite the differences of culture and physical milieu of school settings in both countries. Though the standing behavior of juvenile delinquents may differ in different cultures and milieus, our hypothesis is that the perception of classroom environment of juveniles in both Japanese and U.S. groups may yield similar patterns. The present study examines, using the Correctional Institution Environmental Scale developed by Moos (1987), student perception of classroom environments in school settings other than those surveyed in the previous study, namely, correctional institutions in Japan and the United States.

METHOD

Correctional Institution Environment Scale (CIES)

Moos (1987) developed the CIES by constructing nine subscales around three central dimensions: Relationship, Personal Growth, and System Maintenance and Change. The relationship dimension consists of Involvement, Support, and Expressiveness. The personal growth dimension is measured by Autonomy, Practical Orientation, and Personal Problem Orientation. The System Maintenance and Change dimension is measured on three subscales: Order & Organization, Clarity, and Staff Control. More detailed definitions are provided in Moos's (1987) manual.

The CIES was translated into Japanese by a bilingual translator and then translated back into English by a native-speaking English translator. Adjustments were then made to get the most natural Japanese wordings and the instrument was pretested on Japanese subjects.

For the high school students, some terms used in the original CIES (e.g., institute and staff) were replaced with words common to high school students (i.e., school, teacher).

SUBJECTS

Moos (1987) standardized the CIES for juveniles on 3,651 residents and 858 staff from representative correctional institutions. These normative samples are composed of 112 units for residents and 96 units for staff. They are state training schools and reception centers, county juvenile halls, county- and state-managed ranches and camps, privately managed vocational training schools and work release programs.

The Japanese juvenile samples consisted of 276 male and 60 female residents detained in the correctional institutions under the administration of the Ministry of Justice. Male samples are drawn from five male institutions and the female sample from one female institution. The control group consisted of 163 male and 154 female university freshmen. A modified CIES was administered on the first day of

TABLE 18.1. CIES Dimension Scores Deviation-Japanese High School Students and Correctional Institution Residents, Japan and the United States

Dimension	Japanese high school students (N = 319)		Japanese residents (N = 336)		U.S. residents (N = 713)	
	Mean	SD	Mean	SD	Mean	SD
Relationship	15.5	5.6	14.3	5.5	13.8	6.5
Personal Growth	16.6	4.8	14.1	5.0	14.6	6.2
System Maintenance and Change	14.2	5.4	15.0	5.0	15.7	6.4

their admission to university. The sample consisted of 184 subjects graduated from private schools and 133 subjects from public or national high schools.

RESULTS

Residents in Correctional Institutions in Japan and the United States

The summed mean scores for each dimension for residents in Japan and the United States, and also for the control group (Japanese high school students), are shown in Table 18.1. The subscales for the three dimensions are compared in Figures 18.1-18.3.

Taking the summed mean score for each of the three dimensions and testing for differences between the residents of correctional institutions in Japan and United States, results showed no significant differences (see Figures 18.1 and 18.2). This result and the results of a previous study of Hirata (1994), which showed similar perception of classroom environment among the nonattendant students of Japan and the United States, tend to support our hypothesis. Despite the difference of culture and physical milieu, the Japanese and United States students who show deviant standing behavior (e.g., refusal to attend school or delinquent behavior) show similar perceptions of their classroom environment assessed on Moos's three dimensions of social climate (see Table 18.2).

Table 18.2 shows the means and standard deviations of the nine subscales of the CIES for both Japanese and U.S. residents, and these groups are compared in Figures 18.1 and 18.2. All means and standard deviations for U.S. residents are quoted from Moos's (1987) manual.

Taking the mean subscale scores and testing for differences, the findings are significant ($p < .05$) for five subscales (marked with asterisks in Figures 18.1 and 18.2), that is, Expressiveness, Autonomy, Practical Orientation, Order & Organization, and Staff Control. Figure 18.4 shows the mean subscale scores in the CIES Relationship dimension for residents in Japan and the United States. A significant difference ($p < .01$) is seen only for the subscale of expressiveness, with U.S. residents scoring considerably less, 4.3 vs. 3.9, showing the more negative attitudes

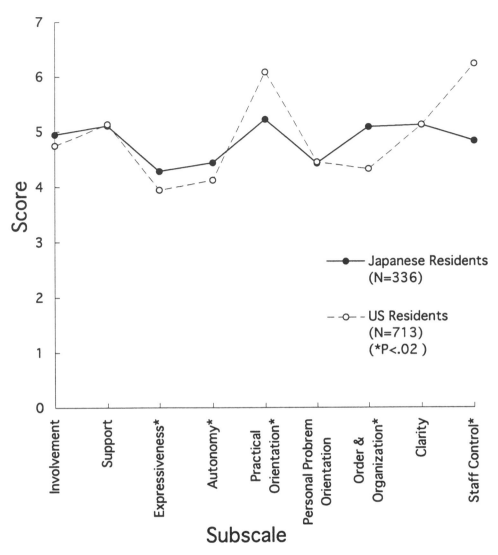

FIGURE 18.1. CIES subscale profiles, Japanese and US residents.

of U.S. residents toward self-expression. The remaining two subscales show no significant differences.

Mean subscale scores on the Personal Growth dimension are compared in Figure 18.5. No significant difference was seen for Personal Problem Orientation, but on the subscales of Autonomy and Practical Orientation, the mean subscale scores between Japan and the United States were found to be significantly different ($p < .05$). Japanese residents (4.4) feel more encouraged to take the initiative in planning activities and to take initiative in the unit than do U.S. residents (4.1). On Practical Orientation, U.S. residents (6.1) perceive more opportunities for learning practical skills than do Japanese residents (5.2).

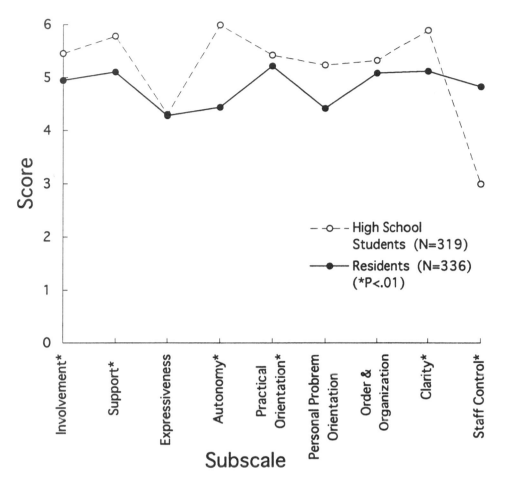

FIGURE 18.2. CIES Subscale profiles, Japanese high school students and residents.

Figure 18.6 compares the three subscales in the System Maintenance and Change dimension for Japan and the United States. Order & Organization and Staff Control showed significant differences ($p < .01$), lower scores for U.S. residents on Order & Organization, 5.08 (Japan) versus 4.32 (U.S.). Japanese residents emphasized importance and order in the program of the institute. For Staff Control, the direction of the difference is reversed, with U.S. residents feeling more pressure under necessary control 4.82 (Japan) versus 6.22 (U.S.).

High School Students and Residents in Japan

Means and standard deviations of the three dimensions for Japanese high school students and residents are shown in Table 18.1 and are compared in Figure 18.7. Student's t-analysis of the mean difference between high school students and residents is significant for each of the three dimensions. There were higher scores for the high school students on the Relationship and Personal Growth dimensions

TABLE 18.2. CIES Subscale Score Japanese High School Students and Correctional
Institution Residents, Japan and the United States

Subscales (number of items)	Japanese high school students (N = 319)		Japanese) residents (N = 336)		U.S. residents (N = 713)	
	Mean	SD	Mean	SD	Mean	SD
Involvement (10)	5.5	2.4	4.9	2.2	4.7	2.5
Support (10)	5.8	2.3	5.1	2.8	5.1	2.3
Expressiveness(9)	4.3	1.9	4.3	1.7	4.0	1.8
Autonomy (9)	6.0	1.7	4.4	1.8	4.1	2.1
Practical Orientation (10)	5.4	2.2	5.2	3.0	6.1	2.3
Personal Problem Orientation (9)	5.2	1.9	4.4	1.7	4.4	1.9
Order and Organization (10)	5.3	2.0	5.1	2.5	4.3	2.4
Clarity (10)	6.0	1.7	5.1	1.9	5.1	2.0
Staff Control(9)	3.0	1.7	4.9	2.1	6.2	2.1

($p < .05$). With respect to System Maintenance and Change, residents showed
higher scale scores than high school students ($p < .001$). Figure 18.2 shows nine
subscale means for both groups.

Figure 18.8 shows the CIES Relationship dimension mean scales cores of
Japanese high school students and Japanese residents. As to the Relationship dimension, high school students view their relations with peers and teachers better
than do the residents. Significant differences are seen on the Involvement and
Support subscales, with high school students showing higher scores than residents.

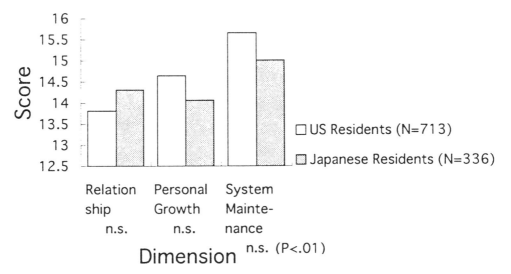

FIGURE 18.3. CIES dimension mean scale scores, Japanese and US residents.

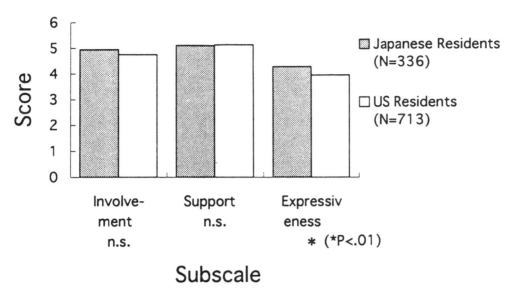

Figure 18.4. CIES Relationship dimension, mean scale scores, Japanese and US residents.

These results suggest that high school students feel more involved in school activities and enjoy the support of their peers.

Figure 18.9 shows mean subscale scores on the Personal Growth dimension. Significant differences are seen between both groups on this dimension, with high school students scoring higher than resident groups. Assessment of their schools

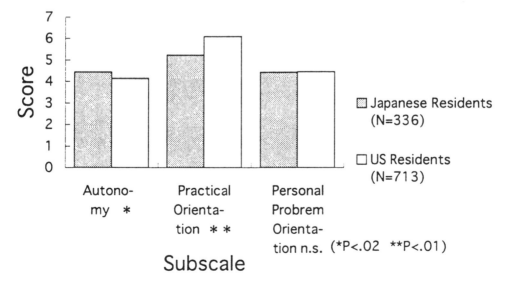

Figure 18.5. CIES Personal Growth dimension, mean scale scores, Japanese and US residents.

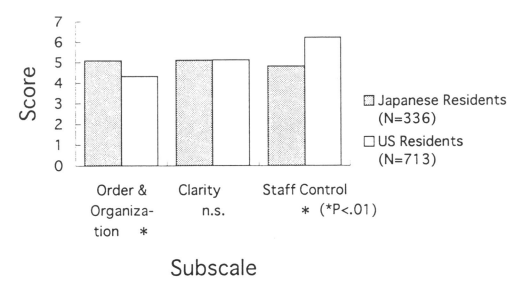

FIGURE 18.6.   CIES System Maintenance dimension, mean scale scores, Japanese and US residents

(institutions) as important environments for personal growth is higher among the high school students than residents in the institutions. Chi square analyses showed significant differences on both subscales for Autonomy and Personal Problem Orientation ($p < .001$).

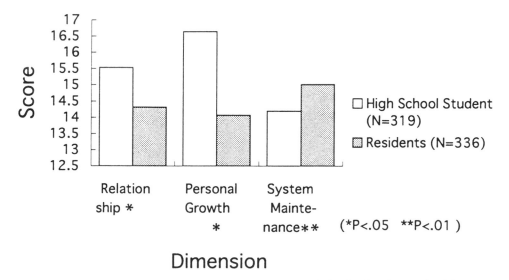

FIGURE 18.7.   CIES dimensions mean scale scores, Japanese high school students and residents.

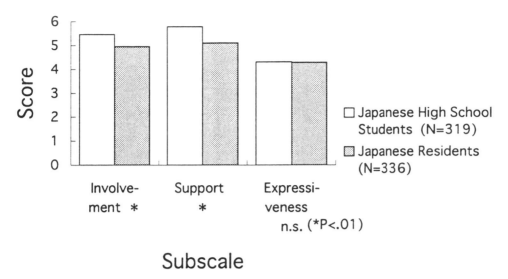

FIGURE 18.8. CIES Relationship dimension, mean scale scores, Japanese high school students and residents.

Mean scale scores of the residents on System Maintenance and Change, in contrast to the other two dimensions, are higher than those of the high school students (*p* < .001). Higher assessment for system maintenance of the classroom environment is seen among the residents than among high school students. Figure 18.10 shows the three mean subscale scores of the System Maintenance and Change dimension. Significant differences are seen on Clarity and Staff Control. Clarity of

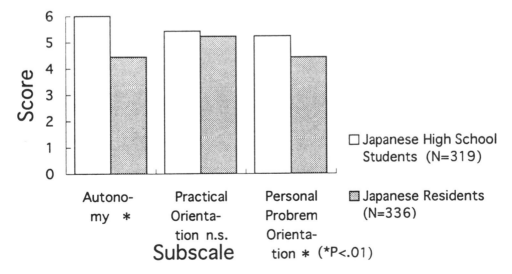

FIGURE 18.9. CIES Personal Growth dimension, mean scale scores, Japanese high school students and residents.

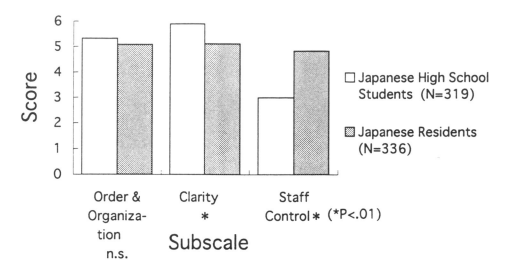

FIGURE 18.10. CIES System Maintenance dimension, mean scale scores, Japanese high school students and residents.

school rules, award and punishment, and self-planning are rated higher among the high school students, but residents feel more constrained by control of the staff than do high school students.

High school students showed higher scores than correctional institution residents on every subscale except the reversed score on Staff Control. This may suggest that the institution residents feel more oppressed in positive involvement with their environment and feel strong control by their staff. The result also may mean that the environment, as seen by the institution residents, is more restrained and closed than the classroom environment as perceived by high school students.

Normal distribution of scale scores on almost all of the nine scales is seen among the high school students, whereas a bimodal or rectangular distribution is observed only in the resident group.

DISCUSSION

When Japanese and U.S. institution residents and a group of Japanese high school students were compared, significant differences on the subscales of Involvement, Support, Personal Problem Orientation, and Clarity were found between Japanese institution residents and high school students, showing that perception of the classroom environment of these two groups is different. However, the difference between the institution residents in Japan and the United States is not significant on these subscales. These two groups showed similar perceptions of their classroom (institutional) environments, despite cultural differences between the two countries.

However, a set of results in which significant differences exist on the three subscales (i.e., Expressiveness, Practical Orientation, and Order & Organization), showed lower mean scale scores in the U.S. institution resident group, but a nonsignificant difference between the Japanese high school students and the resident groups. This may reflect cultural differences between Japan and the United States in the perception of the classroom environment. Similar tendencies in the domain of self-expressiveness were seen in the cross-cultural study of the Work environment Scale (WES) between groups of American and Japanese workers (Asai & Bechtel, 1990). In the present study, the U.S. institution residents showed the highest subscale scores of the three groups on Autonomy and Staff Control. Together, these results may support our hypothesis that similar perceptual patterns of sociopsychological classroom environments may elicit similar standing behaviors despite differences in culture and physical milieus.

REFERENCES

Asai, M., & Bechtel, R. B. (1990). A comparison of some Japanese and US workers on the WES. In Y. Yoshitake, R. B. Bechtel, T. Takahashi, & M. Asai (Eds.), *Current issues in environment-behavior research* (pp. 1–9). Tokyo: University of Tokyo.

Gump, P. V. (1980). The school as a social situation. *Annual Review of Psychology, 31* 553–582.

Hirata, S. (1994). Perception of classroom environment by junior high school student. Some characteristics of nonattendant student. Master's Thesis submitted to Waseda Univ. (In Japanese)

Moos, R. H. (1987). *Correctional institutions environment scale manual* (2nd ed.). California: Consulting Psychologist Press.

Moos, R. (1974). *School environment scale manual*. California: Consulting Psychologist Press.

Chapter 19

"BIG SCHOOL, SMALL SCHOOL" REVISITED

A Case Study of a Large-Scale Comprehensive High School Based on the Campus Plan

Toshihiko Sako

The psychological effect of high school size was first studied by Barker and his colleagues (Barker & Gump, 1964), who conducted a survey of students' participation in extracurricular activities. In schools whose size varied from 50 to 2,300 students, they conducted behavior setting surveys. Behavior settings have certain time space boundaries and standing patterns of behavior. Behavior settings are environmental units that connect teachers and students with their school as a whole. Comparisons between big and small schools showed that the number of settings did not have a simple proportional relation to school size. This means that even small schools have essential settings as a school in spite of a low level of structure and differentiation. As for the use of settings, small schools stimulate students to participate in settings responsively so that they have varied experiences. On the other hand, big schools seem to offer their students only nominal participation. School size relates positively to setting size. This means that a small school's small setting is in an "understaffed condition," which allows all participants to be substantial performers even though their performances are relatively poor. On the other hand, a big school's big setting yields some excellent performers but many

Toshihiko Sako • Department of Human Health Sciences, School of Human Sciences, Waseda University, Saitama 359, Japan.

Handbook of Japan–United States Environment–Behavior Research: Toward a Transactional Approach, edited by Seymour Wapner, Jack Demick, Takiji Yamamoto, and Takashi Takahashi. Plenum Press, New York, 1996.

anonymous and nominal participants. In the latter case, individuals deviate from the interpersonal control that might function in a small group and become unable to enjoy the useful activities provided by small settings. The question is: which school, big or small, is desirable? The answer depends on which educational objectives are considered favorable: essential experience or excellent performance.

Many researchers have examined the effects of school size since the "Big school, small school" study. Baird (1969) studied the relationship between achievement in extracurricular activities and school size and found small school size effects similar to those in the above mentioned study. Schoggen (1989) reviewed relevant studies and referred to the advantage of small schools. Wicker (1979) proposed a "staffing theory" based on his school and church size studies. His theory proposes an optimal size of settings, but he is prudent about generalizing across organizations because of the different effects of size on members' position.

Barker and Gump (1964) referred to "the campus plan" as an option to cope with the negative effects of a big school should its construction be unavoidable. Most campus schools, according to Barker and Gump, have a number of small schools, divided on the basis of subject education, with the big school as a whole sharing the large facilities available for extracurricular activities. They suggested that a big school, including its facilities, should be divided, because the social value of small schools was found to be prominent in the domain of extracurricular section activities rather than in subject education.

Gump (1987) reviewed high school size studies in the United States and pointed out that small schools would be effective if their sizes were from 500 to 700 students. He referred to the campus plan (specifically, the schools-within-a-school plan), reiterating that small schools should not share the entire domain of the big school for extracurricular activities because the "understaffed" behavior settings of small schools facilitated interpersonal contact between teachers and students. Based on findings such as these, it seems advisable to divide a big school into smaller units or small schools. However, as yet, there has been no study of "the schools-within-a-school plan" from the viewpoint of environmental psychology.

Ina-gakuen Comprehensive Upper Secondary School, located in Ina town, established in 1984 by Saitama Prefecture, is the first and only large-scale comprehensive high school in Japan. Ina-gakuen is an innovative high school model based on the campus plan. One important feature of this school is that it has adopted a comprehensive elective system for various subjects through unifying three schools as one. However, teachers and students have to belong to one of six houses (campuses or small schools) that together constitute one whole school. This big school comprises several connected buildings so as to form a circle with one wide corridor called the mall on the second floor of each building (Figure 1).

The first and second floors of each of the six buildings are for each house. On the first floor are a hall, eight home rooms (four rooms on each side of the hall), two restrooms, and an entrance. On the second floor are four home rooms, a teachers' room, two rest rooms, a student council room, and a storeroom. The third floor of each building is allotted for a particular subject, and contains classrooms, utility rooms, and a study room for teachers, including an allotted supervisory teacher. These rooms are located on both sides of the learning center that corre-

HOUSE 1: JAPANESE
HOUSE 2: FOREIGN LANGUAGE
HOUSE 3: SOCIAL STUDIES
HOUSE 4: MATHEMATICS
HOUSE 5: PHYSICAL SCIENCES
HOUSE 6: PHYSICAL SCIENCES

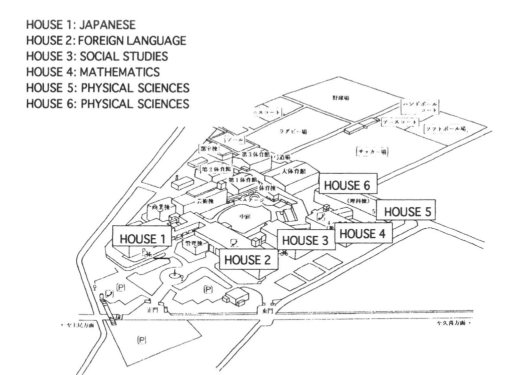

FIGURE 19.1. A bird's eye view of Ina-gakuen high school.

spond to the hall downstairs, except for the natural science subject rooms in the
fifth and sixth houses. In addition, the school has one building for art and home-
making and some facilities for physical education linked with the mall. Each house,
as a small school, has its own activities and events. Home rooms constitute a small
school unit and are maintained as such for 3 years after students' admission,
keeping the same teachers and students. Students attend classes for common
required subjects in their home room. In addition, they have to move about to
attend classes for elective subjects in each subject building. Students share in club
activities for the whole school. Most of the curriculum and all of the extracurricular
activities are carried out in the big school. The number of students is about 3,300
in all and each house has 550 students. This numerical allocation is in the range at
which the above-mentioned small school effects are reported to be successful
(Gump, 1987).

We have already reported some of the problems relating to teachers' movements and communication caused by extremely large-scale architecture such as this
(Sako, 1993). This chapter presents some results on teacher student relationships
in this large school based on the campus plan. The purpose of the study is to
examine differences between human relations in a big school and in a small school.
A preliminary survey of teachers indicated that most felt dissatisfaction rather than
satisfaction in their relationship to students and that the feeling of anonymity

annoyed them. Teachers frequently complained that they could not identify students and that they did not have enough personal contact with them. Consequently, we had to clarify in which part of the school the anonymity arose: in the big school or in the small schools. In doing so, we can examine to what extent the campus plan is effective in resolving big school problems. We report here perceived teacher student relationships in big and small school settings.

METHOD

We examined teachers' and students' perceptions of the teacher student relationship in the following settings: "home room" and "house" as small school settings; and "elective subject classes," "club activities," and "school as a whole" as big school settings. When a big school is partly divided based on the campus plan, there exist small school settings as well as big school settings. Ina-gakuen high school has about 220 staff and 3,300 students. The area of the school site is 160,000 m². It has 12 major buildings. Each of the six houses or small schools has 12 home rooms. Each house has its own student council that performs various activities. The school provides about 182 rooms (home rooms, classrooms, laboratories, and studios) for subject education. About 160 elective subject classes are taught annually. As for extracurricular activities, 150 clubs exist regularly or temporarily. An annual school festival, a field day, and a school excursion are held as a whole school. A whole school student council is steered by representatives from each house. Thus, the school has an extremely large number of settings. Teachers and students move around according to their own schedule within a big school. Therefore, we adopted generic names that represented various sociophysical settings in this school. *Home rooms* and *houses* refer to small school settings. *Elective subject classes*, *clubs*, and the *whole school* represent big school settings. Settings occur in places suitable for their respective activities.

In this study, we prepared a set of six bipolar adjective rating scales that seemed adequate to measure anonymity in human relations, on the basis of the above-mentioned preliminary survey of teachers. They were: "familiar-distant," "loose-tight," "delicate-rough," "active-passive," "individual-uniform," and "deep-shallow." Each scale had seven categories. The center of the scale was either "middle of the scale," "equally applicable to both adjectives," or "completely inapplicable to either adjective." For both adjectives, we provided the following categories: "a little applicable," "fairly applicable," and "very applicable."

Using these scales, we asked teachers to rate teacher student relationships in the above-mentioned settings in October 1992. We asked them to indicate three ratings on each scale at each setting: "present state," "optimal state," and "other schools' state in general." Twenty-eight teachers responded, comprising 4 female and 24 male teachers. Next, we asked third-grade students to indicate the same ratings (except for "other schools") in April 1993. Fifty-three students responded, comprising 22 females and 31 males. We selected club participants only because we wanted to evaluate the comparative status of club activities among other settings. The rate of club participation had been, usually, 70 to 75%.

TABLE 19.1. Rating of "Familiar-Distant" (1–7)

	Students (n = 53)					Teachers (n = 28)				
	Home room	House	Elective	Club	Whole school	Home room	House	Elective	Club	Whole school
Present state										
M	3.19	4.62	3.96	2.79	4.47	2.93	4.39	4.29	2.79	4.96
SD	1.08	1.36	1.57	1.68	1.64	1.18	1.23	1.54	1.40	1.26
Optimal state										
M	2.93	3.36	2.77	2.62	3.11	2.54	3.04	2.79	2.07	2.75
SD	1.07	1.16	1.25	1.35	1.50	0.96	1.11	1.52	0.90	1.35

RESULTS

Tables 1 to 6 show the results of the six scales. The seven categories of each scale correspond to a numerical value from 1–7. For example, 1–7 for "familiar-distant" is interpreted as follows: 1 is "very applicable," to the left of the adjective "familiar," and 7 to the right side, "distant."

According to Table 1, for the present state of settings, both students and teachers perceive "club" to be the most "familiar," and "home room" the next most "familiar." This result was expected since the sample of students was drawn from club participants only. The other settings are relatively neutral, except for teachers' perception of "school." Optimal state ratings all tend to be more "familiar." Differences between present and optimal states are larger in "school," "elective," and "house" than in "club" and "home room."

Rating of "loose-tight" (Table 2) shows that the "home room" is presently the "loosest" setting for both teachers and students. Teachers perceive all settings to be loose. Students' optimal state ratings are "looser" than teachers' ratings. Students like a looser state and teachers prefer a less loose state.

Table 3 shows that both teachers' and students' present state ratings are generally neutral. The optimal state ratings of both groups are for more "delicate" ratings, but teachers rate "delicate" more importantly than students do.

On the rating of "active-passive" (Table 4), the present state of settings seems neutral, except for "club," which is "active." Ratings of the optimal state are more

TABLE 19.2. Rating of "Loose-Tight" (1–7)

	Students (n = 53)					Teachers (n = 28)				
	Home room	House	Elective	Club	Whole school	Home room	House	Elective	Club	Whole school
Present state										
M	2.59	3.77	3.70	3.93	3.40	2.68	2.79	3.32	3.75	2.89
SD	0.95	1.17	1.31	2.03	1.18	0.95	1.29	1.06	1.65	1.07
Optimal state										
M	2.70	3.13	3.15	3.28	3.04	3.43	3.46	4.00	3.86	3.64
SD	1.07	1.11	1.17	1.73	1.22	1.20	0.96	1.19	1.27	0.95

TABLE 19.3. Rating of "Delicate-Rough" (1–7)

	Students (n = 53)					Teachers (n = 28)				
	Home room	House	Elective	Club	Whole school	Home room	House	Elective	Club	Whole school
Present state										
M	4.04	4.28	3.83	3.51	4.23	3.96	4.25	4.00	3.68	4.64
SD	0.92	0.95	1.11	1.63	1.34	1.50	1.14	1.54	1.39	1.28
Optimal state										
M	3.66	3.89	3.38	3.32	3.47	3.11	3.43	2.71	2.79	3.14
SD	1.06	0.89	1.15	1.38	1.34	1.29	1.29	1.08	1.32	1.33

"active" for both teachers and students. Teachers rate optimal settings to be more "active" than students do.

Table 5 shows that students perceive present settings to be relatively "individual," except for "house." Students' optimal ratings seem not to differ from their present ratings. Teachers' ratings are similar to students', but teachers rate optimal settings as more "individual."

Ratings of "deep-shallow" (Table 6) indicate that the present state of settings is perceived as being almost the same for both teachers and students. Ratings are similar and neutral, except for "club." Students rate optimal "elective," "house," and "school" settings "deeper" than present ones, while teachers rate all settings "deeper."

We applied an analysis of variance to each scale. Position (teacher and student) was considered a between-subjects variable and setting ("home room," "house," "elective," "club," and "school") and state (present and optimal) as within-subject variables.

The effect of "position" was not significant on any of the scales. Perceptions of teachers and students were similar across state and setting.

The effect of "state" was significant on five scales, but not significant on "loose-tight": "familiar-distant," $F(1, 79) = 73.63$, $p < .001$; "delicate-rough," $F(1, 79) = 40.20$, $p < .001$; "active-passive," $F(1, 79) = 60.21$, $p < .001$; "individual-uniform," $F(1, 79) = 40.24$, $p < .001$; or "deep-shallow," $F(1, 79) = 52.25$, $p < .001$. Both teachers and students perceived the optimal state of settings to be more "familiar," "delicate," "individual," and "deep" than the present state.

TABLE 19.4. Rating of "Active-Passive" (1–7)

	Students (n = 53)					Teachers (n = 28)				
	Home room	House	Elective	Club	Whole school	Home room	House	Elective	Club	Whole school
Present state										
M	3.60	4.28	3.66	2.87	3.94	3.46	4.14	3.82	3.29	4.14
SD	1.03	1.15	1.37	1.37	1.55	1.26	1.11	1.31	1.30	1.24
Optimal state										
M	3.06	3.21	3.09	2.64	3.15	2.50	2.93	2.46	2.11	2.54
SD	1.06	1.08	1.42	1.24	1.42	1.23	1.05	1.07	1.03	1.20

TABLE 19.5. Rating of "Individual-Uniform" (1–7)

	Students (n = 53)					Teachers (n = 28)				
	Home room	House	Elective	Club	Whole school	Home room	House	Elective	Club	Whole school
Present state										
M	3.09	3.60	2.93	2.93	3.32	3.36	3.71	3.71	2.96	3.46
SD	1.17	1.03	1.25	1.40	1.30	1.39	1.08	1.24	1.14	1.11
Optimal state										
M	3.02	3.13	2.62	2.77	2.89	2.64	2.93	2.46	2.39	2.50
SD	1.26	1.08	1.18	1.22	1.20	1.13	1.25	0.92	1.20	1.00

The effect of "setting" was significant on all scales: "familiar-distant," $F(4, 316)$ = 27.65, $p < .001$; "loose-tight," $F(4, 316) = 8.50$, $p < .001$; "delicate-rough," $F(4, 316) = 7.46$, $p < .001$; "active-passive," $F(4, 316) = 13.69$, $p < .001$; "individual-uniform," $F(4, 316) = 5.33$, $p < .001$; and "deep-shallow," $F(4, 316) = 13.80$, $p < .001$. Considered holistically, "club" was distinctively "familiar," "active," "individual," and "deep," "home room" was fairly "loose" and "familiar," and "elective" was "individual." Other settings tended to be rather neutral.

The interaction between "position" and "state" was significant on five scales, except for "familiar-distant": "loose-tight," $F(1, 79) = 27.02$, $p < .001$; "delicate-rough," $F(1, 79) = 7.21$, $p < .01$; "active-passive," $F(1, 79) = 6.43$, $p < .05$; "individual-uniform," $F(1, 79) - 10.00$, $p < .01$; and "deep-shallow," $F(1, 79) = 5.50$, $p < .05$. Both teachers and students rated the optimal state of settings to be more "delicate," "active," "individual," and "deep" than the present state of settings, whereas teachers rated "delicate," "active," "individual," and "deep" as being more important than students did. As for "loose," however, students rated the optimal state "looser," but teachers' ratings were less "loose." The interaction between "position" and "setting" was not significant on any of the scales.

The interaction between "state" and "setting" was significant on all scales: "familiar-distant," $F(4, 316) = 20.70$, $p < .001$; "loose-tight," $F(4, 316) = 5.10$, $p < .001$; "delicate-rough," $F(4, 316) = 3.27$, $p < .05$; "active-passive," $F(4, 316) = 3.34$, $p < .05$; "individual-uniform," $F(4, 316) = 3.07$, $p < .05$; and "deep-shallow," $F(4, 316) = 7.72$, $p < .001$. Noteworthy tendencies were as follows: Though the present

TABLE 19.6. Rating of "Deep-Shallow" (1–7)

	Students (n = 53)					Teachers (n = 28)				
	Home room	House	Elective	Club	Whole school	Home room	House	Elective	Club	Whole school
Present state										
M	3.76	4.28	4.04	3.40	4.19	3.75	4.39	4.35	3.00	4.68
SD	1.11	0.97	1.45	1.47	1.36	1.14	1.17	1.22	1.41	0.95
Optimal state										
M	3.42	3.28	3.19	3.13	3.49	3.00	3.18	2.79	2.07	2.96
SD	1.18	1.08	1.23	1.35	1.28	1.41	1.44	1.40	1.25	1.23

state of "house" and "school" was slightly "distant" and "elective" was neutral, the optimal state inclined toward "familiar." As for "loose-tight," differences were not so clear. This may be related to the role of "position," which showed significant interaction with "state" on this scale. The optimal state of "school" and "elective" was more "delicate" than the others. The optimal state of "house" and "school" was more "active" than the others. The optimal state of "house," "school," and "elective" was more "individual."

The optimal state of all settings was "deeper" than the present state. However, since the present state of "house," "elective," and "school" was in the direction of "shallow," differences between present and optimal states were greater than for the other settings.

Finally, the interaction between "position," "state," and "setting" was not significant on any of the scales.

DISCUSSION

Two surveys of teachers and students revealed that big school settings were not necessarily anonymous and that teachers and students responded to anonymity differently.

First, as for big school settings, teacher student relationships in clubs were "familiar," "active," "individual," and "deep." Clubs did not show anonymity. On the other hand, the school as a whole was a little "distant" and "shallow." Elective subject classes were perceived as a little "shallow" or relatively neutral but at the same time "individual." These big school settings result in somewhat anonymous relations between teachers and students. Home rooms and houses are small school settings. Unexpectedly, houses manifest the same characteristics as the whole school or the big school. On the contrary, home rooms were perceived as being "loose" and "familiar." They showed the positive properties of small school settings. The major features of home room settings are long term and fixed.

Setting utilization is a plausible explanation for this contradiction. Students come into their own home rooms everyday. They participate in club activities voluntarily. However, house settings and school settings are not necessarily daily settings. Official house meetings are held once a month. Whole school meetings are held at similar intervals, though some of them are conducted as school events.

The "loose" property of home rooms seems peculiar because this setting is one of the most permanent and is perceived as being "familiar." Nojima (1993) asked students to list familiar teachers and students in different settings and concluded that home rooms were the most important settings in "familiarity" development. Since teachers perceived house and school as well as the home room as being "loose," this might be related to the influence of big school anonymity.

Accordingly, we cannot simply say "small is beautiful" or "big is dangerous." However, both teachers and students hoped that teacher student relationships in both house and school would become more "familiar," "delicate," "active," "individual," and "deep." They saw present house and school settings as being "anonymous." We might advise improving house and school settings to make them more attractive. The campus plan, however, involves an inevitable trade-off between small schools and a big school. In this study, home rooms were effective as stable bases of school life. The role of a house as a small school remains ambiguous. However, it seems to have a certain effect on teachers and students that the present research could not clarify. Because the house is one of the membership groups for both teachers and students and is an extra group that other schools do not have, it must offer students something meaningful.

Second, there was an interesting difference between teachers' and students' perceptions. In general, teachers tended to perceive the present state as being more negative and the optimal state as being more idealistic than students did. This may be because of their awareness of responsibility. At the same time, they may regard themselves as being administrators or guardians. They saw the present state as being "loose" and wished the state to be less "loose." In a sense, the big school may not be so preferable for teachers. Differences between present and optimal states were larger in teachers' than in students' perceptions. On the other hand, it seems that students might enjoy big school anonymity because of their own firm home room foundations.

Finally, a comparison between Ina-gakuen and other schools (general size) showed that Ina-gakuen's elective subject classes and the whole school were relatively "distant" and "shallow." This represents features of a big school. However, "loose" and "individual" were unique properties of teacher student relationships in the settings of Ina-gakuen. The success or failure of the campus plan seems to revolve around the subtle balance between the small schools and the big school. Again, the question is: Which is desirable; a big school, a small school, or a campus school? The answer depends on the educational objectives that educators prefer. Environmental psychology is helpful for evaluating school models, such as this case study.

NOTES

Daniel Stokols suggested that anchor point theory and staffing theory should be distinguished. Home rooms and clubs might work as anchor point settings among various overstaffed big school settings in the campus plan.

Robert Bechtel recommended counting setting participation. According to archival data, house (small school) settings do not occur frequently, even though participating in them is required. Attending elective subject classes is voluntary, but most participants are unfamiliar with each other. In this case, setting size seems not so influential.

REFERENCES

Baird, L. L. (1969). Big school, small school: A critical examination of the hypothesis. *Journal of Educational Psychology, 60*, 253–260.

Barker, R. G., & Gump, P. V. (1964). *Big school, small school: High school size and student behavior.* Stanford. CA: Stanford University Press.

Gump, P. V. (1987). School and classroom environment. In D. Stokols & I. Altman (Eds.), *Handbook of environmental psychology: vol.1.* (pp. 691–732). New York: Wiley.

Nojima, E. (1993). The effectiveness of the "house system" in a comprehensive high school: A case study of Ina-gakuen comprehensive high school. *Journal of Human Sciences, 6*(1), 13–22. (In Japanese)

Sako, T. (1993). Teachers' movement patterns in the physical setting of a large-scale comprehensive high school based on the campus plan. *Journal of Human Sciences, 6*(1), 2–12. (In Japanese)

Schoggen, P. (1989). Behavior settings: A revision and extension of Roger G. Barker's ecological psychology. Stanford, CA: Stanford University Press.

Wicker, A. W. (1979). *An introduction to ecological psychology.* Monterey, CA: Brooks/Cole (Republished in 1983 by New York: Cambridge University Press).

Chapter **20**

SOJOURN IN A NEW CULTURE
Japanese Students in American Universities and American Students in Japanese Universities

Seymour Wapner, Junko Fujimoto, Tomoaki Imamichi, Yuko Inoue, and Karl Toews

INTRODUCTION

Attending a university in a country different from one's own has a potent impact on a person (see Church, 1982, whose review and summary lists more than 40 references on this topic; Klineberg & Hull, 1979). Despite the accumulation of a vast amount of information concerning the more general problem of a sojourn to a different culture and the more specific case of a sojourn to a university in a different culture, there still remains "the need for a multicausal model in which home and university factors are influential" (Fisher & Hood, 1987, p. 439). Or as Shaver, Furman, and Buhrmeister (1985) in their study on transition to college see it, "The field needs a broad social and developmental perspective which can coherently encompass the burgeoning, and not well-integrated literature on personal relationships" (p. 219).

It is the goal of the present study to take some beginning steps in meeting this challenge by utilizing a broad theoretical perspective—the holistic, developmental,

Seymour Wapner • Heinz Werner Institute, Clark University, Worcester, Massachusetts 01610. **Junko Fujimoto** • 2-13-7 Kitahorie, Nishi-Ku, Osaka 550, Japan. **Tomoaki Imamichi** • 2-7-13 Shroganedai, Minato-ku, Tokyo 108, Japan. **Yuko Inoue** • 1-34-20 Negao Higashimachi, Hirakata, Osaka 573-01, Japan. **Karl Toews** • 1558 Stonemill Road, Lancaster, Pennsylvania 17603.

Handbook of Japan–United States Environment–Behavior Research: Toward a Transactional Approach, edited by Seymour Wapner, Jack Demick, Takiji Yamamoto, and Takashi Takahashi. Plenum Press, New York, 1997.

systems-oriented approach (Wapner, 1977, 1981, 1987)—to analyze *experience* and *action* of the sojourn of students to a university in a culture different from their own, in particular, Japanese students who attend universities in the United States and American students who attend universities in Japan. Hopefully, such research will throw light not only on the general nature of the transition but also on the similarities and differences of the student sojourner coming from or going to the same two cultures, Japan and the United States.

To accomplish this task, we shall first briefly review some of the central assumptions of the perspective. Second, we shall review some of the variety of studies that have dealt with cultural relocation generally as well as with the specific problem addressed here of attending a foreign university. Third, we shall present four studies that have sought evidence on the transition to a university in a different country from the student's homeland, three of which deal with Japanese students in American universities and a fourth study that is concerned with American students in Japanese universities.

THE PERSPECTIVE

Basic to the approach is the *transactionalist* assumption that the *person-in-environment* is the *unit to be analyzed*. Given this unit of analysis, three aspects of the person and three aspects of the environment are considered. With respect to the person: *physical/biological* (e.g., health), *psychological/intrapersonal* (e.g., mental status), and *sociocultural* (e.g., role) aspects are examined. With respect to the environment: *physical* (natural and built), *interpersonal* (friends, relatives, etc.), and *sociocultural* (rules, regulations, mores, etc.) aspects are examined.

Transactions of the person with the environment which include both *action* and *experience* (knowing—encompassing sensory-motor functioning, perceiving, thinking, etc.—feeling, and valuing) are *teleologically* directed; that is, the person transacts with the environment in terms of *ends or goals*, which are achieved by a variety of *instrumentalities*.

Closely linked to *teleological directedness* is the notion of *planning* and *expectation*. This involves identification of required environmental resources, the nature of anticipated barriers, definition of goals and the means for accomplishing them, and a motivated basis for adhering to relevant regimens of action.

Person-in-environment system states are assumed to be *developmentally orderable* in terms of the *orthogenetic principle* (Kaplan, 1966; Wapner, 1987; Werner, 1957a, b; Werner & Kaplan, 1956), which defines development in terms of the degree of organization of the system: development proceeds from a *dedifferentiated* to a *differentiated and hierarchically integrated* person-in-environment system state. The developmental ideal is characterized by *control* over self world relations, *subordinated* rather than *interfused* functions, *discrete* rather that *syncretic* mental phenomena, *articulate* rather than *diffuse* structures, *stable* rather than *labile*, and *flexible* rather than *rigid* modes of coping (cf. Kaplan, 1966). Moreover, "such movement is assumed to involve (1) greater salience of positive affective states; (2) diminution of isolation, anonymity, helplessness, depersonalization, and entrapment; (3) coordination of long-term and short-term planning; and (4) optimal use of available instrumentalities to accomplish personal goals. In short, the developmental ideal involves movement toward an integrated unity of the person's overt and covert actions (Wapner, 1987, pp. 1444–1445).

A final feature of the perspective to be described here concerns the *relation between experience and action*. This is the problem of whether the person in fact *"does"* what he or she *"would like to do."*. This, of course, is closely related to *planning*. For example, if prior to a transition to a university in another country, the individual has plans about how to cope with difficulties that arise, does he or she, in fact, carry out those plans? Further, what are the general factors and precipitating events that make for the appropriate change from thought to action?

How do these general assumptions help categorize and explain the various findings in the literature concerning the general problem of transitions to other cultures and the more specific sojourn of the student to a university in another cultural context? Let us review the main findings of such research in terms of the categories and developmental modes of analysis combined in the holistic, developmental, systems-oriented perspective.

GENERAL RELOCATION

Person-in-Environment System State

The general difficulties, problems, and experience encountered when relocating to a new culture, as described by a variety of authors, are summarized in Table 20.1, according to the three aspects of the person and the three aspects of the environment encompassed by the person-in-environment system state. These apply to general relocation, whether it be *migration*, involving a long-term move, or a sojourn that is limited to a short stay.

Developmental Analysis

Both migration and a sojourn may be analyzed developmentally as indicated by the form of acculturation adopted by the individual involved. For example, there may be *assimilation* and *identification* of self with the new culture (e.g., "melting pot"); there may be *isolation* from and/or *conflict* with the culture of the host environment (e.g., ethnicity); or there may be flexible maintenance of both the values of the culture one leaves as well as the culture one enters (e.g., *biculturalism*; see Ramirez & Casteneda, 1974; Wapner, 1988). These alternatives formally parallel the self world relationships implied in the orthogenetic principle as ranging from lesser to more advanced developmental status, that is, from (1) dedifferentiated to (2) differentiated and isolated, to (3) differentiated and in conflict, to (4) differentiated and hierarchically integrated person-in-environment system states (Wapner, Ciottone, Hornstein, McNeil, & Pacheco, 1983, p. 122; Wapner, 1987).

TABLE 20.1. Some of the Literature on General Factors Affecting Adaptation of Migrants and Sojourners

	General factors	References
Person		
Biological	At risk for such disorders as cardiovascular disease.	Wolff (1952)
Psychological	"Culture shock" involving strain in adapting; loss from uprooting; rejection by host; surprise; discomfort; indignation; disgust; impotence linked to unfamiliarity with culture and lack of capacity to cope.	Oberg (1960) Taft (1977)
	Excessive occupation with water, food, minor pains; excessive fears of being cheated or robbed; anger toward avoidance of local people; longing to be with fellow nationals.	Clark-Oropeza, Fitzgibbons, & Baron (1991) Church (1982)
Sociocultural	Role; adjustment as foreigners; loss of personal status.	Cormack (1968) Bochner (1972) Church (1982)
Environment		
Physical	Need for adaptation to climate, food.	Church (1982)
Interpersonal	Communication with members of host culture; differences in language (verbal & nonverbal); link of favorable attitudes, judgement, & satisfaction to increased social interaction.	Hall (1959) Porter (1972) Ekman (1972) Klineberg & Hull (1979)
Sociocultural	Stress linked to sensing structure or rules of society; how to cope with relevant reinforcers, aversive & discriminative stimuli.	Schutz (1964) Triandis (1975) Triandis, Vassiliou, Tanaka, & Shanmugam (1972)

SOJOURNER TO A FOREIGN UNIVERSITY

There are many differences between migration and a sojourn to a foreign university. For example, in addition to a restriction in time of stay for the latter, there are two perturbations involved, namely, entering college and a different culture. Both of these features, the subculture of the college and the new culture of the country, must be taken into account in the analysis.

Since individuals involved in a sojourn to a university plan a relatively temporary stay as compared with the migrant who plans to remain in the new country, students may not exhibit all of the developmental stages described in the orthogenetic principle. Dedifferentiation in the sense of "melting" into the culture is unlikely to occur. Moreover, for American students in Japanese universities, this is exaggerated because foreigners in the homogeneous Japanese society are differentiated by appearance, by language, and by cultural barriers.

The specific problem of transition to a university has been studied extensively. Church (1982) has summarized the commonly mentioned problems of foreign students under the following categories: *academic problems*, *personal problems*, and *sociocultural problems*. From our perspective, Church's categorization has the disadvantage that any of these categories may encompass or fail to differentiate

TABLE 20.2. Commonly Mentioned Problems of Foreign Students (Church, 1982)
Categorized According to the Holistic, Developmental, Systems-Oriented Perspective

Person	
Biological	Somatic complaints
	Adjusting to food
Psychological	Inadequate prior preparation
	Inadequate academic orientation, advice
	Adjustment to new educational system: Frequent exams and assignments; classroom and professor/student informality; competitiveness; grading methods: credit system.
	Loneliness
	Homesickness
	Depression
	Arrival confusion
	Maintaining self-esteem
	Lack of personal guidance and counseling
	Overambitious goals to succeed
	Inappropriate motivation for overseas study
	Concern over employment opportunities on return home
	Understanding lectures
	Decision to stay or return home
	Staying current with events at home
Sociocultural	Defining role as a foreign student
	Religious problems
Environment	
Physical	Adjusting to food, climate
	Housing difficulties (cost, noise, privacy)
Interpersonal	Family problems or loss of loved ones
	Problems with verbal and nonverbal communication: need to delay educational goals while studying host language; written and oral reports
	Culture-bound professional vocabulary
	Getting along with roommates
	Superficial American friendships
	Racial discrimination
	Dating and sexual problems
	Difficulty making social contacts
Sociocultural	Adjusting to social customs and norms
	Contrasting or conflicting values and assumptions
	Balancing simultaneous culture group memberships
	Visa, immigration problems
	Difficulties obtaining employment
	Financial, employment problems
	Ignorance of host nationals about home culture
	Political upheaval at home
	Inadequate placement or credit for previous coursework
	Selection of institution and coursework

between specific features of experience that can fit more functional, meaningful categories of our general perspective. Accordingly, it seemed useful to reclassify the problems Church (1982) drew from more than 50 sources in terms of the category system of our theoretical perspective. Such a categorization is presented in Table 20.2.

Examination of the table shows that the specific examples of Church's three categories can be meaningfully incorporated within the person-in-environment system (including its three levels of person and environment) that serves as the unit of analysis of our theoretical perspective. Moreover, the differential categorization not only systematizes the areas of concern and findings therein, but also shows the gaps (e.g., cognitive, affective, and valuative aspects of experience; development of interpersonal relationships; relations between expectation and actuality) that might be filled for a more complete analysis of the problem of the student sojourner to a culture different from his or her own.

The four studies reported here were directed at accumulating information that could fill these gaps as well as replicate some of the information already known about the sojourn to a foreign university.

STUDY I: THE AMERICAN CULTURE AS VIEWED BY STUDENTS ENROLLED IN A UNIVERSITY IN THE UNITED STATES: A PILOT STUDY

The aim of this preliminary investigation was to obtain information, through an in-depth interview, about the general character of the transactions (experience and action) of the sojourner enrolled in an American university, utilizing the category system of the holistic, developmental, systems-oriented perspective (Wapner, 1987).

METHOD

Subjects

The subjects were 15 Japanese students (8 males and 7 females) enrolled in a university in the northeastern United States.

Procedure

Volunteers completed an Informed Consent Form, and participated in an audiotaped interview that contained the open-ended question, "Compared to Japan, what was the USA like when you first came here to college?" and answered a series of specific questions concerning their experience with respect to all six aspects of the person-in-environment system, as well as information about the discrepancies between their expectations and actual experience in the new university environment. Participants were characterized as to their fit in the developmental categories mentioned earlier, namely, dedifferentiation ("melting pot notion"), differentiation and isolation, differentiation and conflict (both of these modes are concerned with maintaining one's ethnicity), and differentiation and hierarchic integration (striving toward multiculturalism).

Analysis

The recorded interviews were transcribed, and case studies were constructed according to all six aspects of the person-in-environment system and the relations among them. Moreover, participants' initial experience on arrival and experience at the time of the interview were described. Where available, data were statistically analyzed by chi-square and the sign test.

RESULTS

Physical Features of the Environment

A variety of negative features of the new environment were spontaneously mentioned, including "transportation system bad," "food bad," "things rough and broken," "weather cold," "polluted," "boring," "dangerous." Considering both initial and later impressions, all 15 subjects mentioned one or more of these negative items. Large size, which is regarded neither negatively nor positively, was commented upon frequently by 7 subjects when describing feelings upon initial arrival, as compared with none commenting on large size only on a later stay and one person commenting on large size on both occasions ($p < .05$, sign test). The positive physical features mentioned (e.g., "buildings are nice," "less polluted," and "inexpensive") are reported very infrequently, with only three persons giving such a response on both initial and later descriptions.

Interpersonal Features of the Environment

A variety of negative features of Americans were mentioned spontaneously, including "superficial," "rude," "less organized," "ignorant about Japanese culture," "other international students easier to communicate with [than Americans]," "not open to other cultures," "prejudiced," "selfish," "short temper[ed]," "not oriented toward own group," not friendly," "not punctual," "less communicative." For the initial and later descriptions pooled, a high proportion, 13/15, gave negative impressions, with no evidence of difference between the two sets of descriptions. There was some evidence of gender differences depending on time: A higher proportion of men (6/8) than women (1/7) reported negative impressions on arrival (X^2 (1, $n = 15$) = 4.06, $p < .05$), whereas there was a higher proportion of women (7/7) than men (2/8) doing so on their later stay (X^2 (1, $n = 15$) = 12.23, $p < .001$). The positive features reported included "easygoing," "friendly/nice," "open/direct," "teachers are dedicated," "welcome others," "hypocritical," "helpful." Whereas nine participants give more positive responses on arrival than after a longer stay, only one participant gave the opposite response ($p < .05$ by sign test).

Sociocultural Features of the Environment

The positive features reported included "many races exist," "provide responsibility," "provide chances [opportunities]," "religious belief is strong," "American education is good," "less restriction," "American society consists of relativism and patriotism," "respect other's opinion," "values homosexuality." The negative features included were as follows: "drinking age is strict," "drugs are easy to obtain," "economy is bad," "teenager values," "allowed to carry shot-guns," "manners are bad," "system of problem solving is forceful and inefficient," "less restriction," "influence [operates]," "politics is important," "different system [with respect to] taxes and tips."

Physical Features of the Person

There was a nonsignificant tendency for more reports of "unhealthy" occurring on arrival (6/15) than after a stay in the environment (1/15).

Psychological Features of the Person

The intrapsychological features of the person reported can be divided into cognitive, affective, and valuative categories.

Cognitive

The responses fitting the cognitive category include "surprise," "learned from new environment," "gained motivation to study," "aware of differences in thinking," "became critical [in thinking]," "influenced as a liberal," "became a new person," "became independent." A greater proportion of subjects (12/15) report these for the time of their arrival only, and only one subject reports these after a longer stay ($p < .05$ by sign test).

Affective

The negative experiences reported include "annoyed," "frustrated," "insecure," "disappointed," "bored," "worried," "uncomfortable," "lonely/homesick," "hurt," "depressed," "mad," "sad," "scared." One or more of these 13 negative features were present in the reports of participants (15/15) regarding their initial contact, and these responses were more frequent than those given after a longer stay (10), with the reverse occurring for only one subject and four showing an equal number of responses for both occasions ($p < .05$ by sign test). Four additional negative features ("no trust," "isolated," "depressed,""bothered," "shy") were added regarding the inquiry concerning later impressions. When these are added to the original features there is an increase to 13/15 subjects reporting negative features. The positive features reported include "comfortable," "fun/happy," "free," "funny," with "satisfactory" added on the second inquiry. There is no evidence of a tendency for a difference between initial and later reports.

Valuative

The items reported with no trends on initial compared to a longer stay include "awareness of prejudice toward him as a foreigner," "change in the way of contacting [others]," "aware of identity," "became a new person," "prejudice toward her as a foreigner," "awareness of difference in culture," "awareness of strong prejudice toward her as a foreigner." Most noticeable were "awareness of differences in values-conflict" (5/15) and "prejudice toward subject as a foreigner" (4/15).

Adjustment Characterized Developmentally

The frequency of subjects falling in the developmental categories (1) dedifferentiation, (2) differentiation and isolation, (3) differentiation and conflict, and (4) differentiation and hierarchic integration were respectively: 4, 4, 5, and 2 for reports on initial contact, and 0, 3, 2, and 10 for reports on adjustment after a longer stay. Thus, there is a tendency for participants to shift in the direction of striving toward multiculturalism (differentiation and hierarchic integration with a longer stay).

SUMMARY

Despite the small size of the sample, some interesting trends emerged. For example, with respect to the physical features of the environment, all subjects mentioned a broad variety of negative items for both their initial and later impressions, whereas there were fewer subjects giving positive impressions. The trend for women to give a higher proportion of negative reports concerning physical features of the environment suggests that it is important to assess gender differences. Though comment about physical features of the person were restricted to reports of "unhealthy" on initial contact, responses concerning the intrapsychological features of the person were large and varied enough to be divided into cognitive, affective, and valuative categories. The responses fitting the cognitive category tend to decrease with length of stay. Those in the affective category that were negative did not differ with respect to duration of stay and there is only a slight increase for the descriptions falling under positive affect. Among the value-oriented responses are those concerned with prejudice against a foreigner.

STUDY II: EXPERIENCE AND ACTION OF JAPANESE STUDENTS IN A VARIETY OF UNIVERSITIES IN THE UNITED STATES

The various examples of experience and action for the three aspects of the person (physical, psychological, and sociocultural) and for the three aspects of the environment (physical, interpersonal, sociocultural) obtained in the pilot study reported above provided a basis for constructing an anonymous questionnaire that could be used in obtaining data on adaptation to American culture from a larger sample of Japanese students in American universities.

METHOD

Subjects

By means of an anonymous questionnaire, data were collected from 60 Japanese student volunteers (48 female and 12 male) who were born and raised in

Japan and were studying at various American universities in 26 of the United States. All had more than 7 years of training in the English language.

Procedure

Volunteer participants were obtained by approaching them personally or through a letter of introduction and an anonymous questionnaire mailed to Japanese students attending a variety of universities in the United States. Potential participants were instructed to complete the questionnaire, not to write their names or any identifying marks on the questionnaire, and to mail it in an unidentifiable postage-paid envelope.

Questionnaire

The questionnaire obtained information on demographics (e.g., gender, age, previous experience in the United States, friendships, ability in English), along with reports on expectations and actual experiences in the United States. The expectations and actual experiences of the subjects were identified through qualitative, open-ended questions to which the participants provided written responses. In addition, participants were requested to answer more specific questions and rating statements (by circling a number on a scale ranging from 1 to 7) concerning their experience and action in the new environment.

Analysis

Ordinal data (rating scales) were analyzed by ANOVA, and the nominal data were analyzed by chi-square tests. Expectations and experience/action were largely assessed within the six categories of the person-in-environment unit of analysis.

RESULTS

Expectations: Qualitative Data

The findings listed separately for the six person-in-environment categories are listed with frequency of occurrence in Table 20.3. Here, mention will be made only of the most frequent responses for the various categories.

To the general question "What were your expectations about America?" the most frequent responses were that: America would be a free society; Americans would be friendly; and America would be racially diverse and dangerous.

Physical Aspects of the Person

More than half expected to gain weight, and thought American food would be greasy and heavy.

TABLE 20.3. Expectations Concerning the United States: Qualitative Data

Category	Responses	Frequency
General question	America would be a free society	21/60
	Americans would be friendly	18/60
	America would be racially diverse	11/60
	America would be dangerous	10/60
Person		
Physical	Expected to gain weight	34/60
	American food would be greasy, heavy	15/60
Psychological	American culture would make their mind freer, more open	12/60
	Become more independent	5/60
	More outgoing	4/60
	Worried about homesickness	3/60
	Single responses: expectations of aggressiveness, loneliness, confusion, isolation, irritation, optimism.	
Sociocultural	Wanted role as representatives of Japanese culture to Americans	10/60
	Be perceived as foreigners	4/60
	Be perceived as minority	4/60
	Expected to experience prejudice	3/60
Environment		
Physical	Broad land, open space	4/60
	"Green" nature	4/60
	Large size	2/60
	Single responses: air dry; too hot.	
Interpersonal	Americans friendly	8/60
	Cheerful, aggressive, & loud	3/60
	Single responses: positive—jolly, smart, active, smiling; negative—unkind, lazy, selfish, superficial, strange, forward, express feelings directly.	
Sociocultural	America as melting pot	
	Free	9/60
	Open	7/60
	Single responses: informal, noisy, cheerful, religious, commercial, innovative, pragmatic, same as American movies.	

Psychological Aspects of the Person

The most common response was that the American culture would make their mind freer and more open, including such comments as "set my mind free," and "make me more open-minded."

Sociocultural Aspects of the Person

The most common response was that they wanted their role to be representing Japanese culture to Americans.

Physical Aspects of the Environment

The expectations expressed were sparse and included: a lot of broad land and open space.

Interpersonal Aspects of the Environment

More than half of the respondents expected Americans to be friendly.

Sociocultural Aspects of the Environment

Expectations included such characteristics of the culture as "free" and open.

Actual Experience: Qualitative Data

The responses concerning the actual experience of the Japanese students in the American environment are presented in Table 20.4. Again, the most frequent responses are noted below.

In response to the *general open-ended question* about experiences in the American environment, the most frequent spontaneous answer concerned size, for example, America was "big," "huge," "large," with respect to objects, space, culture, "everything." Next, was the racial diversity of people.

Physical Aspects of the Person

When answering the open-ended question about health, the most frequent report was a complaint about difficulty with American food. For example, one student reported "I try to avoid oily and sweet food, it makes me sick," and a number of students mentioned gaining weight, sometimes in conjunction with the nature of the food.

Psychological Aspects of the Person

In response to the open-ended question concerning their feelings, the most frequent response was a feeling of being a "foreigner."

Sociocultural Aspects of the Person

When asked about actual experience with respect to their roles and values in America, the most frequent response was that they felt the same as before, yet when the varied responses suggesting change are added together over half the subjects report some kind of change.

Physical Aspects of the Environment

In response to the open-ended question, the most common response was the ineffectiveness of the transportation system. Five said a car is necessary to get around: "you can't survive without a car."

Interpersonal Aspects of the Environment

The most common answer was that the Americans are friendly, next that they are superficial: for example, "they do not talk from the heart."

TABLE 20.4. Actual Experience of Japanese Students in the United States: Qualitative Data

Category	Responses	f
General question	Americas large size (big, huge, large with respect to objects, space, everything	28/60
	Racial diversity of people	14/60
	Friendly	6/60
	Kind	6/60
	Natural, beautiful, paradise	4/60
	Not much difference between U.S. & Japan	4/60
	Felt like Asian minority	2/60
	Single responses: Americans live in cars; use a lot of materials and throw them away.	
Person		
Physical	Difficulty with American food	15/60
	Gaining weight	14/60
	Feeling pretty good	8/60
	Sick with flu	6/60
	Feeling very good, excellent	5/60
	Feeling ill/tired because of schoolwork	3/60
	Dry skin	2/60
	Stomach aches	2/60
	Single response: afraid of AIDS.	
Psychological	Feeling of being a "foreigner"	7/60
	Embarrassed by level of ability to speak English	6/60
	Identity with other Asians	3/60
	Feelings of isolation & loneliness	3/60
	Stress from environment	3/60
	Feeling free	3/60
	Enjoying American culture	3/60
	Single responses: frustrated, annoyed, anxious, tired, bored, insecure.	
Sociocultural	*Single responses*: realization of self as foreigner; self as "yellow"; women have more freedom; women and men equal; cannot disregard Japanese collectivism though liked individualism.	
Environment		
Physical	Ineffective transportation system	8/60
	America is big, huge, spacious	6/60
	Need car to get around	5/60
	Environment is dangerous	5/60
	Beauty of nature	3/60
	U.S. not sanitary	2/60
	Single responses: long, straight, roads going on and on; less pollution than in Japan; weather too cold in winter; so much interesting stuff in USA environment.	
Interpersonal	Americans are friendly	13/60
	Americans are superficial	10/60
	Single responses: Americans keep eye contact while talking; don't speak to foreigners; self-centered; don't care about others; some kind, gentle; some extremely rude.	
Sociocultural	Separation of races	6/60
	Racial discrimination	3/60
	Much religion	3/60
	Drinking age should be lowered	3/60
	Lack of respect for older people	2/60
	Single responses: more liberty and rigidly governed by law; respect others; open for everything; accept new things; more crime; women treated equally; guys cook & wash clothes.	

Sociocultural Aspects of the Environment

The most prevalent answer to the open-ended question with respect to American culture referred to the apparent separation of the races, "It seems that white and black people are still separated."

Actual Experience: Ratings

Findings on participants' ratings (ranging from 1 to 7) regarding specific questions of their actual experience with respect to all six aspects of the person-in-environment system are presented in Table 20.5 in terms of mean rating, standard deviation, mode, and frequency of the modal value of the distribution. The items included are restricted to those that had a mean rating of 3 or below, 5 or above, or had a modal value at the extremes (1,2 or 6,7), where the former represented agreement and the latter disagreement with a statement. Examination of Table 20.5 shows the main findings were as follows:

Person

With regard to the *physical* aspect of the student's experience, there was a strong tendency to feel that they gained weight after arriving in the United States. For the *psychological* aspect of the person, students reported not feeling angry when in America; enjoying the new culture; and feeling like a foreigner. For the *sociocultural* aspect, Japanese students reported that being in the United States helped them to learn what it is to be Japanese; and that they have no political power in their new university.

Environment

The questions on the *physical* aspects of the environment received the greatest number of similar responses. Japanese students reported that the distances between American cities is too far; the United States is very spacious; the transportation system is bad; American food is tasteless; it was difficult to buy Japanese food; cars, houses, and buildings appear to be larger in America when compared with Japan; and they enjoyed American architecture.

Students showed less agreement with respect to the *interpersonal* environment. There was a tendency to agree that American teachers are friendlier; Americans are ignorant of Japanese culture, and show more feelings to others than Japanese.

With regard to the *sociocultural* aspect of the environment, students agreed that one can obtain a driver's license more easily in the United States than in Japan; the United States has poor gun control laws; Americans are more open about sex and pornography; Americans are more able to talk about difficult issues such as child abuse, teenage pregnancy, and homosexuality; and Americans are ignorant with regard to Japanese culture.

TABLE 20.5. Experience and Actions of Japanese Students at American Universities: Ratings

Category	Item	Mean	Standard deviation	Mode	Frequency of mode	Percent (mode/ total)
Person						
Physical	Gained weight in the U.S.	5.1	2.3	7	27	45
Psychological	Enjoy new culture	5.6	1.6	7	20	33
	Being in America does not make me angry	5.4	1.8	7	21	35
	Feel like a foreigner	5.1	1.8	7	17	28
Sociocultural	Do not have political power	6.3	1.3	7	41	68
	Has helped to learn what is Japanese	5.7	1.6	7	24	40
Environment						
Physical	Many trees and grassy areas	6.4	.9	7	40	67
	Transportation system is bad	6.4	1.1	7	38	63
	Distances between cities too far	6.3	1.3	7	39	65
	U.S. is spacious	6.3	1.2	7	43	71
	Houses and cars are bigger	6.2	1.3	7	33	55
	Difficult to buy Japanese food	5.6	1.6	7	25	42
	Enjoy architecture	5.5	1.6	7	21	35
	American food is tasteless	5.4	1.7	7	24	40
Interpersonal	American teachers are friendlier	5.8	1.3	7	22	37
	Americans show more feelings	5.7	1.7	7	26	43
	Americans ignorant of Japanese culture	5.3	1.4	7	19	32
Sociocultural	The U.S. has bad gun control laws	5.9	1.5	7	26	43
	Can easily obtain driver's license	5.8	1.5	7	33	55
	Americans more open about sex	5.7	1.4	7	24	40
	U.S. education gives one a chance to think	5.6	1.7	7	23	38
	Americans talk more about difficult issues	5.4	1.6	7	19	32
	Age is not as important as in Japan	5.3	2.1	7	23	38

Moderator Variables

Relative to Japanese men, (1) Japanese women rated themselves as feeling more unhealthy (M_m = 4.5; M_f = 4.6; F = 4.58, 1,58 df, $p < .05$); (2) Japanese women reported feeling more comfortable and happy (M_m = 2.9; M_f = 4.3; F = 5.90; 1, 58df, $p < .05$); and (3) Japanese women tended not to rate themselves as feeling less like foreigners (M_m = 4.8; M_f = 4.3; F = 3.30, 1,58df, $p = .07$).

Those with prior experience in the United States rated themselves as (1) more tired (M_e = 4.9) than those without prior experience (M_{we} = 3.8) F = 4.16; 1, 58 df; $p < .05$) and (2) tended to feel more unhealthy (M_e = 4.9) than those without prior experience(M_{we} = 3.9) (F = 3.25; 1,58 df; $p = .08$).

Relative to those Japanese students not expecting people to be friendly, those expecting people to be friendly (1) had fewer headaches and stomachaches (M_f = 2.2; M_{nf} = 3.6; F = 6.50; 1, 58 df; $p = .01$) and (2) tended to regard Americans as superficial (M_f = 4.9; M_{nf} = 4.2; F = 3.19; 1, 58 df; $p = .08$).

SUMMARY

A number of trends were evident. For example, being in the American culture helped the Japanese students define and understand what it means to be Japanese; there was considerable consensus in the modes of experiencing aspects of the physical environment as compared with the physical, psychological and sociocultural aspects of the person; Japanese students found America to be a freer society, and Americans to be ignorant of Japanese culture.

In addition, a number of interesting findings were obtained with respect to moderator variables, including the effect of prior experience on present experience in the United States (e.g., those students with prior experience felt more unhealthy and more tired); effect of gender (e.g., female students felt more unhealthy, but were happier and more comfortable than male students); and expectations that Americans were friendly (e.g., those students expecting Americans to be friendly reported having less headaches and stomachaches).

STUDY III: DEVELOPMENT OF INTERPERSONAL RELATIONSHIPS OF JAPANESE STUDENTS IN AN AMERICAN UNIVERSITY

The cultural characteristics of group formation in Japanese society are reinforced in young generations through various features of the society, for example, through the organization of business and through the educational system. Instead of working individually, one is taught to work with a group. Those who work cooperatively are rewarded, and those who work independently are not. Beginning early in childhood, as cooperation and harmony are stressed as important, as is being part of a group in Japanese society.

While Japanese society places importance on the cultural characteristics of group formation, there remains the question of the value of group formation when in another culture, such as when Japanese students are in universities in the United States. The possibility that Japanese students in the United States restrict themselves to remain part of a group of Japanese in the United States, including friendships, could be regarded negatively since such group allegiance isolated from the general cultural context would inhibit the learning of English and obtaining information and understanding of the American culture.

Accordingly, the present study aimed to assess the formation, changes, and basis for the interpersonal relations (friends, family, professors, etc.) of Japanese students attending an American university. Both first-year and advanced students are included to assess the nature of changes in interpersonal relations depending on the duration of stay in the United States.

METHOD

Subjects

Participants included 16 Japanese students drawn from two groups: a first-year group (first year at university; short stay) consisting of 6 students (3 males and 3 females); and an advanced group (longer stay) consisting of 10 students (6 males and 4 females).

Procedure

The subjects participated in three sessions throughout the fall semester. At the first session, in the middle of October, they filled out questionnaires concerning background information (e.g., age, sex, duration of stay in the United States). They then completed a so-called Psychological Distance Map (PDM) for people, an instrument designed to obtain information about people important in the world of the subject (e.g., Minami, 1985; Wapner, 1977, 1978, 1981).

The subject is presented with a blank piece of paper except for a small circle placed in its center that contains the word "ME," which represents the subject. The task is to add circles on the paper representing people at distances from "ME" according to how close the subject felt to the person they added. The closer the person entered to "ME" the closer the subject felt to that person. The name or initials of the person as well as the sequence of entry was noted within or next to the added circle.

After completing the PDM, subjects completed an Interpersonal Network Questionnaire that provided information about the people entered on the PDM and the subject's relationship to their entries. A second session in which the PDM and the Interpersonal Network Questionnaire was completed took place in November. A week following the second session, an interview was conducted that consisted of 28 open-ended questions about the subjects' experience in developing friendships in the United States and Japan.

RESULTS

Entries on PDM from Old and New Environment: First-Year Group

Comparison of the number of people important to the Japanese student from the old versus the new environment is restricted to the first-year group. It was found that people listed on the PDM map from the old environment (M_o = 12.0) were significantly greater than the mean number of people listed from the new environment (M_n = 1.1) ($p < .05$ by sign test). The relationship is in the same direction on the second test occasion (M_o = 10.0; M_n = 4.7) but is no longer significant ($p > .05$).

Further, for the first-year students, the mean number of people from the new environment entered on the PDM (first occasion) (M = 1.1) significantly increases

(M = 4.7) on the second test occasion ($p < .05$, sign test). However, there is no significant difference for number of people from the old environment on the first compared with the second test occasion.

Entries on the PDM from New Environment: First Year versus Advanced Students

First test occasion. On the first test occasion, the mean of the total number of university students entered for the first-year group (M = 1.1) was significantly smaller than the mean (M = 6.5) for the advanced group ($t = 2.15$; 14 df; $p = .05$).

Second test occasion. On the second test occasion, the mean of the total number of students entered on the PDM for the first year group (M=4.2) was significantly smaller than the mean (M = 7.8) for the advanced group ($t = 3.60$; 14 df; $p <. 01$).

Differences in Origins (Japanese, International, American) of Entries on the PDM

Differences, computed separately for entries of people with different origins (Japanese, International, American), in the effect of duration on campus, for each test occasion as well as for extreme groups (first-year students on first occasion versus advanced students on second occasion), yielded no evidence of statistical significance.

Nature of Relationship to Subject of People Entered on PDM

In these analyses, the relationship of the people entered on the PDM to the subject was assessed using the Interpersonal Network Questionnaire that included five types of relationships: social interaction; emotional support; informational guidance; tangible assistance; and conflict and stress. Each part had five questions to be answered "yes" or "no" for each person entered on the PDM.

Americans versus Japanese and others entered on the PDM. There was a nonsignificant tendency for the Japanese students entered on the PDM to show a decrease from the first to the second occasion on the four parts of the Network Questionnaire that refers to positive features. The opposite tends to hold for the American students entered on the PDM.

Qualitative Analysis: Interview

The interviews, though not providing definitive data, suggest that various aspects of the experience of the Japanese students are worthy of further exploration. For example, in response to the question, "Is there any difference in your relations with those friends you have met in the United States compared with your friends in Japan," 44% (7/16) said there was a difference. They suggested that there was a cultural difference. For example, one respondent said: "I talk something, like, what happened here, not about myself or family thing, private thing, we just talk about common things, general things. In Japan I talk with them a little bit more of my feelings. Probably, its because of different culture. It's kind of embarrassing to talk about those private or self things. I don't feel uncomfortable or embarrassing when I speak to Asian or international students. . . . I don't think they [American students] understand, I don't expect them to understand."

In response to the question, "Do you try to avoid being with other Japanese at [name of university]?", 75% (12/16) said they did not. However, they did say that they once did try to avoid being with Japanese, because it would not help them learn English. One participant said: "I'm not trying to avoid them now, but first when I came here, I was consciously trying to avoid to involve in Japanese community. . . I know its comfortable to be with Japanese because you don't have to speak English and [having the] same background. That's not what I wanted to do here. I wanted to interact with those with different background."

Many Japanese students said they do not avoid other Japanese, but also they do not stay together. For example, one student replied: "...it doesn't mean if they are Japanese they are okay. That's definitely untrue. Basically, they are Japanese people. There are American people that I wouldn't get along with, of course, and, of course, I have Japanese people who I don't prefer hanging around with."

In response to the question, "Do you try to avoid being with other Japanese at _____ University?" 75% (12/16) said they did not.

Finally, to the question "Do you have many international friends other than Americans, 75% (12/16) said they do. Many suggested this was the case because they were in a similar situation with respect to language skills and similarity of culture. One participant stated: "I can feel more comfortable being with Asian people, in terms of skin and color. We have more similar interest[s] than Americans, sometimes."

Another participant replied: "There are many things that we [with international students] can communicate. With Americans, because of cultural differences, they sometimes do unbelievable things, but with international students, the reasons why we are here are similar...with oriental, we have the same kind of sense towards things, so with them I can reveal myself totally."

Finally, another student answered: "Sometimes, we make fun of American students together, to see how they understand, to see how they see things in America, and compare how I see America."

SUMMARY

In general there was evidence, albeit limited, that the shorter the stay of Japanese students in the new college environment, the greater the number of people in their personal networks from the prior Japanese environment in which they lived; in time, the first-year Japanese students increase the extent of their personal networks among American, International, and other Japanese students; those Japanese students on campus for more than a year had larger personal networks than the first-year students.

There was a tendency for Japanese students to have more international than American friends. This might be a consequence of Japanese students being better able to understand each other because of cultural similarities as compared with Americans and because of the Japanese students' lack of confidence in their English skills. Only 25% (4/16) of the Japanese students answered "yes" to the question of whether they were confident in their language skills.

STUDY IV. EXPERIENCE AND ACTION OF AMERICAN STUDENTS IN JAPAN

In contrast to the first three studies, this fourth study deals with American students attending Japanese universities. This study is of interest because undergraduates from Western countries studying in Japan must cope with a very different situation than students from one Western country to another or indeed for Japanese students studying in the United States. American students are usually less knowledgeable about the Japanese language since they do not study the language until they enter college; in contrast, the Japanese study English beginning at an early age. To be fully literate in Japanese requires knowledge of about 2,000 signs, which takes a long time to acquire. Another complication concerns the organization of the Japanese universities. Some have different campuses for international divisions; others have separate departments exclusively for international students; and only a few others have Japanese and international students in the same setting where classes are taught in Japanese. Another difficulty concerns the culture. The difficulty that foreign students have to adjusting may be because Japan, though on the surface it may appear Westernized, is viewed by Westerners as a closed society. In general, since American society is not as homogeneous as Japanese society, it seems easier to integrate into American than into Japanese society.

There are, however, factors that facilitate Americans' study in Japan. For example, a study by Tanaka, Takai, Kohyama, and Fujihara (1991) suggests that lesser language ability can work to an advantage. Though this may be surprising, Tanaka et al. (1991) suggest that being competent in the Japanese language could imply that the Japanese hosts would be less likely to perceive the student as a "guest" or "visitor"; thus, he or she would no longer receive the special welcome typical of that given to a *gaijin* (foreigner). Moreover, once a foreigner is competent in the Japanese language there is also the expectation that such a person would be knowledgeable of Japanese unwritten rules; however, this does not necessarily follow from having competence in the language.

While there may be some commonalities in the difficulties faced by both Japanese students in America and American students in Japan there are likely to be differences, such as the Japanese student in America being concerned about crime, while the American student in Japan concerned with strictness of rules.

The aim of this study was to explore the nature of the transactions (experience and action) of American students in Japan utilizing a variety of aspects of the holistic, developmental systems-oriented approach (Wapner, 1987) that was employed in the previous studies of the Japanese students in American universities.

Hopefully, the analysis would point to similarities and differences concerning sojourns in the two cultures by natives from the other culture.

METHOD

Subjects

The subjects were 26 American citizens (13 males and 13 females), ranging in age from 19 to 26 years, who were attending Japanese universities at the time they volunteered to participate. Seventeen were Caucasian Americans, six half Asian half white, two black, and one Asian American. Their mean length of stay in Japan was 20 months, with a range of from 3 months to 8 years.

Sixteen participants were obtained from Sophia University at Ichigaya, where the International Division was separate and had a rather small campus in urban Tokyo; five were from the International Christian University, which has a large campus in suburban Tokyo; and five came from Waseda University at Takaanobaba, a medium sized campus in urban Tokyo.

Potential subjects, who had to be American students attending a Japanese university at that time, were approached by the experimenter informally on campus with a request to volunteer to participate. After informing the potential participants about the nature of the study and after they showed interest in participating they were given a consent form they were asked to sign. This was followed by the audiotaped interview that contained open-ended and specific questions concerning their experiences, including relations between their expectations and actual experiences, using a questionnaire with a Likert rating scale. Finally, a Psychological Distance Map (PDM) for people and places (e.g., Minami, 1985; Wapner, 1977) was administered using the same procedures described earlier for Study III. The overall adjustment was assessed by a MANOVA of the four ratings of the degree to which they feel part of the Japanese environment (dedifferentiated—going along completely with the environment; differentiated and isolated—feeling isolated; differentiated and in conflict; and differentiated and integrated—feeling integrated with the culture).

RESULTS

Quantitative Findings

Gender Differences

There were only two significant gender differences: Women rated themselves as significantly more troubled by crowds ($M_w = 5.4$) than men ($M_m = 3.8$) ($t = 2.0$; 24 df; $p = .05$); and women watched Japanese TV more frequently ($M_w = 4.8$) than men ($M_m = 3.6$) ($t = 2.4$; 24 df; $p < .05$). Since there were only two gender differences, it seems appropriate to review the findings pooled for gender to obtain a general description of the findings.

Ratings on Environmental Experience

Table 20.6 presents findings, pooled for gender. The most striking findings are as follows: with respect to the *physical* environment, students experienced public transportation as very effective.

As to the *interpersonal* environment they report liking the Japanese; having many American friends; finding the Japanese attractive; finding Americans attractive; that they do not write Japanese well; and that they do not read Japanese well.

With respect to the sociocultural aspects of the Japanese environment the participants indicated that they would like to know more about Japan; felt very adept at using chopsticks; were interested in hearing about American news; liked shashimi; and liked Japan. In general, they also report that they feel good about themselves.

Qualitative Findings

The qualitative findings (12 men, 12 women) derived from the interview are summarized in Table 20.7 which presents questions and frequency of responses of occurrence according to the questions posed to the participants. The items for which there was consensus by more than one-third of the group, that is, eight or more participants gave the same response, included participants who reported that: they came to Japan to study language and culture; they felt that Japan was an important country for their field of concentration (business); they could not imagine what Japan was like; they found the Japanese shy, reserved, and less expressive; they liked the features of safety and high quality of transportation that characterized Japan; they disliked the high prices; they felt the Japanese were kind, friendly, and polite; they felt there were more rules and greater adherence to them; they would describe the United States to Japanese as dangerous, and diverse; they felt there was difficulty with language; they were more reserved and less outgoing when with Japanese; they did not avoid being with other Americans at their Japanese college; they had many international friends other than Japanese; they largely communicated by use of mixed English and Japanese or by use of English alone; and if they were lost they would ask how to get where they were going.

Detailed examination of Table 20.7 also reveals many other interesting features of the experience and action of the American students in Japanese universities even though there is no marked consensus with respect to these items.

Adjustment Characterized Developmentally

The overall adjustment, indicated by the mean ratings (1 = not at all fit the category, to 7 = completely fit the category) assigned to the four developmentally ordered person-in-environment system state categories was as follows: (1) *dedifferentiated* (that is, going along with the Japanese cultural context) M = 4.1; SD = 1.45; (2) *differentiated and isolated* (that is, isolated from the Japanese cultural context) M = 4.0; (3) *differentiated and in conflict* (that is, in conflict with the

TABLE 20.6. Quantitative Data: Experience of and Action in the
Japanese Environment by American Students (Pooled for Gender)
Attending Japanese Universities[a]

Environment		Mean	Standard deviation
Physical			
	Effectiveness of transportation	6.3	1.7
	Tolerate climate	3.6	1.9
Interpersonal			
	Like Japanese	5.5	1.2
	Have American friends	5.4	1.3
	Find Japanese attractive	5.3	1.2
	Find Americans attractive	5.0	1.6
	Japanese like me	4.9	1.6
	Have Japanese friends	4.7	1.5
	Troubled with crowds	4.6	2.0
	Understand Japanese	4.4	1.1
	Speak Japanese	3.8	1.0
	Read Japanese	3.1	.8
	Write Japanese	2.7	.8
Sociocultural			
	Like to know more Japanese	6.0	1.4
	Adapted to chopsticks	6.0	1.6
	Interest in U.S. news	5.3	1.6
	Like sashimi	5.2	2.1
	Like Japan	5.0	1.5
	Go along with rules	4.9	1.8
	Want to go back to U.S.	4.7	1.9
	Interest in Japanese news	4.7	1.6
	Proud of U.S.	4.2	1.6
	Watch Japanese TV	4.2	1.5
	Know Japan well	3.7	1.2
	Comfortable with hierarchy	3.5	1.8
	Feel good about self	5.5	1.1
	Feel American	4.9	1.8
	Feel foreign	4.4	1.9

[a]Rating of 1=not at all; 7=very much or many.

Japanese cultural context) M = 3.5, SD = 1.70; and (4) *differentiated and integrated* (that is, integrated with the Japanese cultural context) M = 3.7, SD = 1.7.

The differences among the ratings for the four categories were assessed by a MANOVA utilizing the variables of gender and developmental mode of adjustment. No evidence was found of significant differences between gender, between adjustment categories, or for the interaction of gender by adjustment.

Psychological Distance Map (*PDM*)

PDM for People

Analysis of the entries on the PDM for people were in keeping with expectations, that a significantly greater number of friends (19/22) of American background

TABLE 20.7. Qualitative Data: Experience of and Action in the Japanese Environment by American Students (Pooled for Gender) Attending Japanese Universities

Question	Responses (frequency)
What has your experience of Japan been like?	Experience was educational (5/24); overall experience positive (5/24); difficulty in adjustment (4/24)
Why did you come to Japan?	To study language & culture (10/24); to study in their special field, e.g., business (7/24); family connections (4/24)
What did you imagine Japan would be like?	Not able to imagine anything (5/24); crowded (4/24)
How did Japan differ from how you imagined it?	More modern (3/24); More crowded (2/24); more expensive (2/24); not as conservative (2/24)
What are the most significant differences: Physical environment?	More crowded (6/24); less space (5/24); less green (4/24); smaller (3/24); cleaner (3/24); more convenient (2/24); dirtier (1/24); less convenient (1/24)
Interpersonal environment?	Japanese more shy, reserved, and less expressive (9/24); hierarchic relations (2/24); traditional conservatism greater in Japan (2/24)
Sociocultural environment?	More rules (6/24); greater adherence to rules (6/24); less crime (5/24);
What do you like about Japan?	Safety (16/24); convenience & high quality of transportation (8/24); Positive feeling about culture (3/24); their high valuation of the family (3/24)
What do you dislike about Japan?	High prices (8/24); crowds (4/24); status of women (4/24); lack of individuality (3/24); unquestioning attitude (2/24); narrow mindedness (2/24); isolationism (2/24)
How do you feel about the Japanese?	Kind, friendly, polite (9/24); like them (5/24); "people are people" (4/24); respect them (3/24)
How do you think other Americans feel about the Japanese?	Japanese unfair in trade (5/24); they engage in Japanese "bashing" (5/24); Americans blame them (scapegoats) for decline in American trade (2/24); generally dislike them (2/24); Japanese work too much (2/24); Japanese are rich (2/24)
How do you think the Japanese feel about you?	They see Americans as foreigners (7/24); uncertain about me (6/24); Japanese are curious & interested (4/24); Japanese like me (4/24)
How do you think Japanese feel about Americans?	Loud (7/24); lazy (3/24); positive (3/24); stupid (1/24); scary (1/24)
Do you feel you have changed since you came to Japan?	Yes, in my views and perspectives (6/24); more tolerance & patience (4/24); more mature (2/24); adopted Japanese qualities (2/24); no (1/24)
How would you describe Japan to an American who did go to Japan?	Crowded (4/24); stressful & busy (3/24); good transportation (3/24); expensive (3/24); different from U.S. (3/24); not that different (2/24); unique (2/24); in some ways like the U.S. (2/24)
How would you describe the U.S. to a Japanese who has not been in the U.S.?	Dangerous (12/24); Diverse (10/24); big (3/24); not that dangerous (2/24)
Are there other aspects of concern in everyday living (language, etc.) that have not been covered?	Difficulty with language (9/24); none (5/24)
Do you vary in behavior depending on whether you are with Americans or Japanese?	More reserved, less outgoing with Japanese (10/24); acting like myself (5/24); loss of humor when with Japanese (3/24)

TABLE 20.7. (continued)

Question	Responses (frequency)
How do you handle yourself in mixed groups of Americans and Japanese?	Act as myself (5/24); same as with Americans (4/24); depends on group or situation (4/24); treat both the same (3/24); more shy & reserved (2/24); try to be helpful (2/24)
Are there any differences in relations with friends met in Japan compared to the U.S.?	Better friends with Americans (5/24); no differences (3/24); friends in japan are polite (2/24); friends in japan are temporary (2/24); friends are friends (2/24); closer with friends in Japan (2/24); foster bonds that are made (2/24); takes longer to make friends (1/24)
Do you avoid being with other Americans at your college in Japan?	No (12/24); kind of (2/24)
Do you have many international friends other than Japanese?	Yes (13/24); not many (5/24); some (4/24)
How do you communicate most of the time?	Mixed Japanese & English (12/24); English (9/24); Japanese (1/24)
If you were lost, how would you go about finding your way to your destination?	Ask (12/24); go to police box (Koban) (6/24); ask in Japanese (3/24); ask in English (2/24)
What general expectations about going to college in Japan did you have?	Easy classes (6/24); hard, demanding classes (6/24); learn a lot of Japanese (5/24); Few (2/24)
In what ways has your actual experience been less than expected?	Classes are less challenging (5/24); learned less Japanese (5/24); did not learn much in school (4/24); horrible teachers (3/24); did not get to know as many Japanese (3/24); expensive (3/24)
In what ways has your actual experience been better than expected?	Made more friends (7/24); learned about culture (4/24); adapted well (2/24); learned about self (2/24); good faculty (1/24); better classes (1/24)
Did you have differences between expectation and actuality with respect to your health?	Lost weight (6/24); smoked cigarettes (5/24); none (5/24); healthier (4/24); walk more (2/24); bad air (2/24); less healthy (1/24); gained weight (1/24)
Did you have differences between expectations and actuality with respect to your psychological state?	Same (5/24); going "crazy" (4/24); culture shock (2/24); more positive (2/24); ups and downs (2/24)
Did you have differences between expectation and actuality with respect to your status as a student?	Doing less work (5/24); same (5/24); studies harder (2/24); segregated from the rest of Japanese students (2/24)
Did you have differences between expectation and actuality with respect to your physical environment?	More crowded (4/24); smaller (3/24); same (2/24); lack of space (2/24); less green (2/24); dirtier (2/24); cleaner (1/24)
Did you have differences between expectation and actuality with respect to your interpersonal environment?	Better (5/24); no (4/24); less friends (4/24); less social (2/24)

would be entered than Japanese friends ($p < .01$ by sign test). The first entries of people were largely restricted to family members, (17/25) but this was not significantly greater than other first entries, including Japanese friends (2/25), American friends (3/25), and friends from other countries (3/25).

PDM for Places

The analysis of the PDM for places was not significant with respect to country mentioned, though there was a tendency for more subjects to mention places in Japan (12/17) than in the USA (5/17).

SUMMARY

In general, there was evidence that all three aspects of the environment physical, interpersonal, sociocultural played a role in the participant's experience of the Japanese cultural context. Students experienced public transportation as very effective, they adapted to the use of chopsticks, they wanted to know more about Japan, and that they liked the Japanese but were concerned about their own lack of Japanese reading and writing ability. While students felt relatively isolated from the environment, they were neither in conflict with it nor highly integrated with it.

DISCUSSION

The literature reviewed and the four studies reported speak to the potential of the holistic, developmental, systems-oriented perspective to serve as a model for encompassing the broad literature and findings on the experience and action of the student sojourner to a university in a country other than his or her own. This is evident in a number of ways. For example: (1) the category system of the approach that considers three aspects of the person (physical/biological, intrapsychological, sociocultural) and three aspects of the environment (physical; interpersonal, sociocultural) operates as a means for integrating the broad diversity of findings and literature in this problem area; (2) the general approach shapes new methodologies for exploring the problems of experience and action of the sojourner; and (3) the approach has heuristic value for opening new questions and problems that arc worthy of investigation.

The utility of the *category system* is evident not only in its handling of the problems of the sojourner posed by Church's (1982), broad summary of the literature, but also because it aided in designing relevant questions that yielded significant findings concerning the general problems posed by the student sojourner to a new culture.

The three studies (I–III) of Japanese students in American universities and the study (IV) of American students in Japanese universities revealed the impact of the sojourn on all aspects of the person, of the environment, and the relations among them. There were similarities and differences in the experience and action of the student sojourners in the two countries. While many of the Japanese students found the United States to be big, inexpensive, ineffective in transportation, dangerous, and free compared to Japan, the American students found Japan to be small, expensive, effective in transportation, safe, and regulated. Both groups experienced an awareness of being a foreigner.

Generally, the Japanese students in the United States were more satisfied with academics than their American counterparts in Japan. A noteworthy difference between Japanese students in the United States and American students in Japan was that most Japanese students came to study for the full 4 years, while for most American students it was only 1 year of study abroad. Both the Japanese students and the American students experienced language barriers although the Japanese students were better prepared in English than the American students were in Japanese.

Moreover, the special *methodologies* developed as a function of the nature of the perspective, for example, the special *methodologies* developed as a function of the *Psychological Distance Map (PDM)* and the *developmental approach* to characterizing the sojourner's mode of adjusting to and coping with the new cultural context, have shown strong beginnings in helping analyze significant issues in the general problem area.

The PDM that is directed toward uncovering information about people and places important in the sojourner's world has helped make inroads on the general character of, and the changes important to, the sojourner during his or her stay in the new cultural context.

For example, the PDM was effective (Experiment III) in uncovering, even with a small group of sojourners at an American university, some important features of change in their interpersonal environment. Comparison of PDM's involving duration of exposure (first- vs. fourth-year student; first vs. second test occasion) showed that with time there is a shift from a significantly greater number of people from home to a significant increase in the number of people from the new environment. It is of further interest that there was a tendency for Japanese students at an American university to have more international than American friends.

In addition, the developmental approach appears to be useful in assessing the sojourner's means of adapting to and coping with the new cultural context in which she or he is located. Assessment of the sojourner's mode of adjustment developmentally yielded a suggestion (Experiment I) that the Japanese at an American university showed a tendency of change toward "multiculturism" (differentiated and hierarchically integrated person-in-environment system state) with time. In contrast, the American sojourners at Japanese universities on the one occasion assessed showed no such tendency.

This points to the need for more systematic, extensive study of change in experience and action *during the course of exposure* of the sojourner to the new culture. It is interesting to speculate about the possibility that a definitive finding of differences in mode of adjustment with a greater trend toward multiculturalism can be found for the Japanese than for the American sojourners. Should that be the case, it might be a function of one or both of the factors mentioned in the introduction, namely, (1) extensive preparation in English by the Japanese sojourners and little or no preparation in Japanese by the American sojourners and (2) the homogeneity of Japanese society relative to the diverse ethnic groups characterizing the United States.

In other future research it may be useful to include more items that identify the difficulties as well as the means of coping with those difficulties. A few of the important variables to consider are the amount of time spent in the new country that impacts the experience and action of the sojourner and the particular environmental context that a given American or Japanese student comes from, as well as the particular environmental context (both university and geographic area) in which he or she is a sojourner. In addition to the environmental context, the characteristics of the sojourner (e.g., gender) are worthy of continued and further study. For example, in what ways are the experiences of Americans who are a minority similar to or different from Americans who are part of the majority when they attend universities in Japan? How do the American natives perceive Japanese students in the United States and how do the Japanese natives perceive American students in Japan? How do the natives characterize the adjustment of the sojourner compared to the sojourner's own experience of adjustment?

In addition to the theoretical significance of the approach and of the studies reported here, the findings have practical implications. For example, some of the findings in the four studies reported may be useful to prepare potential students who will spend a year or more in a college abroad by informing them about possible difficulties they may experience in the new culture, both in the college as well as the broader environmental context. Moreover, on the side of the university administration, it might be of value to identify the strengths and weaknesses of their programs from the perspective of the foreign student. Finally, on the side of the government, it might be of value to know how their country is perceived by foreign student sojourners.

Hopefully, with accumulation of findings in these areas the holistic, developmental, systems-oriented perspective will advance—in keeping with the needs expressed by Fisher and Hood (1987), and Shaver et al., (1985)—as a model that encompasses the personal, interpersonal, home, university, and cultural factors that play a significant role for students sojourning at a foreign university.

NOTE

The four studies reported here were conducted by students (Study I—Yuko Inoue; II—Karl Toews; III—Junko Fujimoto; and IV—Tomoaki D. Imamichi) in collaboration with S. Wapner as part of his research course on Developmental Aspects of Transactions of Persons-in-Environments in 1993 and 1994.

REFERENCES

Bochner, S. (1972). Problems in culture learning. In S. Bochner & P. Wicks (Eds.), *Overseas students in Australia*. Randwick, N.S.W.: New South Wales University press.
Church, A. T. (1982). Sojourner adjustment. *Psychological Bulletin, 91*(3), 540–572.

Clark-Oropeza, B. A., Fitzgibbons, M., & Baron, A., Jr. (1991). Managing mental health crises of foreign college students. *Journal of Counseling and Development, 69*(3), 280-284.

Cormack, M. (1968). *The wandering scholar.* International Educational & Cultural Exchange,3(4),44–51.

Ekman, P. (1972). Universals and cultural differences in facial expressions of emotion. In J. K. Cole (Ed.) Nebraska *Symposium on Motivation: Vol. 19.* Lincoln: University of Nebraska Press.

Fisher, S., & Hood, B. (1987). The stress of the transition to university: A longitudinal study of psychological disturbance, absent-mindedness, and vulnerability to homesickness. *British Journal of Psychology, 78,* 425–441.

Hall, E. T. (1959). *The silent language.* New York: Doubleday.

Kaplan, B. (1966). The comparative developmental approach and its application to symbolization and language in psychopathology. In S. Arieti (Ed.), *American handbook of psychiatry: Vol. 3.* New York: Basic Books.

Klineberg, O., & Hull, W. F. (1979). *At a foreign university: An international study of adaptation and coping.* New York: Praeger.

Minami, H. (1985). *Establishment and transformation of personal networks during the first year of college: A developmental study.* Doctoral dissertation, Clark University, Worcester, MA (Dissertation Abstracts International DA 8608772).

Oberg, K. (1960). Cultural shock: Adjustment to new cultural environments. *Practical Anthropology, 2,* 177–182.

Porter, R. E. (1972). An overview of intercultural communication. In L. A. Samovar & R. E. Porter (Eds.), *Intercultural communication: A reader.* Belmont, CA: Wadsworth.

Ramirez, M., III, & Casteneda, A. (1974). *Cultural democracy bicognitive development and education.* New York: Academic Press.

Schutz, A. (1964). The stranger. In A. Boderson (Ed.), Collected papers II:*Studies in social theory* (pp. 91–105). The Hague: Martinus Nijhof.

Shaver, P., Furman, W., & Buhrmeister, D. (1985). Transition to college: Network changes, social skills and loneliness. In S. Duch & D. Perlaman (Eds.), *Understanding personal relationships: An interdisciplinary approach* (pp. 193–219). London: Sage.

Taft, R. (1977). Coping with unfamiliar cultures. In N. Warren (Ed.), *Studies in cross-cultural psychology: Vol. 1* (pp. 121–153). London: Academic Press.

Tanaka, T., Takai, J., Kohyama, T., & Fujihara, T. (1991). *Social skills performance of foreign students in Japan.* Paper presented at the 21st International Congress of Psychology, Brussels, Belgium.

Triandis, H. C. (1975). Culture training, cognitive complexity, and interpersonal attitudes. In R. Brislin, S. Bochner, & W. J. Lonner (Eds.), *Cross-cultural perspectives on learning.* New York: Wiley.

Triandis, H. C., Vassiliou, V., Tanaka, Y., & Shanmugam, A. (Eds.) (1972). *The analysis of subjective culture.* New York: Wiley

Wapner, S. (1977). Environmental transition: A research paradigm deriving from the organismic-developmental systems approach. In L. van Ryzin (Ed.), *Wisconsin Conference on Research Methods in Behavior Environment Studies Proceedings* (pp. 1–9). Madison: Univ of Wisconsin Press.

Wapner, S. (1981). Transactions of persons-in-environments: Some critical transitions. *Journal of Environmental Psychology, 1,* 223–239.

Wapner, S. (1987). A holistic, developmental, systems-oriented environmental psychology. In D. Stokols & I. Altman (Eds.), *Handbook of environmental psychology* (pp. 1433–1465). New York: Wiley.

Wapner, S. (1988). The experience of environmental change in relation to action. *The Journal of Architectural and Planning Research, 5*(3), 237 256.

Wapner, S., Ciottone, R., Hornstein, G. A., McNeil, O., & Pacheco, A. (1983). An examination of studies of critical transitions through the life cycle. In S. Wapner & B. Kaplan (Eds.), *Toward a holistic developmental psychology* (pp. 111–132). Hillsdale, NJ: Erlbaum.

Werner, H. (1957a). *Comparative psychology of mental development.* New York: International Universities Press. (1st edition, New York: Harper, 1940; 2nd edition, Chicago, Follett, 1948).

Werner, H. (1957b). The concept of development from a comparative and organismic point of view. In D. Harris (Ed.), *The concept of development* (pp. 125–148). Minneapolis: University of Minnesota Press

Werner, H., & Kaplan, B.(1956). The developmental approach to cognition: Its relevance to the psychological interpretation of anthropological and ethnolinguistic data. *American Anthropologist, 58*, 866–880.

Wolff, H. G. (1952). *Stress and disease*. Springfield, IL: Thomas.

Chapter **21**

PUBLIC SPACE AS ART AND COMMODITY
The Spanish American Plaza

Setha M. Low

INTRODUCTION

This chapter tackles the problem of the relationship of architecture and visual culture to "unstable" cultural meanings by providing an analysis of a designed urban form—the Spanish American plaza. The examination of the public space as art and commodity provides a glimpse of the contradictions between the artistic and often idealized representational purposes of the urban plaza and its political and economic base. Bringing these contradictions to light helps to demystify visual culture and highlights the ways in which architecture and urban design are deeply ideological both in artistic style and political purpose.

Further, by reconsidering a designed public space as a commodity, its planning, design, construction, or refurbishing takes on new economic meaning. A public space that is valued ostensibly as a place for people to sit, read, and gather becomes a financial strategy for revitalizing a declining city center, a way to maintain real estate values, and/or a means of attracting new investments and venture capital. In Mexico, a central space such as the *zocalo* can even become the battleground for representational dominance and the re-presentation of a nation's history. In Japan, public spaces in front of important shrines (Figure 21.1) or in front of busy railway stations may play the same role, whereas in New York City public spaces such as

Setha M. Low • Department of Environmental Psychology and Department of Anthropology, Graduate School and University Center of the City University of New York, New York, New York 10036.

Handbook of Japan–United States Environment–Behavior Research: Toward a Transactional Approach, edited by Seymour Wapner, Jack Demick, Takiji Yamamoto, and Takashi Takahashi. Plenum Press, New York, 1997.

Figure 21.1. Shrine public space in Tokyo, Japan.

Bryant Park behind the New York Public Library and the planned renovation of Times Square carry the same symbolic burden of artistic and economic intentions.

One of the best ways to analyze a visual/cultural artifact is as a moment in a particular historical time and cultural place. The analysis should consist of visual and spatial as well as social strategies. Yet even in the clearest of place narratives, it is difficult to separate out what is design and what is commodification. The interpenetration of artistic, political, and economic intent and interpretation is part of the mystification process by which art/architecture serves ideological and economic rather than simple artistic ends. Certainly the position of the viewer—socially, politically, physically—also influences what can be and is seen. Thus, the place narratives presented are but one perspective on how the tension between the art/commodity analyses uncovers meanings that cannot be seen "at first sight."

THEORETICAL ASSUMPTIONS

This research is part of a larger project concerned with developing a method for studying human–environment relations and the built environment that incorporates cultural meanings and social relations in a significant way. This approach contextualizes space both in terms of the political economic conditions of the social production of the built environment as well as the personal meanings and experience (or "social construction") of space. Theoretically, this work draws upon anthropological, geographical, and sociological understandings of the social world, rather than upon a strictly environmental psychological one. Further, the analysis attempts to move away from a structural/functional definition of "culture" that reduces culture to behavior and activities, but instead deals with culture as a complex, multifaceted, and ever-changing set of symbolic and meaning-centered strategies for ordering and giving meaning to social life.

In this type of analysis the relationship of problem, theory, and method is loose at best, following the conventions of anthropologically based ethnographic research. The problem emerges from lengthy "fieldwork" in a place, and the analysis of field notes written while participating and observing social life. The method is the participant observation, that is, the observing and participating that one does when living in the field. Participant observation is supplemented with interviews, counts of users, maps, historical documents, and whatever other methods are necessary to collect data that bear upon the selected problem. Theory is brought to the problem in order to understand the phenomena that are observed and felt by the researcher. In some cases the theory is grounded, that is generated by the data. In other cases, such as in this work, the theoretical orientation to the problem is borrowed from the work of others who are working on similar ethnographic problems.

The notion that a plaza can be analyzed as a commodity as well as art is drawn from the work of Sylvia Rodriquez, who is concerned with the way that painting "mystifies" or covers the economic or political objectives of the production of the painting. It seems that the design and building of public plazas serve these same purposes, thus this chapter is an exploration of this theoretical idea applied to the situation of the Spanish American plaza in Costa Rica.

METHOD

Methodology

The ethnographic descriptions presented in the discussion are based on long-term fieldwork in San José, Costa Rica—1972 to 1974, the summers of 1976 and 1979—and three intensive fieldwork periods focusing only on the urban plazas—the first from May through September 1986, the second from December 1986 through February 1987, and the third from November through December 1993. I was able to observe differing behaviors during these visits since they were made at different times: the summer months coincide with Costa Rica's "rainy" season, while the winter months are the "dry" season. The rainy season is characterized by late-afternoon rain that, when heavy, interrupts plaza life for at least an hour. During these periods I would stand with other plaza occupants under the closest available shelter or visit one of the local cafés to wait for the rain to stop. During Christmas and New Year's, in the dry season, plaza use is most intense. The weather is clear, sunny, and cooler and the plazas become informal markets for seasonal gifts, nuts, and fruits.

Observation

I was concerned that participant observation in a public space might not capture all the activities that occur, so the fieldwork utilized three different observational strategies:

1. Each plaza was observed by sector and everything that occurred in that sector was recorded for a designated period of time. This time–space sampling provided a system for nonsequentially observing all the sites throughout the day on both weekends and weekdays. A series of behavioral maps locating activities and counts of people by location, sex, and age were also collected.
2. After the first month of time–space sampling, a map of activity locations had emerged, so a second set of observations concentrated on documenting those activities and the people engaged in them throughout the plazas.
3. During the third phase of participant observation, I carried a camera and map, spoke to people, and became more involved in everyday plaza life. By this time plaza users were quite used to seeing me with my clipboard and pen, and were delighted that I was now taking photographs and involving them in my up until then seemingly clandestine task. The camera gave many people an excuse to talk to me and to ask what I was doing. I began to make friends and hang out with some of the plaza occupants, even visiting them at home or joining them when they went for a drink or a meal.

Interviews and Historical Documentation

At the conclusion of these observations I collected a series of interviews with a variety of plaza users, using questions that had emerged during the observational period. A systematic series of interviews with the managers, owners, and directors of the relevant local institutions located on or near the plazas were completed; and blueprints, design guidelines, and plans for the design of the plazas were collected. A series of interviews with local historians and archival work in the National Library and at the Universidad de Costa Rica provided documentation of the oral histories of Parque Central. Interviews with the current and previous Ministers of Culture (including President Oscar Arias) and the head of Urban Planning, as well as with the architects involved in the design of the Plaza de la Cultura, provided contextual data for the ethnographic descriptions and documentation of the processes of material production. Finally, novels, newspapers, television presentations, and conversations with friends and neighbors provided data on the broader cultural context of public life. These phases of the research project were repeated during each of the three field visits.

FIGURE 21.2. Pargue Central, San José, Costa Rica.

Analysis

This methodology was quite effective in providing various kinds of data that could be compared and analyzed. The content analysis of the field notes, interviews, maps, and historical documents generated a series of themes and theoretical insights into the cultural underpinnings of plaza design and use. The observations, interviews, archival documentation, and spatial and architectural maps and drawings provided distinct "texts" that were read in relation to each other in the search for breakdowns and incoherence that would uncover areas of cultural conflict and contestation (Agar, 1986; Geertz, 1973).

Setting

The plazas of San José, Costa Rica, are public spaces, planned, designed, and maintained by the municipal government and paid for by national funds. They are symbolically important places as a focus of national pride and generally represent the most notable artistic achievements of their respective periods.

Parque Central, the original Plaza Major, is the oldest plaza in San Josè, and represents Costa Rica's Spanish colonial history in its location, spatial form, and surrounding buildings. Its relatively long history spans the colonial, republican, and modern periods. Historical photographs, paintings, and written descriptions of the successive stages of its design and social life document the changes in its built form and social use. Today, Parque Central is densely covered with trees, benches,

FIGURE 21.3. Plaza de la Cultura, San José, Costa Rica.

arbors, and grass divided into rough parterres by pathways reinforced with iron-link post and chain fences (Figure 21.2). It remains a vibrant center of traditional Costa Rican culture inhabited by a variety of workers, pensioners, preachers and healers, tourists, shoppers, sex workers and gamblers, and people who just want to sit and watch the action.

Plaza de la Cultura, a contemporary plaza only two blocks west of Parque Central, is a recently designed public space heralded to become the new city center and emblem of the "new Costa Rican culture." Because it was completed in 1982, it was possible to interview almost all of the individuals involved in its design and planning, while at the same it could be studied as a well-established place.

The Plaza de la Cultura proved to be an intriguing contrast to Parque Central in terms of its style of design, spatial configuration, surrounding buildings and institutions, activities, and daily users and visitors. Its design derives from the famous public space of colorful pipes and abstract spaces in front of the Pompidou Museum in Paris. Plaza de la Cultura is a site of modern consumption filled with young Costa Ricans, tourists, and families who bring small children to chase the pigeons or play in the fountain. North American culture is consumed by Costa Rican teenagers carrying radios blaring rap music, and North American tourists "consume" Costa Rican culture by buying souvenirs, snacks, theater tickets, and artworks as well as the sexual favors and companionship of young Costa Ricans.

RESULTS

Parque Central

There are three definable historical periods in the design and redesign of Parque Central. In each of these, urban design elements seemingly represent only the values and concerns of those who were in power—the Spanish colonists (and their class and ethnicity-based social order), the coffee-based landed elite, and the urban middle-class. However, in each case there is visual evidence that contradicts the designers' values and intentions.

Parque Central was planned and built by Spanish colonists who came to the central plateau of Costa Rica from the cacao-growing lowlands to claim territory and establish a new town in 1751. It was designed as a grassy, tree covered rectangular *Plaza Mayor* placed at the center of the growing town and was oriented as a square city block with north–south and east–west roads as its boundaries. On the east side of the plaza a cathedral was built; on the north side were the military barracks; the *Botica Francesa* and a small hotel were constructed on the south edge, and private residences and small businesses soon filled the remaining building sites. The earliest Parque Central was a gathering place and weekend marketplace.

The urban design of Parque Central is part of the town-planning tradition of the Spanish American colonial empire codified in the 1573 *Orders for Discovery and Settlement*. These edicts were modified in the building and designing of towns throughout Latin America, but always retained elements of the colonial intent to recognize and redefine the administrative and political authority of urban power. The result was a strictly hierarchical organization of space with a gradual progression outwards from the central grid of the *Plaza Mayor* extending in every direction. Each square or rectangular lot had its function and value assigned to it in inverse relationship to its distance from the central square (Lefebvre 1991:151).

The centralized and hierarchical organization of space of the Spanish American plaza and gridplan town is actually a syncretic urban design form derived from European architectural traditions of medieval *bastides* and the Mesoamerican plaza temple complex and urban plans of the cities that the Spanish encountered during the conquest of the New World (Low 1993). Some of the earliest Spanish American plazas were built directly on top of the ruins of their Aztec or Maya antecedents. Both European and Mesoamerican plazas were designed to display military conquest and market domination by the conquering rulers, whether those rulers were the Aztec, Maya, or Spanish. Therefore, although the Spanish American plaza was used as a means of spatial domination and colonial control, its form also derived from indigenous forms of political and economic control expressed in the Mesoamerican plaza temple complex. Since the spatial relations of plaza to buildings, hierarchy of spaces, and functions of the plaza remain somewhat the same, the symbolism (artistic representation) retains aspects of both cultural histories. Returning to the *zocalo* of Mexico city, the juxtaposition of the Great Temple and plaza spaces with the colonial Cathedral and plaza illustrates the kind of symbolic tension and complex cultural meanings generated in these plazas. I am not sure if there is a comparable Japanese example in which a public space retains symbolic aspects of

FIGURE 21.4. Buddhist shrine public space, Japan.

conflicting and different histories, but this could be a focus of further research and discussion.

Parque Central retained its colonial form and meaning well into the nineteenth century when the plaza was redesigned and refurbished with all the trappings of European bourgeois elegance—an iron fence with gates, a grand fountain imported from England, and a wooded Victorian kiosk on which the military band would play for the Sunday *retreta*. It was from this historical period that I began to find evidence of increasing class-based social control of urban public space. The accumulated wealth of coffee growers and a republican government created a politically and socially powerful landed elite who began to impose their class-based conception of appropriate use and behavior in public places.

Historical texts, retrospective interviews, and diaries describe Parque Central as a place where the elite would gather and stroll in the evening or meet after church on Sunday to listen to the military band. The gates began to be locked, and the plaza patrolled at night. Nonetheless, photographs of this period show workers in open shirts sitting and barefoot boys playing in the Parque Central, and novels include references to street children and homeless people who could be found living there.

This contestation of the image of Parque Central as an elite plaza or as a heterogeneous urban center has continued in the recent public debate about the proposed replacement of the current 1944 modern cement kiosk with a model of the wooden Victorian structure. In spring 1992, a group of citizens brought a petition to the Legislative Assembly to tear down the cement structure and reconstruct the original Victorian one. The citizens that are attempting to reconstitute the Parque Central in its elite, turn-of-the-century image are not the daily users, but are professional and middle-class *Josefinos*. The conflict over the architectural form of the kiosk is a struggle about control of the artistic style and the appropriate use of the Parque Central, in which the architectural furnishings represent broader social and class-based meanings.

My second task in discussing Parque Central is to comment on how it can be understood as a commodity, or good. The transformation of Parque Central from a social gathering place to a commodifed center of commerce is not very clear. For one thing, except for the period of bourgeois control, the Parque Central has always been some kind of marketplace. But the dramatic economic changes of the 1950s and 1960s resulted in an increasingly dense, dirty, and heterogeneous city center. The remaining upper- and middle-class residents chose to move to the western sections of the city or the suburbs, abandoning Parque Central to the poor and working classes. These residents were replaced with architectural symbols of a new kind of economy based on debt and banking controls, and dependent on foreign capital. These modern steel and glass office buildings provide a new economic and visual context for Parque Central.

During the 1970s and 1980s the growth of the service sector and increase in unemployment because of the decline in agricultural exports has produced a proliferation of informal workers in Parque Central. Shoeshine men who control the northeast corner, ambulatory vendors on sidewalks and pathways, salesmen who use the benches as offices, construction workers who wait for pickups under the arbor, sex workers who stand in the kiosk, and gamblers who move among the crowd selling stolen goods are all examples of how the plaza is used as an urban place of commerce. Benches have become a kind of urban real estate in that salesmen who have lost their offices moved to these new "open-air" locations. There is growing competition for space to work or sell as the increasing numbers of marginally employed move to the Parque Central to open shop. Conflicts have developed between the shoeshinemen who defend their inherited spaces and newcomers trying to find a place to work. In the context of these economic conditions the debate about whether a Victorian kiosk would improve the space seems ironic, at least.

Plaza de la Cultura

The recent design of the Plaza de la Cultura does not provide the historical depth of the Parque Central, but it provides a more detailed analysis of the underlying motives and meanings of its contemporary landscape design. Constructed over a subterranean museum with bright yellow ventilation pipes, the Plaza de la Cultura and has a shallow pool containing three water jets, metal pipe benches, and only a few trees. The plaza is bordered on the south by the turn-of-the-century National Theater, on the west by the Gran Hotel, the major tourist hotel for North

American visitors, and on the north and east, by busy shopping streets lined by McDonald's and Sears as well as local businesses. The few trees in planters line the western edge alongside the hotel shops that include a jewelry store, a clothing store, a shop that sells the renowned Costa Rican ice cream, "Pops," and a newspaper stand that carries the Miami Herald.

The idea to build the Plaza de la Cultura is said to have been the inspiration of the Minister of Culture in 1976. The head of the bank of Costa Rica had gotten the legislative assembly to allocate the funds to build a museum to display the world famous collection of pre-Columbian gold artifacts that were housed on the second floor of the central bank. The museum was said to represent pride in indigenous Costa Rican culture and was supported by the Liberation party, a political party representing a coalition of professional, middle-class, and working-class Costa Ricans as opposed to the landed gentry and coffee-growing elite. The land around the National Theater was selected by the Minister of Planning and the head of the Bank of Costa Rica as the site that would easily accommodate tourists and that would represent a new center of culture in San José.

According to the Minister of Culture, when he saw the National Theater sitting in the open space created by the destruction of the surrounding buildings, he realized that it would be a powerful visual image to have a plaza with the gold museum underground so that the "architectural jewel" of Costa Rica, the National Theater, could be seen. Thus the plans for the original gold museum were scrapped and a new design for a subterranean museum and plaza began.

Here, it seems we have an excellent example of the role of landscape design in the creative destruction of forms of society (Harvey 1989, Zukin 1995). A residential and small-scale commercial neighborhood was transformed into an advertisement for Costa Rican culture. At the same time, this transformation generated new investment opportunities for foreign capital to expand their interests in tourism and tourist-related activities. The disguise for this commodification of a public space was the sociopolitical ideology for the Liberation party. The leadership of a new professional class wanted to represent Costa Rican culture as modern, drawing upon modern European idioms of design, but also as indigenous, based on the pre-Columbian past. North American capital was already fueling the Costa Rican economy and had influenced the siting of the plaza—placing it next to the major tourist hotel and in the center of North American businesses (i.e., McDonald's, Sears) and tourist activity. Thus, the siting, spatial form, and ultimately the design of the Plaza de la Cultura came from a combination of ideological and economic forces.

A team of three architects worked together and won the competition to design the plaza. Each of the architects had a different vision of the kind of place it should be. One architect imagined it to be a plaza where men could watch women walk by and designed a vast paved open space that, according to him, provided the longest sight line available for women-watching in San José. Another architect saw the plaza as a meeting place that was to be symbolically linked to other plazas in the city by making a second grid, a double grid with pedestrian walkways and trees. He imagined young men leaning on the outside rails of the perimeter piping and put a rail just where a man's foot might rest. The third architect was concerned that the new plaza be a significant open space; "Costa Ricans have their gardens (*jardinci-tos*) and their parks, and they have their special places, but they do not have a center for jugglers, music, political meetings, and large gatherings like in New York." He wanted public performances: "but we did not want a huge dry space, so we put in trees along the edges."

These images and imaginings created a rather eclectic plaza with a spatial form and style that many Costa Ricans did not like or understand. When the plaza first opened in 1982 there were demonstrations by people who tore out the plantings, started fires in trash cans, and tried to destroy as much as possible. It is not entirely clear from the reports who the people were or exactly what they were protesting, but the media interpreted this demonstration as a protest against the plaza's stark modernity.

The unusual urban space produced by the ideological, economic, and professional forces has now been appropriated by a group of users who did not have a place in the city before this plaza was constructed. The vast open space and view is used by street performers, religious singing groups, political speakers, and teenagers break-dancing or playing soccer. These activities are not accommodated by the parklike atmosphere of Parque Central. The small plazas created by the design in front of the National Theater and Gran Hotel are used by officially licensed vendors who have semipermanent stands from which they sell local crafts to tourists. These stands have proliferated and by late 1992 lined the entire edge of the plaza. The Gran Hotel generates a seemingly endless stream of tourists who sit on the edges of the plaza watching people from the safety of the hotel's sidewalk cafe. Women and families bring their children so that they can run after pigeons or play in the fountain.

Based on interviews with key informants and conversations with users and friends, one learns that this tranquility is contested by a number of illicit activities that make it perceived as an unsafe and unpleasant place to be. The newspapers regularly run articles criticizing the municipal government's management of the plaza by reporting any mishap or transgression that might occur there. The hotel bouncer remains posted at the edge of the plaza ready to protect his guests from the sight of beggars or homeless people looking for a place to rest. Official uniformed police stand outside the National Theater and refuse entrance to anyone who looks unsuitable. When a young man ran by and grabbed a gold chain from the neck of a young girl, there were police everywhere within seconds. The intensity of social and spatial control appears even greater here than in Parque Central, more intensely contested, and as yet unresolved (somewhat reminiscent of Bryant Park in New York City). This recently designed urban plaza represents and accommodates "modern" spatial practices based on youth, foreign capital, tourism, and an ideology of liberal democracy, contested by the public discourse about the danger and uncomfortable quality of the public space.

CONCLUSIONS

Design, like art, can be seen as a case study in mystification: urban public spaces that are said to be designed for the common good are often designed to accommodate activities that will exclude some people and benefit others. Further, the economic motives for the design of urban public space often have more to do with increasing the value and attractiveness of the surrounding property, rather than with increasing the comfort of the daily users. Analyses of design as art and commodity, however, allow for some degree of demystification of the ideological, political, and economic bases of public urban design. In the analysis of two urban plazas in Costa Rica, the artistic and economic goals of the Plaza de la Cultura do not meet the needs of traditional Costa Rican plaza users, and instead accommodate tourists and teenagers who previously did not have a public space. The plaza enhances the value of the Gran Hotel that sits on its edge and the associated tourist-related spaces more than it provides a quiet retreat for local users. On the other hand, in Parque Central there is conflict about the image of the plaza and design of the kiosk between the working users who are concerned about function and the middle-class residents who want to re-create an image of the past.

Landscape design and the reorganization of space is part of the creative destruction of forms of society, replacing traditional forms with new capitalist/modern relationships. Global economic forces are influencing both the production and construction of these new spaces. At the same time, spatial forms such as public plazas are systems of representation and social products whose style is a result, rather than a cause, of social differentiation. So the designs that are produced are at some level simply reflections of social changes that have already occurred. Nonetheless, their analysis highlights the cultural conflict and contestation that is ongoing as economic forces restructure the public space of the city.

NOTE

The author would like to thank Laurel Wilson, Joel Lefkowitz, Bob Rotenberg, and Gary McDonogh for their comments on this chapter. The author would also like to acknowledge the support provided by a Fulbright Fellowship, NEH Fellowship at the John Carter Brown Library, and by the Wenner-Gren Foundation for Anthropological Research that made this research possible.

REFERENCES

Agar, M. (1986). *Speaking of ethnography*. Beverly Hills: Sage.
Geertz, C. (1973). *The interpretation of culture*. New York: Basic Books.
Harvey, D. (1989). *The condition of postmodernity*. Cambridge: Blackwell.
Lefebvre, H. (1991). *The production of space* (Nicholson-Smith, trans.). Oxford: Basil Blackwell.
Low, S. (1993). Cultural meaning of the plaza. In B. Rotenberg & G. McDonogh (Eds.), *The cultural meaning of urban space* (pp. 75–94). Westport, CT: Bergin and Garvey.
Zukin, S. (1995). *The cultures of cities*. Cambridge: Blackwell.

Chapter 22

URBANIZATION AND QUALITY OF LIFE IN ASIA

Kanae Tanigawa and Kunio Tanaka

INTRODUCTION

Many cities in Asia, as in many other parts of the world, find their cities' level and rate of urbanization excessive and are unable to provide necessary services and facilities to the residents (United Nations, 1989). This, in turn, produces various urban problems in housing, employment, and social services, crime, traffic congestion, pollution, etc. A recent study found that, of the 100 largest cities in the world, 34 are located in Asia. Except for three cities in Japan, most cities are located in the developing countries in Asia (Population Crisis Committee, 1990). This reality breaks the historical connection between city size and levels of economic development. Some of the cities in Asia, in part from rapid population growth, find their population size unmanageable, and lose their strength and vitality.

In this chapter, we will try to understand the impact of urbanization on various urban conditions through urban administrators' perceptions of how the level of urbanization affects people's quality of life.

There have been arguments about city size and the level of urban conditions, sometimes known as the optimum size of cities (Henderson, 1986; Thomas, 1978). Does a larger city enjoy a better quality of urban life than a smaller one? There are a number of reasons that a larger city is more beneficial than a smaller city. The concentration of the population, for example, may attract business and capable professionals. Or there may be more potential opportunity for employment in a larger city than in a smaller city. In this chapter, we can not attempt to arrive at

Kanae Tanigawa • Bell Road Rokho #204, 14-14 Teraguchi-cho, Nada-ku, Kobe 657, Japan. Kunio Tanaka •1579-4 Mikage-cho, Kishimoto, Higashinada-ku, Kobe, 658, Japan.

Handbook of Japan–United States Environment–Behavior Research: Toward a Transactional Approach, edited by Seymour Wapner, Jack Demick, Takiji Yamamoto, and Takashi Takahashi. Plenum Press, New York, 1996.

definitive conclusions. We can, however, explore conditions of urban life through urban administrators' perceptions. There are a number of ways to measure the conditions of urban life, and administrators' perception is one of them since they are the frontline people who manage and try to offer comfortable living for the people of the city.

METHOD

Data for this study were collected by the Asian Urban Information Center of Kobe. The inquiry was carried out through mailed self-administered questionnaires.

For large countries, the inquiry focuses on cities with a population of 100,000 or more. For the smaller countries, where this cut-off point would eliminate most cities, the inquiry is directed to all state or provincial capitals.

This inquiry was designed to obtain from urban administrators information on the conditions that they perceived to be major problems or major advantages in their cities—measurement of the quality of urban conditions.

The results that we report came from 128 cities in eight countries in Asia: India (15 cities); Indonesia (30); Japan (26); Republic of Korea (31); Malaysia (5); Nepal (5); the Philippines (12); and Thailand (4).

To assess urban conditions, administrators were asked to judge each of 39 specific conditions, indicating for each whether this was an *urgent major* problem, a *serious* problem, or merely a *minor* problem. For each condition, they could also indicate that it was not a problem but rather a *satisfactory* condition or even an *advantage* for the city. Each condition was scored from 1, for an urgent major problem, to 5, for an advantage. This permits us to examine the overall score for each city and each country and the overall score for all the cities together. We can also examine the extent to which any specific condition or group of similar conditions constitutes a problem or advantage for the city.

The 39 conditions were grouped under 12 major categories as follows: 1. *General* (health, educational level); 2. *Utilities* (water, sewage, garbage disposal); 3. *Transportation* (public transportation, traffic volume, traffic flows); 4. *Housing* (population without shelter, low-cost housing, middle-income housing, high-income housing); 5. *Employment* (general unemployment, male unemployment, female unemployment, child labor); 6. *Health and Family Planning* (primary health care, hospital care, family planning services, social welfare services); 7. *Education* (primary education, secondary education, vocational education, tertiary education); 8. *City Personnel* (quality, quantity); 9. *City Revenues* (size of revenue base, control of revenues); 10. *Crime* (violent crime, property crime, prostitution, organized crime, drug abuse); 11. *Pollution* (industrial waste, sewage, automobile exhaust, noise pollution); 12. *Industrial Change* (rapid industrial growth, manufacturing decline).

RESULTS AND DISCUSSION

We can begin with the overall assessment of the urban conditions in each country and then move to the 12 categories of problems. To examine the overall

TABLE 1. Mean Scores for 12 Categories of Urban Conditions in 8 Asian Countries and Japan[a]

Condition	India	Indonesia	Korea	Malaysia	Nepal	Philippines	Thailand	All	Japan
General	3.5	3.0	3.1	3.9	2.5	3.1	3.3	3.2	3.3
Public utilities	2.4	1.9	2.7	3.3	2.2	2.1	2.6	2.3	2.7
Transportation	2.7	2.4	2.5	2.9	2.7	2.4	1.5	2.5	2.6
Housing	2.9	2.8	2.8	4.0	2.8	2.5	2.6	2.8	3.3
Employment	2.2	2.1	3.0	3.5	3.0	2.3	2.3	2.6	3.4
Health/family planning	3.7	3.2	3.4	4.0	3.0	3.3	3.4	3.4	3.1
Education	3.6	3.0	3.3	4.1	2.9	3.5	3.6	3.3	—
Personal	3.4	**	3.0	3.4	2.0	3.1	4.0	3.0	3.1
Revenue	2.5	**	2.2	3.0	3.8	3.6	2.3	2.5	3.0
Crime	3.3	2.9	2.8	3.5	3.0	2.8	2.4	2.9	—
Pollution	2.5	2.7	2.5	3.3	2.3	2.5	2.1	2.6	3.0
Industry	3.1	2.8	2.5	4.1	3.4	3.0	2.7	2.9	—
Total	2.96	2.66	2.84	3.58	2.78	2.80	2.70	2.83	3.05

[a]Scored from 1 to 5: 1 = urgent major problem; 2 = serious problem; 3 = minor problem; 4 = satisfactory condition; 5 = advantage for city.
**Omitted since the scores are based on only two responses.

condition, we can simply take the average of all the scores of the 39 conditions. Recall that a score of 1 is given if the condition is considered an urgent major problem and a score of 5 is given if the condition is considered an advantage for the city. This gives us a positive score such that the higher the score the better the condition is perceived to be.

Table 1 shows the mean scores for each of the 12 categories of conditions, both for all cities and for each country. This allows us to see which types of problems are most serious and which conditions are most favorable for the cities. Since the Japanese figures are not directly comparable with those for the other seven countries, it will be necessary to separate the groups of countries somewhat. For our 102 cities other than in Japan, the overall mean for all 39 conditions was 2.8. The range was from a low of 1.5 (Haydin, Thailand) to a high of 4.0 (Penang Island, Malaysia).

Country Scores

Malaysia had the highest overall mean (3.6), followed by India (3.0), Korea (2.8), the Philippines (2.8), and Nepal (2.8), with Thailand (2.7) and Indonesia (2.7) having the lowest scores or the most serious urban problems. The overall mean score for Japan was 3.1 with a range from 2.2 to 3.9. This overall ranking is probably best explained by a combination of wealth and urbanization. Japan is by far the most wealthy and has the capacity to solve many of the urban problems that other countries find so pressing. Malaysia is among the wealthiest of the less developed countries. It is less wealthy than Korea, but it has a strong rural development program and a far slower growth of urbanization; thus, it has less objective urban pressures. Malaysia's small size, lower density, slower urbanization, and greater wealth therefore insulate it from some of the most serious urban problems we find in the less developed countries.

Problem Area Scores

From column 9, which provides mean scores for each problem area for all countries excluding Japan, we can see which problems were the most serious. With an overall mean score of 2.8, we find six conditions *below* the mean: Public Utilities, Transportation, Revenues, Employment, Pollution, and Housing in ascending order. These tend to be among the most serious problems that reflect poverty plus development and rapid urbanization or the conditions that most strain urban infrastructure and services. For Japan, the lowest score is for Transportation followed by Public Utilities, Pollution, and Revenues, which are all below Japan's overall mean score. Like the poorer countries, Japan has urban problems of transportation, public utilities, and pollution. It does not, however, face the serious problem of poverty—low employment and lack of housing—that plague its less wealthy neighbors.

At the other end, on the side of more advantageous or less serious conditions, we have social services. For countries other than Japan, Health (welfare and family planning) and Education plus City Personnel are the conditions that are most favorably assessed by the urban administrators overall. These conditions also ran very high for all countries, though they are not the highest for all. For India, Indonesia, Korea, and Malaysia they are the highest categories. Thailand ranks the quality and quantity of urban personnel above its health and education services. For the Philippines, education is the highest followed by its revenue conditions with health advantageous conditions followed closely by housing and general educational and health conditions and personnel.

Individual Problem Scores

When we turn from the 12 broad categories to the 39 individual conditions in countries other than Japan, the broad pattern described above remains generally stable, but it also shows some interesting trends. The conditions that received the lowest scores included garbage disposal (2.0), low-cost housing (2.2), sewage (2.3), unemployment (2.3), traffic flows and volume (2.3), and the homeless (2.4). At the other end, the highest mean scores went to family planning (3.7) and primary education (3.7). Then came secondary education (3.5), high-income housing (3.4), and primary health care and middle-income housing (both with 3.3). Not surprisingly, the most serious problems are those associated with poverty and crowding, congested traffic, lack of housing for the poor, inadequacy of the most basic utilities, and unemployment.

There is an important policy implication in this finding. Much of this set of problems could be alleviated by employment, which would provide the income for housing and the revenue base for better utilities. However, the problems would address the serious issue of the physical infrastructure and it would also provide jobs and income for the poor.

Health, Family Planning, and Education

It is interesting to find health, family planning, and education ranking highest in all countries. This is true even in Japan, though note that it is not family planning or education services but the "general" level of health and education among the population. The thrust of economic development over the past few decades has turned attention and resources to human capital. All of these countries, and most Asian countries as well, have made massive headway since roughly 1950 in providing basic education and health services to the great majority of their populations. Everywhere we have seen the decline of mortality and the extension of education as Asian countries have mobilized resources to fulfill the massive demands and the elite dreams that came with independence. Since about 1960, Asia has also led the world in establishing effective national family planning programs. The results are seen in the rise of contraceptive use rates and the decline of fertility. All of this progress is clearly reflected in the urban administrators' assessment of problems and advantages.

The Personnel Paradox and the Problem of Centralization

Finally, we can draw attention to two administrative conditions whose interrelations provide an interesting window to some of the specific problems some administrators face. This concerns the relationship between the quality and quantity of urban government personnel and the revenue base and the control over revenues the city administrators have. We first deal with all countries except Japan since the latter used slightly different questions on these items. The quantity of personnel scores above average (3.0) and the quality of personnel is judged higher (3.1) for the 70 administrators who provided responses. On the other hand, revenue is considered a problem (2.5) and the city's control over revenue is an even more serious problem (2.5). That is, personnel is less a problem than the magnitude and control of financial resources. This pattern is found, however, only in Korea, Malaysia, Thailand, and India. In Indonesia, there were only too responses on the quality and quantity of personnel; thus, the personnel quality score is omitted. For Japan, the question of resource control was omitted. For the Philippines and Nepal, the resource control scores are higher than the personnel scores.

For the four countries with the dominant pattern, there is a reflection of a condition on which many urban specialists of the developing world have commented for some time. The problem lies in central government, which wishes to maintain control over local units. Most specialists agree that this initiative prevents sensitive adaptation of general policies to local conditions. In effect, central control retards the very development of local initiative that both central and local government want. Our urban administrator respondents are saying much the same thing. They are saying that they have good people and would like to get on with the job, but they are restrained both by the lack of revenues and even more by the lack of control over their revenues. Many governments have attempted to correct this problem by promoting administrative decentralization, but few have been really willing to give up control over the local units.

Nepal and the Philippines represent deviant cases in this set of observations. The findings for Nepal suggest that both the quality and quantity of urban personnel represent a serious problem. On the other hand, both revenue base and the control of revenue are considered satisfactory. Nepal's scores not only reverse the general trend, but the difference between the two scores is the largest for all countries. It should be noted, however, that all of the five questionnaires for Nepal were completed by a single administrator in the capital. In the case of many of the objective urban conditions, this may well provide accurate information, though it doesn't obtain the views of the frontline administrators, which was the original plan of the inquiry. In the case of urban personnel and resource control, however, it is quite possible, perhaps even likely, that local and central administrators will have different judgments. It may well be that in the case of Nepal, we are seeing common differences between central and local administrators, rather than gaining the kind of view from the local administrators we have received in the other countries.

For the Philippines, the scores are all quite close. Neither the quality nor quantity of personnel is considered any more than a minor problem (3.1 and 3.2), and the base and control of revenues both get a score of 3.6, almost satisfactory. The Philippines has indeed been promoting administrative decentralization for some time, and it may be that scores we find reflect some success in this movement. At any rate, it would appear that the relationship between administrators' judgments of personnel and resource control would be a fruitful line of research. It is possible that follow-up interviews may well uncover useful suggestions for effective administrative reform that would increase local initiative.

REFERENCES

Henderson, J. V. (1986). Urbanization in a developing country, city size and population composition. *Journal of Development Economics, 22,* 260–293.

Population Crisis Committee (1990). *City-life in the world's 100 largest metropolitan areas.* Washington, DC: Author.

Rosin, K. T., & Resnick, M. (1980). The size distribution of cities: An examination of the pareto law and primacy. *Journal of Urban Economics, 8,* 165–186.

Thomas, V. (1978). *The measurement of spatial differences in poverty: The case of Peril.* World Bank Staff Working Paper No. 273.

United Nations (1989). *Trends in population policy.* New York: United Nations Population Studies, No. 114.

Part **VII**
FUTURE THEORETICAL AND EMPIRICAL DIRECTIONS FOR ENVIRONMENT–BEHAVIOR RESEARCH

DIRECTIONS OF ENVIRONMENTAL PSYCHOLOGY IN THE TWENTY-FIRST CENTURY

Daniel Stokols

OVERVIEW

During the past 30 years, environmental psychology has documented the behavioral significance of the large-scale, sociophysical environment and contributed a variety of new concepts and methods for analyzing people environment transactions (Stokols, 1995). At the same time, effective applications of environment behavior research have been achieved within several community problem-solving arenas. Environmental psychology, as it exists today, spans multiple disciplinary and cultural perspectives, and offers abundant opportunities for collaboration among researchers from different regions of the world. The cross-cultural scope of the field has expanded dramatically in recent years, with the establishment of several international organizations that sponsor yearly conferences on environment–behavior research (e.g., IAPS, EDRA, the Environmental Psychology Section of the International Association of Applied Psychology) and the *Journals of Environmental Psychology (JEP)*, *Architectural and Planning Research (JAPR)*, and *Environment and Behavior (E&B)*, which highlight cross-cultural research and reviews of scientific developments in different regions of the world. The Proceedings of the Fourth Japan USA Seminar on Environment–Behavior Research, from which this book

Daniel Stokols • School of Social Ecology, University of California, Irvine, Irvine, California 92697-7050.

Handbook of Japan–United States Environment–Behavior Research: Toward a Transactional Approach, edited by Seymour Wapner, Jack Demick, Takiji Yamamoto, and Takashi Takahashi. Plenum Press, New York, 1997.

evolved, clearly exemplify the multidisciplinary and international orientation of the field today.

The present chapter examines prospective directions of environmental psychology in relation to both global or "transcultural" forces and culturally specific concerns that are likely to shape the course of research in this field during the 1990s and beyond. These contextual circumstances include (but are not limited to) growing concerns about the rapid depletion of natural resources and adverse changes in the earth's ecosystem; increased opportunities for international scientific collaboration afforded by the Internet, World Wide Web, and other advanced telecommunications technologies; and the geographic, historical, political, and sociocultural forces that are uniquely experienced by researchers living and working in particular countries and cultures. The influence of these global and regional factors on scientists' selection of topics for empirical study, and on the evolution of theoretical perspectives within different national or cultural contexts, is referred to in this chapter as the *ecology of theory development.*

From an ecological perspective, the emergence of scientific theories and research programs can be modeled as a particular category of human–environment transaction, in which scientists' views of the world and their decisions to study particular phenomena are influenced jointly by their personal dispositions, life-span developmental experiences, and a variety of contextual factors that impinge on scientific endeavors at local, regional, and global levels. This ecology-of-theory-development perspective raises several intriguing questions about the future of research on environment and behavior:

First, what are the prospects for achieving greater cross-cultural collaboration and integration of scientific perspectives in this field across diverse national and regional contexts? On the one hand, the unique historical, geographic, sociocultural, and political factors that have influenced the directions of environment behavior research in particular regions may limit the development of transcultural theories and methodologies—those that are both valid and applicable within diverse cultural contexts. As examples of the differential selection and weighting of research foci within various world regions, Scandinavian environmental psychologists have emphasized the effects of interior design features on occupants' mood and social behavior (Kuller, 1987), whereas Latin American researchers have given considerably greater attention to the behavioral consequences of sociocultural and political processes such as poverty, colonization, and class conflict (Low, Chapter 21, this volume; Sanchez, Wiesenfeld, & Cronick, 1987; Wiesenfeld, 1992, 1994).

Similarly, Japanese studies have revealed the substantial influence of climate and natural disasters on individuals' behavior (Hagino, Mochizuki, & Yamamoto, 1987; Kobayashi, et al., Chapter 15, this volume), while French research has devoted greater attention to the impact of sociocultural and political factors on people environment transactions (Jodelet, 1987; Milgram & Jodelet, 1976). The divergent research emphases evident within different countries and cultures raise the possibility that certain environment–behavior theories and methodologies are culturally specific in their scope, validity, and applied utility. Consequently, efforts to consolidate these concepts and methods into transcultural research programs may foster an artificial "homogenization" of environmental psychology that blunts scientific creativity by reducing the diversity of cultural and scientific perspectives.

On the other hand, it is conceivable that certain world wide developments may promote greater consolidation of theoretical perspectives within environmental psychology. Examples of these "unifying" forces in environment–behavior research, noted earlier, include heightened awareness of global environmental problems (e.g., damage to the earth's ozone layer, global warming, depletion of natural resources), and increased access to advanced telecommunications technologies among scientists working in different parts of the world. For example, Pol (1993) has suggested that the emergence of "green environmental psychology" during the 1990s—a scientific paradigm that is explicitly global in its orientation and substantive concerns—is attributable in large measure to growing concerns among researchers worldwide about the recently discovered, rapidly occurring, and potentially catastrophic changes in the earth's ecosphere (cf., Leaf, 1989; Stern, 1992).

In the ensuing discussion, prospective directions of environment–behavior studies are examined in relation to five major societal concerns that have arisen in recent decades and are likely to become even more salient during the 21st century: (1) toxic contamination of environments and rapid changes in the global ecosystem, (2) the spread of violence at regional and international levels, (3) the pervasive impacts of information technologies on work and family life, (4) escalating costs of health care delivery and the growing importance of disease prevention and health promotion strategies, and (5) processes of societal aging in the United States and other regions of the world. An important continuity of environmental psychology since its inception during the mid-1960s has been its long-standing concern with community environmental problems (Craik, Chapter 26, this volume; Stokols, 1995). Thus, the contemporary societal trends considered in this chapter provide a plausible basis for anticipating at least some future directions of the field.

At the same time, it is important to recognize that the salience and expression of these societal concerns are likely to vary across different regions of the world, and that their potential influence on environment behavior research must be considered in the context of culturally specific circumstances such as regional history and geography (Altman, Chapter 29, this volume; Low, Chapter 21, this volume; Saegert, Chapter 27, this volume). Accordingly, the latter sections of the chapter elaborate on the ecology-of-theory-development theme outlined earlier, and consider global and regional societal concerns as one subset of factors among a broader array of contextual circumstances that are likely to influence the future scientific and applied directions of environmental psychology.

RESEARCH AND POLICY CHALLENGES POSED BY CONTEMPORARY SOCIETAL PROBLEMS

Psychological and Behavioral Dimensions of Environmental Pollution and Global Environmental Change

The cumulative toxic effects of agricultural, industrial, and military technologies developed during the 20th century pose a growing threat to population and ecosystem health. Geophysical studies indicate that global environmental changes are occurring at an alarmingly rapid rate (Silver & DeFries, 1990). Ecological research also reveals the direct links between individual and group behaviors toward the environment (e.g., consumption of electricity and fossil fuels, recycling of used materials, corporate ride-sharing programs and efforts to reduce environmental pollution), and the severity and rapidity of atmospheric ozone depletion, global warming, and reduced biodiversity (Stern, Young, & Druckman, 1992). The adverse consequences of these global changes highlight the importance of understanding the circumstances under which individuals and groups make decisions and enact behaviors that affect levels of resource consumption and environmental pollution (Leaf, 1989; Stern, 1992).

Prior research by environmental psychologists and sociologists has examined the correlates of environmentally supportive behavior and has demonstrated the effectiveness of certain interventions (e.g., providing household members with monthly feedback about their energy consumption patterns; implementing corporate policies to encourage recycling and ride-sharing among employees) in promoting ecologically protective actions (Cone & Hayes, 1980; Dunlap, Grieneeks, & Rokeach, 1983; Geller, Winett, & Everett, 1982; Oskamp et al., 1991; Stern & Gardner, 1981). Some studies suggest that people are more likely to enact ecologically supportive behavior when they feel personally and immediately threatened by environmental problems (Baldassare & Katz, 1992; Platt, 1973) and are able to recognize the local implications of global environmental changes (Zube, 1991). These and related studies provide an empirical basis for developing more comprehensive public policies aimed at slowing the pace of environmental deterioration, and promoting higher levels of population and ecosystem health.

Regulatory initiatives designed to protect regional and global environmental quality are already being implemented and evaluated for their effectiveness in Canada, the United States, and worldwide (e.g., Giuliano et al., 1993; Saunders, 1990; World Resources Institute, 1994). Similarly, community-wide coalitions to promote population health have become more prevalent in recent years, and WHO-sponsored programs to encourage the development of "healthy cities" have been organized in several countries (e.g., Ashton et al., 1986; Conner, 1994; Duhl, 1986; Goodman et al., 1993). Considering the time-urgency of current global environmental changes, the development of theory-based policies to ameliorate these problems, and programmatic evaluations of their health and cost benefits, should be a high priority for future research on environment and behavior.

Stemming the Tide of Violence at Regional and International Levels

In the United States between 1981 and 1990, all categories of violent crime (e.g., murder, nonnegligent manslaughter, forcible rape, aggravated assault) increased by nearly 30% among youths under the age of 18 (Federal Bureau of Investigation, 1990). Homicide is the second leading cause of death among U.S. adolescents and young adults, and the leading cause of death among black youths (USDHHS, 1991). The number of reported child abuse incidents in the United States also has increased steadily from 1.7 million in 1984 to 2.4 million in 1990

(Goldstein, 1995), with the homicide rate among children under 4 years of age reaching a 40-year high in 1995 (U.S. Advisory Board on Child Abuse and Neglect, 1995).[1] Moreover, the proliferation of interracial and interethnic violence in many parts of the world (e.g., Bosnia, Rwanda, regions of the former Soviet Union, the Middle East, and several U.S. cities) demonstrates how little progress we have made during the 20th century in curbing violence, terrorism, and war.

Earlier studies suggest that the physical and social environment can influence the occurrence and severity of violence in several ways. For example, physical conditions such as ambient heat may increase the likelihood of violent outbursts among aggression-prone individuals by intensifying their levels of discomfort and annoyance (Baron & Ransberger, 1978). Environmental design features of urban areas may create opportunities for assaultive behavior among already-motivated offenders (Nasar & Fisher, 1993). The interior design and spatial arrangement of homes, the temporal patterning of household activities, and neighborhood transience and incivilities may function as predisposing or constraining factors in the etiology of child abuse (Belsky, 1993; Holman & Stokols, 1994). Frequent portrayals of violent episodes in the mass media may weaken societal norms against aggressive behavior and suggest opportunities for "copy-cat violence" among audience members (Goldstein, 1995; Slaby, 1992). And, a variety of social environmental factors, including historical patterns of intergroup conflict and differential access among community groups to educational and employment opportunities, may increase the likelihood of interracial and interethnic violence (Baldassare, 1994; Merton, 1938). A major challenge for environmentally oriented research is to develop more integrative theoretical and policy perspectives that account for the joint influences of these environmental factors on the etiology of violence (USDHHS, 1993).

The prevalence of intergroup violence in the United States and other countries highlights the importance of documenting the links between environmental design, urban planning, the multicultural structure of society, and community cohesion (Baldassare, 1994; Vila, 1994). An important question in this regard is whether community environments can be designed to provide functional and symbolic supports for diverse lifestyles and cultural identities, while at the same time strengthening collective allegiance to superordinate (or widely shared) goals. Both Riger (1993) and Leavitt and Saegert (1990) emphasize the importance of balancing group empowerment efforts with the cultivation of organizations and settings that foster a strong sense of community. Workable strategies for achieving this goal, however, remain to be developed and tested in future research.

Impacts of Technological Change on Individuals and Groups

The processes by which people create new technologies and are, in turn, transformed by them remain as an unexplored frontier for future theory development and research in environmental psychology (Ittelson, 1986). Technological

---

[1]In addition to these acute and severe forms of child victimization, a vast quantity of "pandemic victimizations" (e.g., physical punishment by parents and non-fatal assaults by siblings and peers) goes unreported each year (Finkelhor & Dziuba-Leatherman, 1994).

innovations such as electronic mail, fax machines, mobile phones, and desktop computing, for example, have fundamentally altered people's work routines, commuting patterns, and social behavior (Bechtel, MacCallum, & Poynter, Chapter 16, this volume; 1995; Handy & Mokhtarian; Meyrowitz, 1985, "Special Report," 1995). The percentages of home-based workers and telecommuters grew during the 1980s and are expected to increase further in the coming years (Rosen, 1991). Yet, little is known about the impacts of telecommuting on organizational effectiveness and social cohesion (Bezold, Carlson, & Peck, 1986; Christensen, 1994). For instance, does telecommuting impair team productivity by reducing face-to-face communication among coworkers? Similarly, are "distance learning" techniques (e.g., teleconferencing, electronic bulletin boards, computer-based courses) educationally inferior to classroom-based instruction, ongoing exchanges between mentors and mentees, and spontaneous discussions among people co-present at the same time and place (Noam, 1995)? And, will computer-based networks exacerbate the tensions between advantaged and disadvantaged groups, by further separating "information-rich" and "information-poor" segments of society? These questions remain to be examined in future research.

Also, in what ways are family dynamics and child-rearing practices being altered by changing work routines and technological innovations? As household structures become more diverse (e.g., single-parent families in which the adult works at home), and as multiple life roles are incorporated within the same environments (e.g., homes that accommodate both work and parenting roles; workplaces that support physical fitness, recreational, and child care needs), innovative design strategies will be needed to help occupants accommodate to these "multifunctional" settings (Christensen, 1994; Franck & Ahrentzen, 1989; Stokols, 1990). The development of design guidelines for multifunctional environments is, therefore, an important direction for future research.

A significant byproduct of urbanization and the rapid deployment of new information technologies is *attentional overload*, a psychological state in which individuals are overwhelmed by higher quantities and faster rates of information than they can manage (Cohen, 1978; Glass & Singer, 1972; Milgram, 1970). An important challenge for future research is to identify environmental resources and behavioral strategies that enable people to cope more effectively with a surfeit of information and stimulation. Certain environments, such as natural and wilderness settings, have the capacity to enhance individuals' recovery from stressful experiences associated with the complexities of urban living and rapid technological change (Hartig, Mang, & Evans, 1991; Kaplan & Kaplan, 1989; Kaplan & Talbot, 1983; Knopf, 1987; Korpela, 1992; Ulrich, 1983). Earlier studies, however, have not examined the ways in which organizational and sociocultural processes affect the restorative value of natural and built environments (Hartig & Stokols, 1994). Settings whose members experience a strong sense of community and attachment to a shared environment may be especially restorative, whereas those in which members feel more detached from their social and physical surroundings may intensify rather than reduce feelings of stress (Stokols, 1990). The social and cultural dimensions of restorative environments remain to be identified in future research.

Environmentally Based Strategies of Community Health Promotion

The costs of health care in the United States grew from about $42 billion in 1965 to approximately $820 billion in 1992—nearly a 20-fold increase (Walsh & Francis, 1992). Health care expenditures accounted for 6% of the Gross National Product in 1966, but they are expected to compose nearly 16% of the GNP by 1998 (O'Donnell, 1994). This dramatic rise in national health costs over the past 30 years highlights the importance of developing more effective disease prevention and health promotion strategies as an adjunct to medical care (Stokols, Pelletier, & Fielding, 1995).

The growing interest in worksite and community health promotion can be expected to open new avenues of research at the interface of environmental and health psychology (Quirk & Wapner, 1995). For example, efforts to develop more comprehensive approaches to community health promotion can combine the focus of health psychology on behavioral and "psychogenic" factors in health and illness (e.g., personality, psychophysiology), with ecological perspectives (Green, 1990; Moos, 1979; Stokols, 1992; Winett, King, & Altman, 1989) that give greater attention to "envirogenic" variables (e.g., geographic, architectural, technological, and sociocultural influences on health status). Similarly, therapeutic strategies that rely on "active interventions" or those requiring voluntary and sustained adherence to prescribed behavioral regimens (e.g., refraining from smoking, maintaining a low-fat diet, using vehicular safety belts) could be fruitfully combined with environmentally based, "passive interventions" (Williams, 1982) that require little or no effort on the part of individuals (e.g., installation of nontoxic furnishings and equipment in work facilities).

In future years, environmental psychological theories of privacy, stress, way-finding, and place attachment, and research-based guidelines for facilities design and management, should play an increasingly important role in the development of comprehensive worksite health promotion programs (Cal/OSHA, 1995; Danko, Eshelman, & Hedge, 1990; Green & Cargo, 1994; Ornstein, 1990; Stokols, in press). The concerns of environmental psychology are also directly relevant to nonoccupational settings and the development of effective community health promotion programs. For example, earlier research on environment and behavior suggests a variety of urban design and planning strategies for improving the quality of residential, neighborhood, school, and health care settings (Cooper-Marcus & Sarkissian, 1986; Moore & Lackney, 1993; Nasar & Fisher, 1993; Ulrich, 1991). These planning guidelines can be used to enhance public health through their incorporation into environmentally based programs for community health promotion (Conner, 1994; Duhl, 1986).

Implications of Societal Aging for Environmental Design and Community Planning

In 1900, people over 65 constituted about 4% of the U.S. population. By 1988, that proportion rose to 12.4%; by 2000 it will be 13% and by 2030, 22%. The most rapid population increase over the next decade will be among those over 85 years of age (USDHHS, 1991). During the past two decades, the population 85 and older has doubled. By 2010, those 75 and older (the "old-old") may constitute more than 40% of the elderly population (Newman, Zais, & Struyk, 1984). Comparable processes of societal aging are occurring in many other countries including Japan (Kose, Chapter 3, this volume; Osada, Chapter 5, this volume).

In view of these trends toward societal aging and the fact that older persons are more burdened by chronic diseases and physical disabilities than younger people, the design of health promotive environments for an aging population becomes increasingly important as a direction for future research (Green, 1990; Pastalan, 1983; Verbrugge, 1990). For instance, because the elderly are disproportionately vulnerable to fatalities from injuries sustained from slipping and falling, the design of stairwells to reduce the likelihood of these events in residential and institutional settings is an important task for future research (Archea, 1985). Also, the design of residential and recreational environments (e.g., physical fitness facilities, neighborhood support groups) to encourage higher levels of physical activity may prove to be an effective strategy for enhancing health status and independent functioning among the elderly (Parmelee & Lawton, 1990; Takahashi, Chapter 4, this volume; USDHHS, 1991; Yamamoto & Hanazato, Chapter 6, this volume). And, because older persons spend more time indoors and are more susceptible to respiratory ailments than younger people, the development of improved ventilation systems and nonsmoking policies to reduce indoor air pollution will become increasingly important tools for promoting well-being among the elderly, and among other age groups as well (Green, 1990; Greenberg, 1986).

Each of the societal concerns noted above has direct implications for environmental design and management. For example, high rates of violence and crime in the United States have altered patterns of public investment in capital projects. In several states (including California, Connecticut, Florida, Massachusetts, Michigan, and Minnesota), more public funds are now allocated to correctional facilities than to colleges and universities ("Prison-Building Binge," 1995). Also, new information technologies are changing the locations and physical designs of residential and occupational environments (Allen, 1977; Becker, 1990; Christensen, 1994). And processes of societal aging are expected to result in higher levels of public and private investment toward the construction of nursing homes, health care facilities, recreational settings, and retirement communities for older adults (Marans, Hunt, & Vakalo, 1984; Kose, Chapter 3, this volume; Newman, Zais, & Struyk, 1984; Parmelee & Lawton, 1990; Takahashi, Chapter 4, this volume). These anticipated changes in the design and construction of new environments, while prompting new areas of applied research, will also pose new theoretical questions and directions for scientific inquiry.

DIRECTIONS FOR THEORY DEVELOPMENT IN ENVIRONMENTAL PSYCHOLOGY

Efforts to ameliorate community problems through applied research often stimulate new theoretical developments as scientists confront the complexities of people environment transactions in naturalistic settings. For instance, the complex realities of global environmental change, diffusion of new technologies, intergroup violence, and societal aging will challenge researchers to bridge previously separate areas of theorizing and research. Thus, the goal of creating health-promotive environments for older adults may stimulate new conceptual links between environmental, developmental, and health psychology (Lawton, 1989). Similarly, the development of programs intended to slow the rising tide of intergroup violence will likely require greater consolidation of psychological, sociological, and environmental design and urban planning perspectives (Goldstein, 1995).

A grand theory of environment and behavior is not likely to emerge in future years due to the enormous diversity and multidisciplinary scope of environmental psychology (Stokols, 1995). However, more modest efforts to consolidate middle-range theories of environment and behavior across different areas of psychological research, and at the interface of psychology and neighboring disciplines, can be expected to continue and, perhaps, become more prevalent over the next several years (McNally, 1992; Merton, 1968; Wapner, 1995). In the remaining sections, I outline some prospective directions for further theoretical development and integration, including opportunities for (1) strengthening the links between the fields of architecture, environmental psychology and urban design; (2) developing contextually broader theories and community problem-solving strategies; and (3) researching the ecology of creativity and theory development—an important and, heretofore, neglected facet of human–environment transaction.

The Expanding Interface between Architecture, Environmental Psychology, and Urban Design

Cross-paradigm research linking the theories and methods of environmental psychology with the fields of architecture, facilities planning, and urban design is likely to expand in future years. Directions for such research include the development of predesign research (PDR) methods (e.g., environmental simulation, programming, and design-review strategies) that are more sensitive to the diverse needs of individuals and multiple user groups occupying common buildings and urban areas (Becker, 1990; Bechtel, 1989; Marans & Stokols, 1993; Mazumdar, 1992; Zeisel, 1981). Also, the design of postoccupancy evaluation (POE) methods that are theoretically anchored and structured to reveal potentially divergent reactions among multiethnic, mixed-gender, multigenerational, and mixed-income groups to shared worksites and public spaces is an important priority for future research (Bechtel, Marans, & Michelson, 1987; Carr, Francis, Rivlin, & Stone, 1992; Zimring, 1989). Moreover, the demographic and economic trends noted earlier relating to increased capital investment in correctional facilities, and in residential, health care, and recreational settings for the elderly, portend greater opportunities for applying environment–behavior theories and research methods toward the design, evaluation, and enhancement of these environments in the coming years (Marans et al., 1984; Parmelee & Lawton, 1990; Wener, Frazier, & Farbstein, 1985).

Attention to individual and group-specific needs in environmental design (based on developmental stage, disabilities, lifestyle, gender, and ethnicity) can be expected to increase as architects and urban designers strive to develop more comprehensive and effective plans for buildings and cities (Altman & Churchman, 1994; Franck & Ahrentzen, 1989; Hubbard, 1992; Michelson, 1985; Preiser, 1988; Sommer, 1983). Scientific interest in the sociocultural aspects of environment and behavior is also likely to grow as advanced telecommunications bring diverse populations and geographic regions into closer contact (Altman & Chemers, 1980; Bechtel et al., Chapter 16, this volume; Meyrowitz, 1985; Rapoport, 1980; Saegert & Winkel, 1990).

Development of Contextually Broader Theories and Community Problem-Solving Strategies

Research in environmental psychology today encompasses both highly focused analyses of individual behavior in particular places, as well as broader formulations of group–environment transactions that span multiple settings in large geographic regions (e.g., neighborhoods, cities), and occur over prolonged periods (e.g., during developmental transitions, relocations). The contextual scope of environmental psychological theories has expanded since the mid-1960s, in that several recent conceptions of environment and behavior subsume spatially, temporally, and socioculturally broader units of analysis (cf. Altman & Rogoff, 1987; Canter & Larkin, 1993; Saegert & Winkel, 1990; Stokols, 1987). This trend toward contextually broader theories and methodological approaches is likely to continue as environmental psychologists delve further into the behavioral underpinnings of global environmental change, design criteria for culturally diverse communities, and the effects of new information technologies on patterns of international communication and scientific collaboration.

At the same time, community intervention strategies based on environmental psychological theories also can be expected to become more integrative and expansive. Previous applications of environment–behavior research to community problem-solving have been targeted primarily toward specific settings and occupant groups, rather than implemented in a more integrative fashion across multiple environments and populations. For example, research on the environmental needs of different age groups suggests design guidelines for enhancing (1) infants' cognitive development in residential settings (Wachs, 1992); (2) the quality of day-care, school, and play environments for children (Moore, 1986; Moore & Lackney, 1993; Noschis, 1992; Susa & Benedict, 1994); (3) the social cohesion of urban neighborhoods (Appleyard, 1981; Newman, 1973; Perkins et al., 1993); (4) the comfort and quality of occupational settings (Becker, 1990; Danko et al., 1990; Sundstrom, 1986; Wineman, 1986); and (5) the design of residential environments for the elderly (Parmelee & Lawton, 1990). What has not been achieved in earlier research, however, is the consolidation of these setting-specific and group-specific guidelines within more comprehensive approaches that address the interdependencies between multiple settings and age groups (Bronfenbrenner, 1979; Friedman & Wachs, in press; Hubbard, 1992; Moen, Elder, & Luscher, 1995)

More comprehensive approaches to community planning and design would (1) consider age, gender, cultural, and ethnic differences in people's response to a wide range of settings; (2) address the relationships that exist among multiple behavior settings (e.g., the social and spatial linkages between residential, child care, and work environments); and (3) incorporate multiple design guidelines for improving the quality of individuals' "overall life situation" (Magnusson, 1981) and the healthfulness of neighborhoods and communities (Ashton, Grey, & Barnard, 1986; Conner, 1994; Duhl, 1986). The importance of developing more comprehensive, research-based guidelines for community planning and health promotion is underscored by the rapidity of recent technological, social, and global environmental changes (Bezold et al., 1986; Dunlop & Kling, 1991; Stern, 1992). The pace and scope of these changes, and the urgency of developing broad-gauged strategies for managing them, will substantially influence the course of research in environmental psychology over the next several years.

The Ecology of Creativity and Theory Development

The preceding sections suggest the value of developing more integrative formulations of environment and behavior—for example, theories, research methods, and community problem-solving strategies that encompass multiple disciplinary perspectives (e.g., architecture, environmental psychology, and urban design) and wider units of contextual analysis (i.e., those that are of broad geographic, historical, and sociocultural scope). The prospects for developing more integrative, comprehensive approaches to the study of environment and behavior are likely to depend on several factors, including (1) the capacity of researchers trained in one discipline to expand their scientific horizons by adopting theoretical and methodological perspectives associated with other fields (Rapoport, Chapter 28, this volume; Wicker, 1985); (2) the extent to which cross-paradigm and multidisciplinary perspectives permit a better understanding of particular research topics, relative to less comprehensive formulations (while avoiding the "diminishing scientific returns" sometimes associated with cumbersome conceptual models and methodologies; cf. Craik, Chapter 26, this volume; Stokols, 1987; Wapner, 1995); and (3) the degree to which historical, geographic, political, and sociocultural circumstances associated with particular regions impose limits on achieving transcultural rather than culturally specific understandings of environment-behavior phenomena (Altman, Chapter 29, this volume; Saegert, Chapter 27, this volume; Wapner, Demick, Yamamoto, & Takahashi, Chapter 1, this volume).

The above-noted circumstances comprise a limited subset of the various intrapersonal and situational factors that influence the development and scope of environment–behavior theories and research programs. The ecology of theory development remains as an intriguing and relatively understudied topic for future research on environment and behavior. This topic subsumes several different issues including (1) the joint influence of psychological and situational factors on the development of an individual's ideas, theories, or artistic contributions and (2) the ways in which historical, geographic, sociocultural, and political circumstances affect the selection of topics for scientific study and the evolution of theoretical perspectives within different national and cultural contexts. Clearly, these different aspects of the ecology of theory development are interrelated, since the development of research programs in different geographic areas reflects the cumulative interests and ideas of numerous individual scientists.

Environmental Salience and Scientific Creativity

Processes of scientific creativity and theory development can be viewed as dynamic transactions between individuals and their sociophysical surroundings (Amabile, 1984; Clitheroe, 1995; Lasswell, 1959; Mead, 1959). To date, little research has been conducted on the ways in which the physical and social features of environments jointly influence scientists' experiences of creativity and the development of their research ideas. One of the ways in which environmental circumstances influence the development of scientific theories is by drawing researchers' attention to highly salient features of situations. Stokols (1986) has distinguished between the *motivational* and *perceptual salience* of environments. The former concerns the extent to which situations or settings are associated with psychologically important needs, whereas the latter category refers to highly noticeable features of the sociophysical environment (cf. Lynch, 1960; Taylor & Fiske, 1978). Motivational salience depends largely on the subjective weightings of importance assigned by individuals to various features of their environments, whereas perceptual salience is determined less by individual or cultural differences and more by objective qualities of the physical and social environment (e.g., the brightness, contrast, novelty, movement, and frequency or prevalence of environmental features).

The concept of environmental salience suggests that those features of the sociophysical environment that are especially prevalent, noticeable, and motivationally significant will disproportionately influence the development of scientists' theories and the directions of their research programs. At least three categories of environmental circumstances appear to be particularly important in this regard: (1) *environmental constraints, deprivations, and challenges* that are prevalent in a specified area (or are commonly found across several different regions); (2) *social and political concerns that are specific to certain cultures and geographic regions*; and (3) *recurring ethical, religious, and social themes that are universally experienced* within all human communities.

Examples of highly salient constraints, deprivations, and challenges include the earthquakes, typhoons, and other natural disasters that frequently occur in Japan (Hagino, Mochizuki, & Yamamoto, 1987; Kobayashi, et al., Chapter 15, this volume); the cyclones, droughts, floods, and brushfires that commonly ravage Australia (Thorne & Hall, 1987); the low levels of natural lighting available in Scandinavia during prolonged periods each year (Kuller, 1987); the high levels of stimulation overload and traffic congestion experienced in large cities (Cohen, 1978; Milgram, 1970; Novaco et al., 1990); and the prevalence of understaffed behavior settings in small, rural communities (Barker, 1968; Barker & Gump, 1964; Barker & Schoggen, 1973). These environmental constraints, deprivations, and challenges have substantially influenced the directions of environment behavior research in different world regions.

At the same time, certain sociocultural and political concerns that have been especially salient in certain regions are clearly revealed in the research emphases of environmental psychology as the field has evolved in those geographic areas. Examples of these concerns include the influence of Marxist–Leninist political ideology on environment–behavior research conducted in the former Soviet Union (Niit, Heidmets, & Kruusvall, 1987); studies of low-income housing design, social-class conflicts, and contested symbolic meanings of urban public spaces in Latin American cities (Low, Chapter 21, this volume; Sanchez, Wiesenfeld, & Cronick, 1987; Wiesenfeld, 1992; Wiesenfeld, 1994); and the emergence of a feminist research perspective in North American and European research on environment and behavior (Ahrentzen, 1990; Altman & Churchman, 1994; Michelson, Chapter 11, this volume; Moore, 1987; Saegert, 1987).

Finally, certain recurring themes of human interpersonal relationships, lifespan development, ethics, and spirituality are universally experienced and motivationally salient within all cultures and communities (Altman & Chemers, 1980; Rapoport, Chapter 28, this volume). Examples of these transcultural themes include human needs for environmental "focal points" (Bechtel, 1977; Takahashi, Chapter 4, this volume; Yamamoto & Hanazato, Chaper 6, this volume) and "anchor points" (Wapner, 1981, 1987), especially during times of developmental and geographic transition; individuals' predilections for establishing strong emotional ties to psychologically significant places (Cooper, 1974; Proshansky, 1978; Proshansky, Fabian, & Kaminoff, 1983; Stokols, 1981); the dialectical tension between people's needs for social contact, on the one hand, and for solitude and privacy, on the other (Altman, 1975); society's treatment of outsiders, minority groups, impoverished individuals, homeless people, and other vulnerable populations (Saegert, 1987; Zimring, Reizenstein-Carpman, & Michelson, 1987); and the ways in which individuals and groups cope with severe environmental stressors and traumatic losses (Baum & Fleming, 1993; Evans & Cohen, 1987; Moos, 1986). These transcultural themes provide a source of continuity in environmental psychology and afford numerous opportunities for future collaborative, cross-cultural research on environment and behavior (Craik, Chapter 26, this volume; Wapner et al., Chapter 1 and 2, this volume).

Opportunities for Cross-Cultural Collaboration in Environment Behavior Research

We return now to a question posed at the outset of the chapter: What are the prospects for achieving greater cross-cultural integration of environment behavior research across diverse national and regional contexts? Considering the important role of salient environmental circumstances in shaping the directions of scientific research, the most immediate opportunities for international collaboration would seem to involve those research topics that pertain to globally significant environmental constraints and challenges (e.g., processes of global environmental change and international conflict; cf. Goldstein, 1995; Leaf, 1989; Pol, 1993; Stern, 1992) and aspects of human–environment transaction that are experienced universally within all countries and cultures (e.g., the diminished physical and mental competencies associated with aging; processes of coping with extreme environmental demands and traumatic life events; and transcultural needs for privacy regulation, territoriality, and environmental security; cf. Altman, 1975; Altman & Chemers, 1980; Lawton & Nahemow, 1973; Moos, 1986; and pertinent chapters in this volume).

At the same time, certain substantive concerns may be more specific to certain cultures and geographic regions and, consequently, less amenable to transcultural theoretical analyses (at least in the short run). Examples of environment–behavior phenomena that have achieved higher levels of salience in certain cultures relative to others include concerns among U.S. women college students about their vulnerability to sexual assaults on campus (Day, 1994; in press); growing concerns among U.S. workers about workplace violence prevention (USDHHS, 1993), and the emergence of a strong feminist research perspective in the environment–behavior field, especially in North America but, heretofore, not in Japan, Scandinavia, and many other regions (cf. Ahrentzen, 1990; Altman & Churchman, 1994; Hagino et al., 1987; Kuller, 1987; Michelson, Chapter 11, this volume; Saegert, 1987). Although transcultural interest in these concerns may expand in the coming years (e.g., through increased global exchange of research ideas via the Internet), they nonetheless appear to be less amenable to cross-cultural investigation at the present time, as compared to other phenomena that have attracted greater interest and attention at international levels.

CONCLUSIONS

The field of environmental psychology has made major strides over the past 30 years in (1) developing novel conceptualizations of people–environment transaction, and (2) applying research concepts, methods, and findings to the analysis and resolution of community problems. Additional research trends include (3) the shift from paradigm-specific to cross-paradigm research, (4) the increasing emphasis on transactional analyses of environment and behavior and (5) on the relationships between groups and their environments, and (6) the expanding international scope of environmental psychology (see Stokols, 1995, for further discussion of these trends).

Future research in environmental psychology will continue to be influenced by societal concerns including (1) global environmental change, (2) intergroup violence and crime, (3) impacts of new information technologies on work and family life, (4) rising health costs and interest in environmental strategies of health promotion, and (5) processes of societal aging. These community concerns will create new opportunities for cross-paradigm research within psychology and between psychology and other disciplines. Examples of these directions include theoretical analyses of individual and subgroup differences in people's reactions to built and natural environments; research on the role of cultural, geographic, and technological factors in creativity and theory development; and the development of contextually broader theories and community problem-solving strategies.

The preceding discussion of research trends and opportunities is undoubtedly incomplete and bounded by the author's own geographic and cultural frame of reference on the environment and behavior field. Additional topics that are likely to receive greater research attention in future years are the design of environments for living and working in outer space (Harris, 1992; Harrison, Clearwater, & McKay, 1991), and the formulation of effective policies for reducing conflicts among industrialized and developing countries related to the contamination of shared environments and the depletion of natural resources (Sommer, 1987). Nonetheless, the summary of research developments and directions for future study, outlined above, provides at least a partial glimpse of environmental psychology's accomplishments and challenges as we approach the 21st century. An especially exciting direction for future study, in this author's view, is to achieve a broader understanding of the ecological circumstances that shape the development and directions of environment–behavior research. Hopefully, the ideas outlined here will stimulate further investigation of the ecology of theory development within the environment and behavior field.

REFERENCES

Ahrentzen, S. B. (1990). Rejuvenating a field that is either "coming of age" or "aging in place": Feminist research contributions to environmental design research. In R. I. Selby, K. H. Anthony, J. Choi, & B. Orland (Eds.), *Coming of age: EDRA 21: Proceedings of the Twenty First Annual Conference of the Environmental Design Research Association* (pp. 11–18). Oklahoma City, OK: EDRA.

Alexander, C., Ishikawa, S., Silverstein, M., Jacobson, M., Fiksdahl-King, I., & Angel, S. (1977). *A pattern language: Towns, buildings, construction*. New York: Oxford University Press.

Allen, T. J. (1977). *Managing the flow of technology*. Cambridge, MA: MIT Press.

Altman, I. (1975). *Environment and social behavior: Privacy, personal space, territory, and crowding*. Monterey, CA: Brooks/Cole.

Altman, I., & Chemers, M. M. (1980). *Culture and environment*. New York: Cambridge University Press.

Altman, I., & Churchman, A. (Eds.), (1994). *Women and the environment. Human behavior and environment: Advances in theory and research: Vol. 13*. New York: Plenum.

Altman, I., & Rogoff, B. (1987). World views in psychology and environmental psychology: Trait, interactional, organismic, and transactional perspectives. In D. Stokols & I. Altman (Eds.), *Handbook of environmental psychology* (pp. 7–40). New York: John Wiley & Sons.

Amabile, T. M. (1984). *The social psychology of creativity*. New York: Springer-Verlag.

Appleyard, D. (1981). *Livable streets*. Berkeley, CA: University of California Press.

Archea, J. C. (1985). Environmental factors associated with stair accidents by the elderly. *Clinics in Geriatric Medicine, 1*, 555–569.

Ashton, J., Grey, P., & Barnard, K. (1986). Healthy cities: WHO's New Public Health Initiative. *Health Promotion, 1*, 319–324.

Baldassare, M. (1994). Introduction. In M. Baldassare (Ed.), *The Los Angeles riots: Lessons for the urban future*. Boulder, CO: Westview Press.

Baldassare, M., & Katz, C. (1992). The personal threat of environmental problems as predictor of environmental practices. *Environment and Behavior, 24*, 602–616.

Barker, R. G. (1968). Ecological psychology: Concepts and methods for studying the environment of human behavior. Stanford, CA: Stanford University Press.

Barker, R. G., & Gump, P. V. (1964). *Big school, small school*. Stanford: Stanford University Press.

Barker, R. G., & Schoggen, P. (1973). *Qualities of community life*. San Francisco: Jossey-Bass.

Baron, R. A., & Ransberger, V. M. (1978). Ambient temperature and the occurrence of collective violence: The "long hot summer" revisited. *Journal of Personality and Social Psychology, 36*, 351–360.

Baum, A., & Fleming, I. (1993). Implications of psychological research on stress and technological accidents. *American Psychologist, 48*, 665–672.

Bechtel, R. B. (1977). *Enclosing behavior*. Stroudsburg, PA: Dowden, Hutchinson, & Ross.

Bechtel, R. B. (1989). Advances in POE methods: An overview. In W. Preiser (Ed.), *Building evaluation* (pp. 199–206). New York: Plenum.

Bechtel, R. B., Marans, R. W., & Michelson, W. (Eds.). (1987). *Methods in environmental and behavioral research.* New York: Van Nostrand Reinhold.

Becker, F. D. (1990). *The total workplace: Facilities management and the elastic organization.* New York: Van Nostrand Reinhold.

Belsky, J. (1993). Etiology of child maltreatment: A developmental-ecological analysis. *Psychological Bulletin, 114,* 413–434.

Bezold, C., Carlson, R. J., & Peck, J. C. (1986). *The future of work and health.* Dover, MA: Auburn House Publishing Company.

Boneau, C. A. (1992). Observations on psychology's past and future. *American Psychologist, 47,* 1586–1596.

Bronfenbrenner, U. (1979). *The ecology of human development.* Cambridge, MA: Harvard University Press.

Cal/OSHA (1995). *Model injury and illness prevention program for workplace security.* Sacramento, CA: State of California, Department of Industrial Relations, Division of Occupational Safety & Health.

Canter, D., & Larkin, P. (1993). The environmental range of serial rapists. *Journal of Environmental Psychology, 13,* 63–69.

Carr, S., Francis, M., Rivlin, L. G., & Stone, A. M. (1992). *Public space.* New York: Cambridge University Press.

Christensen, K. (1994). Working at home: Frameworks of meaning. In I. Altman & A. Churchman (Eds.), *Women and the environment. Human behavior and environment: Advances in theory and research: Vol. 13* (pp. 133–166). New York: Plenum.

Clitheroe, C. (1995). Creativity in context: Defining and researching creativity from an environmental psychology perspective. In J. Nasar, P. Grannis, & K. Hanyu (Eds.), *Proceedings of the twenty-sixth annual conference of the Environmental Design Research Association.* Oklahoma City, OK: EDRA, 163.

Cohen, S. (1978). Environmental load and the allocation of attention. In A. Baum, J. Singer, & S. Valins (Eds.), *Advances in environmental psychology, Vol. 1: The urban environment* (pp. 1–29). Hillsdale, NJ: Lawrence Erlbaum Associates.

Cone, J. D., & Hayes, S. C. (1980). *Environmental problems/Behavioral solutions.* Monterey, CA: Brooks/Cole.

Conner, R. (1994). *Evaluation of the Colorado Healthy Communities Initiative.* Paper presented at the 23rd International Congress of Applied Psychology, July 17–22, Madrid, Spain.

Cooper, C. (1974). The house as symbol of the self. In J. Lang, C. Burnette, W. Moleski, & D. Vachon (Eds.), *Designing for human behavior.* Stroudsburg, PA: Dowden, Hutchinson, & Ross.

Cooper-Marcus, C., & Sarkissian, W. (1986). *Housing as if people mattered: Site design guidelines for medium-density family housing.* Berkeley, CA: University of California Press.

Danko, S., Eshelman, P., & Hedge, A. (1990). A taxonomy of health, safety, and welfare implications of interior design decisions. *Journal of Interior Design Education and Research, 16,* 19–30.

Day, K. (1994). Conceptualizing women's fear of sexual assault on urban college campuses: A review of causes and recommendations for change. *Environment and Behavior, 26,* 742–765.

Day, K. (1995). Assault prevention as social control: Women and sexual assault prevention on urban college campuses. *Journal of Environmental Psychology, 15,* 261 281.

Duhl, L. (1986). The healthy city: Its functions and its future. *Health Promotion, 1,* 55–60.

Dunlop, C., & Kling, R. (Eds.). (1991). Computerization and controversy: Value conflicts and social choices. Boston: Academic Press.

Dunlap, R. E., Grieneeks, J. K., & Rokeach, M. (1983). Human values and pro-environmental behavior. In W. D. Conn (Ed.), *Energy and material resources: Attitudes, values, and public policy.* Boulder, CO: Westview Press.

Evans, G. W., & Cohen, S. (1987). Environmental stress. In D. Stokols & I. Altman (Eds.), *Handbook of environmental psychology, Vol. 1* (pp. 771–610). New York: Plenum.

Federal Bureau of Investigation (1990). *Uniform crime report,* 1989. Washington, DC: U,S. Government Printing Office.

Finkelhor, D., & Dziuba-Leatherman, J. (1994). Victimization of children. *American Psychologist, 49*, 173–183.

Franck, K. A., & Ahrentzen, S. (Eds.). (1989). *New households, new housing*. New York: Van Nostrand Reinhold.

Friedman, S. L., & Wachs, T. D. (Eds.) (in press). *Measurement of environment across the lifespan*. Washington, DC: American Psychological Association.

Geller, E. S., Winett, R. A., & Everett, P. B. (1982). *Preserving the environment: New strategies for behavior change*. New York: Pergamon Press

Giuliano, G., Hwang, K., & Wachs, M. (1993). Employee trip reduction in Southern California: First year results. *Transportation Research A, 27A*, 125–137.

Glass, D. C., & Singer, J. E. (1972). *Urban stress*. New York: Academic Press.

Goldstein, A. P. (1995). *The ecology of aggression*. New York: Plenum.

Goodman, R. M., Burdine, J. N., Meehan, E., & McLeroy, K. R. (Eds.). (1993). Community coalitions for health promotion. *Health Education Research: Theory and Practice, 8*, 305–448.

Green, L. W. (1990). *Community health* (6th edition). St. Louis: Times Mirror/Mosby Publishers.

Green, L. W., & Cargo, M. D. (1994). The changing context of health promotion in the workplace. In M. P. O'Donnell & J. S. Harris (Eds.), *Health promotion in the workplace* (2nd ed., pp. 497–524). Albany, NY: Delmar Publishers.

Greenberg, M. R. (1986). Indoor air quality: Protecting public health through design, planning, and research. *Journal of Architectural and Planning Research, 3*, 253–261.

Hagino, G., Mochizuki, M., & Yamamoto, T. (1987). Environmental psychology in Japan. In D. Stokols & I. Altman (Eds.), *Handbook of environmental psychology, Vol. 2* (pp. 1155–1170). New York: John Wiley & Sons.

Handy, S. L., & Mokhtarian, P. L. (1995). Planning for telecommuting: Measurement and policy issues. *American Planning Association Journal, 61*, 99–111.

Harris, P. R. (1992). Living and working in space: Human behavior, culture, and organization. New York: Ellis Horwood.

Harrison, A. A., Clearwater, Y. A., & McKay, C. P. (1991). *From Antarctica to outer space: Life in isolation and confinement*. New York: Springer-Verlag.

Hartig, T., & Stokols, D. (1994). *Toward an ecology of stress and restoration. Man and nature: Working Paper 54–1994*. Humanities Research Center, Odense University, Odense, Denmark.

Hartig, T., Mang, M., & Evans, G. W. (1991). Restorative effects of natural environment experiences. *Environment and Behavior, 23*, 3–26.

Holman, E. A., & Stokols, D. (1994). The environmental psychology of child sexual abuse. *Journal of Environmental Psychology, 14*, 237–252.

Hubbard, P. J. (1992). Environment-behavior studies and city design: A new agenda for research. *Journal of Environmental Psychology, 12*, 269–279.

Ittelson, W. H. (1986). The psychology of technology. In W. H. Ittelson, M. Asai, & M. Ker (Eds.), *Cross cultural research in environment and behavior* (pp. 145–154). Tucson, AZ: University of Arizona.

Jodelet, D. (1987). The study of people-environment relations in France. In D. Stokols & I. Altman (Eds.), *Handbook of environmental psychology*, Vol. 2 (pp. 1171–1193). New York: John Wiley & Sons.

Kaplan, R., & Kaplan, S. (1989). *The experience of nature: A psychological perspective*. New York: Cambridge University Press.

Kaplan, S., & Talbot, J. F. (1983). Psychological benefits of a wilderness experience. In I. Altman & J. F. Wohlwill (Eds.), *Human behavior and environment: Advances in theory and research, Vol. 6. Behavior and the natural environment* (pp. 163–203). New York: Plenum.

Knopf, R. C. (1987). Human behavior, cognition, and affect in the natural environment. In D. Stokols & I. Altman (Eds.), *Handbook of environmental psychology, Vol. 1* (pp. 783–825) New York: John Wiley & Sons.

Korpela, K. M. (1992). Adolescents' favourite places and environmental self-regulation. *Journal of Environmental Psychology, 12*, 249–258.

Kuller, R. (1987). Environmental psychology from a Swedish perspective. In D. Stokols & I. Altman (Eds.), *Handbook of environmental psychology, Vol. 2* (pp. 1243–1279). New York: John Wiley & Sons.

Lasswell, H. D. (1959). The social setting of creativity. In H. H. Anderson (Ed.), *Creativity and its cultivation* (pp. 203–221). New York: Harper & Row.

Lawton, M. P. (1989). Behavior relevant ecological factors. In K. W. Schaie & C. Schooler (Eds.), *Social structure and aging: Psychological processes* (pp. 57–78). Hillsdale, NJ: Lawrence Erlbaum Associates.

Lawton, M. P., & Nahemow, L. (1973). Ecology and the aging process. In E. Eisdorfer & M. P. Lawton (Eds.), *Psychology of adult development and aging* (pp. 619–674). Washington, DC: American Psychological Association.

Leaf, A. (1989). Potential health effects of global climatic and environmental changes. *New England Journal of Medicine, 321*, 1577–1583.

Leavitt, J., & Saegert, S. (1990). *From abandonment to hope: Community households in Harlem*. New York: Columbia University Press.

Lynch, K. (1960). *The image of the city*. Cambridge, MA: MIT Press.

Magnusson, D. (1981). Wanted: A psychology of situations. In D. Magnusson (Ed.), *Toward a psychology of situations: An interactional perspective* (pp. 9–32). Hillsdale, NJ: Erlbaum.

Marans, R. W., & Stokols, D. (1993). *Environmental simulation: Research and policy issues*. New York: Plenum.

Marans, R. W., Hunt, M. E., & Vakalo, K. L. (1984). Retirement communities. In I. Altman, M. P. Lawton, & J. F. Wohlwill (Eds.), *Elderly people and the environment. Human behavior and environment: Advances in theory and research, Vol. 7* (pp. 57–93). New York: Plenum.

Mazumdar, S. (1992). How programming can become counterproductive: An analysis of approaches to programming. *Journal of Environmental Psychology, 12*, 65–91.

McNally, R. J. (1992). Disunity in psychology: Chaos or speciation? *American Psychologist, 47*, 1054.

Mead, M. (1959). Creativity in cross-cultural perspective. In H. H. Anderson (Ed.), *Creativity and its cultivation* (pp. 222–235). New York: Harper & Row.

Merton, R.K. (1938). Social structure and anomie. *American Sociological Review, 3*, 672–682.

Merton, R. K. (1968). On sociological theories of the middle range. In R.K. Merton, *Social theory and social structure*. New York: The Free Press, 39–71.

Meyrowitz, J. (1985). *No sense of place: The impact of electronic media on social behavior*. New York: Oxford University Press.

Michelson, W. H. (1985). *From sun to sun: Daily obligations and community structure in the lives of employed women and their families*. Totowa, NJ: Rowman & Allenheld.

Milgram, S. (1970). The experience of living in cities. *Science, 167*, 1461–1468.

Milgram, S., & Jodelet, D. (1976). Psychological maps of Paris. In H. M. Proshansky, W. H. Ittelson, & L. G. Rivlin (Eds.), *Environmental psychology: People and their physical settings* (2nd edition, pp. 104–124). New York: Holt, Rinehart & Winston.

Moen, P., Elder, G., & Luscher, K. (Eds.). (1995). *Linking lives and contexts: Perspectives on the ecology of human development*. Washington, DC: American Psychological Association.

Moore, G. T. (1986). Effects of the spatial definition of behavior settings on children's behavior: A quasi-experimental field study. *Journal of Environmental Psychology, 6*, 205–233.

Moore, G. T. (1987). Environment and behavior research in North America: History, developments, and unresolved issues. In D. Stokols & I. Altman (Eds.), *Handbook of environmental psychology, Vol. 2* (pp. 1359–1410). New York: John Wiley & Sons.

Moore, G. T., & Lackney, J. A. (1993). School design: Crisis, educational performance, and design applications. *Children's Environments, 10*, 99–112.

Moos, R. H. (1979). Social ecological perspectives on health. In G. C. Stone, F. Cohen, & N. E. Adler (Eds.), *Health psychology: A handbook* (pp. 523–547). San Francisco: Jossey Bass.

Moos, R. H. (Ed.). (1986). Coping with life crises: An integrated approach. New York: Plenum.

Nasar, J. L., & Fisher, B. (1993). "Hot spots" of fear and crime: A multi-method investigation. *Journal of Environmental Psychology, 13*, 187–206.

Newman, O. (1973). *Defensible space: Crime prevention through urban design*. New York: Macmillan.

Newman, S. J., Zais, J., & Struyk, R. (1984). Housing older America. In I. Altman, M. P. Lawton, & J. F. Wohlwill (Eds.), *Elderly people and the environment. Human behavior and environment: Advances in theory and research, Vol. 7* (pp. 17–55). New York: Plenum.

Niit, T., Heidmets, M., & Kruusvall, J. (1987). Environmental psychology in the Soviet Union. In D. Stokols & I. Altman (Eds.), *Handbook of environmental psychology, Vol. 2* (pp. 1311–1335). New York: John Wiley & Sons.

Noam, E. (1995). Electronics and the dim future of the university. *Science, 270*, 247–249.

Noschis, K. (1992). Child development theory and planning for neighborhood play. *Children's Environments, 9*, 3–9.

Novaco, R.W., Stokols, D., & Milanesi, L. (1990). Objective and subjective dimensions of travel impedance as determinants of commuting stress. *American Journal of Community Psychology, 18*, 231–257.

O'Donnell, M. P. (1994). Employers' financial perspective on health promotion. In M. P. O'Donnell & J. S. Harris (Eds.), *Health promotion in the workplace* (2nd ed., pp. 41–65)). Albany, New York: Delmar Publishers.

Ornstein, S. (1990). Linking environmental and industrial/organizational psychology. In C. L. Cooper & I. T. Robertson (Eds.), *International Review of Industrial and Organizational Psychology, Vol. 5* (pp. 195–228). Chichester, England: Wiley.

Oskamp, S., Harrington, M. J., Edwards, T. C., Sherwood, D. L., Okuda, S. M., & Swanson, D. C. (1991). Factors influencing household recycling behavior. *Environment and Behavior, 23*, 494–519.

Parmelee, P. A., & Lawton, M. P. (1990). The design of special environments for the aged. In J. E. Birren & K. W. Schaie (Eds.), *Handbook of the psychology of aging* (3rd ed., pp. 464–488). New York: Academic Press.

Pastalan, L. A. (1983). Environmental displacement: A literature reflecting old person-environment transactions. In G. D. Rowles & R. J. Ohta (Eds.), *Aging and milieu: Environmental perspectives on growing old* (pp. 189–203). New York: Academic Press.

Perkins, D. D., Wandersman, A., Rich, R. C., & Taylor, R. B. (1993). The physical environment of street crime: Defensible space, territoriality, and incivilities. *Journal of Environmental Psychology, 13*, 29–49.

Platt, J. (1973). Social traps. *American Psychologist, 28*, 641–651.

Pol, E. (1993). *Environmental psychology in Europe: From architectural psychology to green psychology*. Aldershot, UK: Avebury Press.

Preiser, W. F. E. (1988). A combined tactile-electronic guidance system for visually impaired persons in indoor and outdoor spaces. In E. Chigier (Ed.), *Design for disabled persons*. Tel Aviv, Israel: Freund Publishing House Ltd.

Prison-building binge in California casts shadow on higher education. *New York Times,* April 12, National Section, p. 1.

Proshansky, H. M. (1978). The city and self identity. *Environment and Behavior, 10*, 147–169.

Proshansky, H. M., Fabian, A. K., & Kaminoff, R. (1983). Place identity: Physical world socialization of the self. *Journal of Environmental Psychology, 3*, 57–83.

Quirk, M., & Wapner, S. (1995). Environmental psychology and health. *Environment and Behavior, 27*, 90–99.

Rapoport, A. (1980). Cross-cultural aspects of environmental design. In I. Altman, A. Rapoport, & J. F. Wohlwill (Eds.), *Environment and culture. Human behavior and environment: Advances in theory and research, Vol. 4* (pp. 7–26). New York: Plenum.

Riger, S. (1993). What's wrong with empowerment. *American Journal of Community Psychology, 21*, 279–292.

Rosen, R. (with L. Berger) (1991). *The healthy company: Eight strategies to develop people, productivity, and profits*. Los Angeles: J.P. Tarcher.

Saegert, S. (1987). Environmental psychology and social change. In D. Stokols & I. Altman (Eds.), *Handbook of environmental psychology, Vol. 1* (pp. 99–128). New York: John Wiley & Sons.

Saegert, S., & Winkel, G. H. (1990). Environmental psychology. *Annual Review of Psychology, 41*, 441–477.

Sanchez, E., Wiesenfeld, E., & Cronick, K. (1987). Environmental psychology from a Latin American perspective. In D. Stokols & I. Altman (Eds.), *Handbook of environmental psychology, Vol. 2* (pp. 1337–1357). New York: John Wiley & Sons.

Saunders, J. O. (Ed.). (1990). The legal challenge of sustainable development. *Calgary: Canadian Institute of Resources Law*, 15–34.

Silver, C. S., & DeFries, R. S. (1990). *One earth, one future: Our changing global environment*. Washington, DC: National Academy Press.

Slaby, R. (1992). Media influences on violence. In American Psychological Association (Ed.), *Report of the American Psychological Association Commission on Youth Violence*. Washington, DC: Edior.

Sommer, R. (1983). *Social design: Creating buildings with people in mind*. Englewood Cliffs, NJ: Prentice-Hall.

Sommer, R. (1987). Dreams, reality, and the future of environmental psychology. In D. Stokols & I. Altman (Eds.), *Handbook of environmental psychology, Vol. 2* (pp. 1489–1511). New York: John Wiley & Sons.

Special report: Rethinking work. *Business Week*, October 17, pp. 14 93.

Stern, P. C. (1992). Psychological dimensions of global environmental change. *Annual Review of Psychology, 43*, 269–302.

Stern, P. C., & Gardner, G. T. (1981). Psychological research and energy policy. *American Psychologist, 36*, 329–342.

Stern, P. C., Young, O. R., & Druckman, D. (Eds.) (1992). *Global environmental change: Understanding the human dimensions*. Washington, DC: National Academy Press.

Stokols, D. (1981). Group x place transactions: Some neglected issues in psychological research on settings. In D. Magnusson (Ed.), *Toward a psychology of situations: An interactional perspective* (pp. 393 415). Hillsdale, NJ: Erlbaum.

Stokols, D. (1986). A congruence analysis of human stress. In C. D. Spielberger & I. G. Sarason (Eds.), *Stress and anxiety, Vol. 10* (pp. 35–64). New York: John Wiley & Sons.

Stokols, D. (1987). Conceptual strategies of environmental psychology. In D. Stokols and I. Altman (Eds.), *Handbook of environmental psychology* (pp. 41–70). New York: John Wiley & Sons.

Stokols, D. (1990). Instrumental and spiritual views of people-environment relations. *American Psychologist, 45*, 641–646.

Stokols, D. (1992). Establishing and maintaining healthy environments: Toward a social ecology of health promotion. *American Psychologist, 47*, 6–22.

Stokols, D. (1995). The paradox of environmental psychology. *American Psychologist, 50*, 821–837.

Stokols, D. (in press). Environmental design and occupational health. In J.M. Stellman & C. Brabant (Eds.), *ILO Encyclopedia of Occupational Health and Safety, 4th Edition. Section IV, Psychosocial and organizational factors* (S.L. Sauter & L. Levi, Section Co-Editors). Geneva, Switzerland: International Labor Office.

Stokols, D., Pelletier, K., & Fielding, J. E. (1995). Integration of medical care and worksite health promotion. *Journal of the American Medical Association, 273*, 1136–1142.

Sundstrom, E. (1986). Workplaces: The psychology of the physical environment in offices and factories. New York: Cambridge University Press.

Susa, A. M., & Benedict, J. O. (1994). The effects of playground design on pretend play and divergent thinking. *Environment and Behavior, 26*, 560–579.

Taylor, S. E., & Fiske, S. T. (1978). Salience, attention, and attribution: Top of the head phenomena. In L. Berkowitz (Ed.), *Advances in experimental social psychology, Vol. 11* (pp. 249–288). New York: Academic Press.

Thorne, R., & Hall, R. (1987). Environmental psychology in Australia. In D. Stokols & I. Altman (Eds.), *Handbook of environmental psychology, Vol. 2* (pp. 1137–1153). NY: John Wiley.

Ulrich, R. S. (1983). Aesthetic and affective response to natural environment. In I. Altman & J. F. Wohlwill (Eds.), *Behavior and the natural environment. Human behavior and environment: Advances in theory and research, Vol. 6* (pp. 85–125). New York: Plenum.

Ulrich, R. S. (1991). Effects of interior design on wellness: Theory and recent scientific research. *Journal of Health Care Interior Design, 3*, 97–109.

United States Advisory Board on Child Abuse and Neglect (1995). *A nation's shame: Fatal child abuse and neglect in the United States*. Washington, DC: U.S. Government Printing Office.

United States Department of Health and Human Services USDHHS). (1991). *Healthy People 2000: National Health Promotion and Disease Prevention Objectives*. PHHS Publication No. (PHS) 91–50212. Washington, DC: U.S. Government Printing Office.

United States Department of Health and Human Services (USDHHS). (1993). *Preventing homicide in the workplace*. DHHS (NIOSH) Publication No. 93–109. Cincinnati, OH: Centers for Disease Control and Prevention, National Institute of Occupational Safety and Health.

Verbrugge, L. (1990). The iceberg of disability. In M. Stahl (Ed.), *The legacy of longevity* (pp. 55–75). Newbury Park, CA: Sage Publications.

Vila, B. (1994). A general paradigm for understanding criminal behavior: Extending evolutionary ecological theory. *Criminology, 32*, 311–359.

Wachs, T. D. (1992). *The nature of nurture*. Newbury Park, CA: Sage Publications.

Walsh, J., & Francis, S. (1992). *U.S. industrial outlook*. Washington, DC: Health Care Financing Administration, Office of the Actuary, U.S. Department of Commerce.

Wapner, S. (1981). Transactions of persons-in-environments: Some critical transitions. *Journal of Environmental Psychology, 1*, 223–239.

Wapner, S. (1987). *A holistic, developmental, systems-oriented environmental psychology: Some beginnings*. In D. Stokols & I. Altman (Eds.), Handbook of environmental psychology (Vol. 2, pp. 1433 1465). New York: Wiley.

Wapner, S. (1995). Toward integration: Environmental psychology in relation to other subfields of psychology. *Environment and Behavior, 27*, 9–32.

Wener, R., Frazier, W., & Farbstein, J. (1985). Three generations of evaluation and design of correctional facilities. *Environment and Behavior, 17*, 71–95.

Wicker, A. W. (1985). Getting out of our conceptual ruts: Strategies for expanding conceptual frameworks. *American Psychologist, 40*, 1094–1103.

Wiesenfeld, E. (1992). Public housing evaluation in Venezuela: A case study. *Journal of Environmental Psychology, 12*, 213–223.

Wiesenfeld, E. (Ed.). (1994). *Spanish American contributions to environmental psychology*. Caracas, Venezuela: Central University of Venezuela, Faculty of Humanities and Education.

Williams, A. F. (1982). Passive and active measures for controlling disease and injury: The role of health psychologists. *Health Psychology, 1*, 399–409.

Wineman, J. (Ed.). (1986). *Behavioral issues in office design*. New York: Van Nostrand Rinehold.

Winett, R. A., King, A. C., & Altman, D. G. (1989). *Health psychology and public health: An integrative approach*. New York: Pergamon Press.

World Resources Institute. (1994). *World resources, 1994–95*. New York: Oxford Univ. Press.

Zeisel, J. (1981). *Inquiring by design: Tools for environment-behavior rersearch*. New York: Cambridge University Press.

Zimring, C. M. (1989). Post occupancy evaluation and implicit theory: An overview. In W. Preiser (Ed.), *Building evaluation* (pp. 113–125). New York: Plenum.

Zimring, C. M., Reizenstein-Carpman, J., & Michelson, W. (1987). Design for special populations: Mentally retarded persons, children, and hospital visitors. In D. Stokols & I. Altman (Eds.), *Handbook of environmental psychology, Vol. 2* (pp. 919–949). New York: John Wiley & Sons.

Zube, E. H. (1991). Environmental psychology, global issues, and local landscape research. *Journal of Environmental Psychology, 11*, 321–334.

Zube, E. H., & Moore, G. T. (Eds.). (1991). *Advances in environment, behavior, and design, Vol. 3*. New York: Plenum.

Chapter **24**

TRANSACTIONAL PERSPECTIVE, DESIGN, AND "ARCHITECTURAL PLANNING RESEARCH" IN JAPAN

Kunio Funahashi

"ARCHITECTURAL PLANNING RESEARCH" IN JAPAN

People in Western cultures may have some curiosity about the phrase "architectural planning," since the concept of "architecture" has not generally been associated with that of "planning." However, the term "architectural planning" is the literal equivalent of *kenchiku keikaku*, which has been a strong research field in Japanese architecture for some time now. Unlike in Western society, the concept of architecture and its studies in Japan have technological characteristics because kenchiku (architecture) belongs to the field of engineering rather than to art and design.[1] The "Architectural Planning Research" (APR) organization has mainly been developed in the academic circles of the Architectural Institute of Japan (AIJ, estab-

---

[1]The reasons *kenchiku* (architecture) in Japan belongs to "engineering" are supposed as follows: After the Meiji Restoration in 1868, Japan hastened her modernization, even if in the way of the *wakon yosai* (Japanese mind, Western technology)." As to architecture, for example, the policy was directed not to make an "architect" in the Western meaning but a master builder who could build new buildings in European fashion in those days. And Japan has often suffered serious damage from natural disasters such as earthquakes or typhoons, and big fires because of wooden construction and dense habitation. We inevitably paid strong attention to technological aspects of building that directly related to making structures resistant to these physical forces. Then, the system of "architecture" should be constituted so as to include all of the technology required for the building processes.

Kunio Funahashi • Department of Architectural Engineering, Faculty of Engineering, Osaka University, Yamadaoka, Suita 565, Japan.

Handbook of Japan–United States Environment–Behavior Research: Toward a Transactional Approach, edited by Seymour Wapner, Jack Demick, Takiji Yamamoto, and Takashi Takahashi. Plenum Press, New York, 1997.

lished in 1886), and has the same attributes. The term "planning" was preferred to "design" because of the former's scholarly sound. By the terminology of the Western architectural society, the concept of *keikaku* (planning) in Japan seems to be rather similar to "programming."[2]

Today, APR in Japan consists of three fields. One concerns the technology of production, such as design methods, planning of building construction, and management; the other two fields are related, in a broad sense, to the relationships between "architecture and human beings." They include mainstream "life and space" studies classified by building type, based on an amalgam of sociology, home economics, ergonomics, etc.; and issues focusing on fundamental aspects in any building type such as theories of design, human behavior, safety and security, etc., which has been influenced by a gamut of sciences. By the multifaceted nature of architecture itself, the study should naturally include relatively broad fields and issues; but particularly in Japan, however, the reasons for focusing on "life and space" in APR involve historical and societal aspects.

Japan has had at least two periods of rapid and complete change, or "modernization" of society, in the past century, that is, the Meiji Restoration (1868) and the defeat of the Second World War (1945). The imported or imitated modernization, or actually a large degree of confusion from serious change in social system and lifestyle, hastened to establish the new concept as a way of life and social system. In the architectural world as well, the society asked to develop a new building method, technology, and in particular new planning/design theory applicable for the new social demands and/or confusing lifestyle because the conventional wisdom in the traditional building industry would not work well under such total change.

On the other hand, although the field has been called "architectural" planning, most studies were curiously limited to the issue of "plan" configuration. This is one of the interesting features of APR, even though the "plan" is the primary means in architectural thinking. In spite of the name, the APR does not deal with many other aspects of architecture. This bias came from the tendency of scholars in the learned society, such as the AIJ, not to pursue multidisciplinary research but rather to strive by themselves even on sociocultural issues. The reasons for this isolation could be attributed to the generally spreading nonconcern for the physical environment among Japanese social scientists in those days and the persistent notorious sectionalism in Japanese academic circles.

Notable epochs forming the history of the development of APR will now be reviewed.

A Brief Historical Sketch of the Development in "Architectural Planning Research"

The Embryonic Stage

In Japan, "scientific" research in architectural planning seems to have started in the late 1920s. One of the first well-organized studies in those days was the series

---

[2]There is no satisfactory Japanese equivalent for "programming." Same for "design."

of 26 volumes of "Advanced Study on Architecture," which had 11 volumes named *kenchiku keikaku* (architectural planning) and dealt with the planning data and methods of many building types. One volume, subtitled "Principles of Planning," consisted of two parts: fundamental studies on interior environmental engineering such as illumination, acoustics, heating and cooling, and air pollution; and basic general studies on the decision-making of requirements for rooms and of their scales, shapes, and arrangements.

At that time, the modern movement of architecture in Western cultures exerted more influence on Japan. For example, some people went to the Bauhaus and/or studied at Le Corbusier; also many works from European countries were introduced in terms of "modern design" in general. One of the strongest and everlasting influences on plan configuration came from Klein (1927), who proposed an evaluation method for small dwelling units from the economy of movements of inhabitants' points of view (i.e., such as the concept of "Verkehrsweg").

In the middle part of the 1930s, Fujii and Yokoyama started their painstaking work on the anthropometric nature of the human body. Also at that time, the German "Bauentwurfs Lehre" by Neufert (1936) was introduced to Japan. Based on these studies and influences, the AIJ began to edit a new Japanese edition from 1937, which was something akin to the *Architectural Graphic Standard* in the United Staes, and finally published as three volumes of *Compilation of Design Data for Architecture* from 1942 to 1952.

The Work of Nishiyama (1911–1994)

Nishiyama (e.g., 1946) started his studies on dwelling houses for the working class as his dissertation at Kyoto University in 1933. It has been admired as an epoch-making paper, uncompleted, only the beginning of his lasting work about the housing problem and planning in Japan. In this and following papers, he developed a position to study the planning of the dwelling house. The important aspects of his research in terms of current environment behavior studies (EBS) could be summarized as follows:

1. He advocated the ethical position that the architect and the scholar of architecture had to tackle the housing problem of the poor or low-income class, which needed to be solved and improved as soon as possible. Today, this argument probably sounds reasonable but, at that time, the elite architect and/or student of architecture did not have any interest in such a field because there were no jobs working with those classes of people.

2. Not only was the theme original but the research methodology was, in contrast to previous speculative ways, revolutionary, that is, adopting field surveys on the "use" of existing houses through observations, questionnaires, behavioral mapping, interviews, etc. conducted on large samples. This method was named a *"sumai-kata* (way of living) survey."

Through actual surveys, Nishiyama aimed at pointing out the users' needs, the "contradiction" between users' needs and given spaces, and the general developmental tendencies of needs, lifestyles, and latent consciousness. His ideological position was fundamentally based on socialism so he paid positive attention to differences among socioeconomic aspects of phenomena, especially class, locality, and historical background.

Sometimes he suffered attacks from opponents that his work had nothing to do with the design of houses. However, he replied that the understanding of people's lifestyles and the development of new housing policies were the most important primary research issues that should be tackled by scholars.

The Work of Yasumi Yoshitake (1916-)

Yoshitake (e.g., 1956) started his main work just after World War II. He has been, in a sense, a follower of Nishiyama, but it would be more exact to say that he adopted another position. That is, his position on architectural planning was that "research is a (relatively direct) standpoint of design" (p. 1). He introduced more "scientific" methods such as stochastic mathematics of data processing and consecutive investigations into the same type of facility and also expanded his research targets to some public institutional buildings such as schools, hospitals, and libraries, as well as housing.

This most important methodological statement of research is summarized as follows: ". . . in order to grasp the people's needs including the latent aspects too, we noticed that we could access to this critical point by the field survey of existing state of use. There should be a certain lawful tendency in that state, i.e., the interaction between given space and human being. If the tendency could not be maintained, there would appear a certain contradiction which means the difficulty to use..." (p. 1). This has been generally called "*tsukaware-kata* (what the space is used for) studies" (Yoshitake, 1956, p. 1)

Some researchers, including students of Yoshitake, also enlarged their scopes to include the "regional planning" of facilities. This involved issues of the total capacity of each kind of institutional building in a certain neighborhood, of the catchment area of each facility, and of the optimum distribution or site selection of a particular facility, etc.

The extensive results by Yoshitake's school were arranged according to the system of building types. The results were also developed into a kind of design guidelines, which were incorporated in the revisions of "Compilation of Design Data for Architecture" in 1960 and again in 1978. These studies worked well at a time when design guidelines for many programs were hastily required to rebuild physical environments after World War II; they introduced a parsimonious standard because of the relatively lower and less varied level of needs relative to today.

Even though their works have been in the mainstream of APR in Japan until today, the work of Nishiyama, Yoshitake, and their followers was criticized for depending heavily on the survey ("surveyism"); such surveys seemed quite easy at first sight to naive followers. In this methodology, there is the logical contradiction of using results extracted from the survey of existing environments for planning the future world. In spite of these criticisms, the literature contains numerous reports of case studies on many facilities, but there is neither any systematic organization of the information retrieval system nor any theoretical framework that could serve to direct research programs.

The Work of Takashi Adachi (1919–1993)

Adachi (e.g., 1959) has not been paid much attention by other scholars, though his work started around 1950. His original interest was in the philosophy of architecture, but he supported his philosophical speculation with work from the humanities and social sciences. At that time, in Japan and probably in the Western countries, the modern movement of architecture was moving to a new age, that is, to the revival of humanity, the reconsideration of the "functionalist" tradition, etc.

Adachi's position was in "the extension of the functional way of thinking." Specifically, his aim was the comprehensive understanding of the nature of human beings. To access such understandings, he adopted a psychological point of view as a prospective position. Through his critical discussions, two kinds of research trends in APR were identified as follows:

1. Studies from the point of view of the science of labor, ergonomics, human factors analysis, and so on may be a necessity but are too physical in nature (cf. the experiments of the Hawthorne Factory).
2. Studies from the social science point of view, such as is typical in the work of Nishiyama (phenomena or successive changes of social behavior relevant to environments are reported but look superficial; or sometimes the whole phenomena are put down to the contradiction between capital and labor) might be true but we need more in-depth research.

Adachi and his students studied such themes as spatial cognition, meanings of and preference for place, territoriality, crowding, privacy and communication, ecological study of older adults' lives, path choice and way-finding, psychophysiological studies, and so on. Unfortunately, these studies have not necessarily produced concrete and/or voluminous results because many of these studies were pilot studies and not successively followed up. However, through the introduction of EBS to Japan in 1970, Adachi's pioneering work has received some attention from more recent scholars.

AN APPROACH TO DESIGN IN TRANSACTIONAL PERSPECTIVE

Differences between APR and EBS

Based on historical developments and the resulting characteristics of APR in Japan, some differences between APR and EBS can be described as follows.

First, "*tsukaware-kata* (what the space is used for) studies" in APR bear some similarity to the Post occupancy evaluation. However, the former tends to be more generality-oriented, applicable to the type of building, and moderate in direct intervention than the latter, which seems to be facility-oriented toward improving problems found through evaluation processes.

TABLE 24.1. Comparisons between Architectural Planning Research (APR) and
Environment Behavior Studies (EBS)

APR	EBS
"tsukaware-kata (what the space is used for) Studies" are: Generality-oriented, applicable to the type of building, and moderate in direct intervention into a specific case.	Post occupancy evaluation is: Specific facility-oriented.
The concept of "life" is based on: The functionalist tradition, which has focused mainly on the everyday activities in the scope of their types, spatiotemporal structures, and extent of smooth performance.	The concept of "behavior" tends to encompass: Not only functional aspects but also the psychology, perception, cognition, and social interaction needs of people, subcultural differences in lifestyle, and meaning and symbolism of environments.
The "environment" implies: "Architectural" term referred only to "physical" or "spatial" factors, e.g., the shape, dimension, and/or arrangement of any architectural components.	The "environment" implies: The reality of whole environments including sociocultural aspects surrounding us in our everyday life.
The people–environment relation is: Deterministic and/or interactional.	The people–environment relation is: Of the transactional perspective as a leading thought.

Second, "*tsukaware-kata* (what the space is used for) studies," which have
been the mainstream of APR, have been investigating the relations between people's lives and architectural conditions in terms of key propositions such as "architecture and human being" or "life and space." Here, architectural conditions mean,
for example, the shape, dimension, and/or arrangement of any architectural components. The concept of people's lives has been based on the functionalist tradition, namely, with a focus mainly on everyday activities in terms of types,
spatiotemporal structures, and extent of smooth performance. In contrast, the
concept of behavior in EBS, which aims to improve the "quality of life," tends to
encompass not only functional aspects (as in APR), but also the psychology, perception, cognition, and the social interactional needs of people, cultural differences in
lifestyle, and meaning and symbolism of environments. Therefore, between APR
and EBS, there are differences in the purpose of both research and practical intervention to solve problems. We would say that EBS attempts to examine the relation
between human spiritual richness/nobleness and the environment.

Third, the "environment" in APR implies the "architectural" environment,
which has referred only to "physical" or "spatial" factors. Even if it succeeded in
clarifying the effects of such environmental factors on human life, the results would
be limited abstractions of the total reality of our lives. The meanings of physical
factors could be well conceived in terms of the comprehensive consideration of
whole aspects of environments. Or how should we understand the concept of

so-called architectural or physical factors? EBS intends to approach closely the reality of whole environments, including sociocultural aspects that surround us in our everyday lives.

Last, in APR, the relationship between architectural conditions and human life has been typically conceived of as deterministic and/or interactional. In particular, from the standpoint of "architectural creation" in functionalist thought, deterministic attitudes would naturally come from our expectations of the effects of architecture on human life. However, it is obvious that such architectural determinism may have serious limitations in underemphasizing the mutual interdependence of people and environments. Thus, we might introduce the transactional perspective on people environment relations from EBS to APR.

Transactional Perspective

There has been much work on the transactional perspective and several discussions on its generality that have focused on the representativeness of situations, the reliability and validity of measurements, and questions of transcontextual generality (Altman, 1990, 1992; Altman & Rogoff, 1987; Altam et al., 1987; Kaplan, 1987; Saegart & Winkel, 1990; Wapner, 1987). However, it would be sufficient for our purpose here to summarize the nature of the transactional perspective, from Altman (1992) as follows:

1. People and psychological processes are embedded in and inseparable from their physical and social contexts. Thus psychological phenomena are treated as holistic units rather than combinations of separate elements.
2. Time, continuity, and change are intrinsic and central aspects of psychological phenomena.
3. The philosophy of science of transactional perspectives involves the study of unique events along with recurring events, understanding phenomena from the view of different types of observers and participants, and a formal cause approach to understanding.

If the nature of the transactional perspective is understood as above, what implications does this have for APR?

What Is Design and How to Design?

What Is Design?

For EBS, the building of architecture does not have a self-evident purpose because architecture as environment is just one of many aspects for improving quality of life. Further, EBS does not primarily aim to provide data that are "useful to design," but attempts to demonstrate the direction and principles of environmental formation based on comprehensive understanding of people environment relationships including the total development of human beings.

Compared with the interactional and/or deterministic views that are pervasive in the architectural design and urban planning, the transactional perspective stresses people's active interventions in environments not only with respect to physical or functional aspects, but also in terms of the significance or interpretation of environments. Although it is not yet clear what produces and pushes the human toward active intervention in environments, it might be, in Maslow's terms, the human aspiration for self-esteem or self-actualization. If active intervention in environments is possible and human beings consciously act and make their surroundings, then the world is well conceived of as the product, not the cause of, our perception, cognition, and actions. Therefore, the concept of "environment-behavior" *has to be expanded to include "environment, behavior, and design (EBD, by this author),*" *reflecting the fact that humans are the only organisms who flexibly, self-consciously, and massively alter their environment* (Ittelson, 1989).

According to conventional understanding of APR, what is designed is usually referred as to the "architectural or environmental" design. This attitude often views the physical components of settings as independent factors for achieving desired effects on the behavior and well-being of users (Gee, 1994; Stokols, 1988). In the transactional perspective, what is designed means the planning of the total system including everyday human life. Environmental design, which basically means to alter or not to alter the affordance of environment, is or should be, then, the expansion of the possibility of human self-actualization.

With respect to the design processes such as programming, planning, design, and occupancy, EBD would encompass all processes from the very generation of the environment to ceaseless repetitive renewals of it. Now, if these understandings are taken for granted, temporality is important, ceaseless changes and uncertainty of system are essential, and homeostasis, preestablished harmony, or ultimate purpose are not possible. Given people environment relations under everlasting changes and in an uncertain world, is the planning/design of "architecture" possible? In what way? Even though we can distinguish software design from hardware design, which is based on the idea of physically ridden attitudes in APR, architecture as physical existence has a certain firmness and permanency. Thus, in our transactions with such firm material existence, extremes are imagined. On the one hand, we try to make architecture as flexible and weakened as possible toward easy transactions. This is the architecture of $mu,$[3] metaphorically. On the other hand, we have inevitably to change ourselves and/or the meanings given to environments. We could say that, in a sense, such eventual changes imply an aspect of human development. If such changes are conceptualized as continuing development, architecture should be made firmer and more permanent, resisting our attempts to intervene in environments, which foster human development through immutability. This is called the Architecture of being.

Situations and Goals of Design

People have designed their environments. As a result, we live in cultural landscapes (Rapoport, 1990) and have to share rules and schemes of the people environment system in a certain culture (subculture) of time and locus. Any design can only be in it.

Design in the contemporary world can be understood as neither activity in the closed circle of designer and relevant people nor activity separate from the surrounding

---

[3]MU in Japanese which is originated in the Taoism by Lao Tzu and Chung Tzu literally means nothing or nothingness, but still does imply ultimate essential being.

world. Contemporary design activity is situated in the whole world and then becomes meaningful. This is a matter of certainty because all human behavior and knowledge are embedded in the surrounding situation. This condition of design is inevitable for the mutual complex interconnectedness among parts in contemporary society. The situations surrounding design have multidimensionality and multivariety, in particular, when examined against a transactional perspective (Saegert, 1993).

When contemporary design activity is regarded as "situated" in the world and also people's active and emergent intervention to any environments are emphasized, the possibility of a "universal theory of design"[4] is highly improbable and the only "individual design theory" valid for the specific situation and its solution become available. If this is the case, we are compelled to say that only a pragmatic approach remains as our design method, which eclectically employs all prospective means including contingent and emergent phenomena and people happened to be about. This method is consonant with cultural relativism but against the ordinary generality or universality that a theory should have. A "designer," compared even with a researcher who acknowledges effects on the people environment system that has been studied, must be more to the system that is designed because his or her presence and intervention affect inevitable decisions, which are based on his or her specific position about the people-environment system, for example, in terms of the emergence of new aspects of the system during the design intervention (Farbstein & Kantrowitz, 1991). How could we identify what is the "specific situation and its solution" if change in the people environment system of contemporary society is essential and if its speed, forms, and aspects are diverse? How can we limit and describe the situation, in a heuristic but not *post factum* way, which is situated in the whole world? The meanings of "solution" become a big issue as well. The evaluation of "solution" seems to be related to the goals of design. In the situation postulated here, however, what on earth could be the goal of design or even of architecture? Moreover, what could be the criterion of the accomplishment of a goal? Design decision postulates value. Although common values include "improving the quality of life" and the "expansion of self-actualization," who could decide the appropriateness of design at present, under the multivariety and multidimensional situations in which change is essential and human active intervention is rather natural? Can we find any universal normative dimensions such as Lynch (1981) described? As the goal of design, could we accept approaching the state of "good" or "must" that is claimed by "anyone" or "everybody?"

We have been longing for "good" architecture. In what and in which state is the architecture "good?" From the narrow point of view of APR, it means the "fit or noncontradiction" between human behavior and the physical environment. However, what is "fit" or what significance could there be in pursuing the "fit?" We have to ask how and what architecture should be from the viewpoint of rightness, authenticity, and/or sustainability.

Last, who designs, the collaboration or division of work, and the contents and styles of intervention of researcher professionals will be raised. These issues are

---

[4]W. H. Ittelson, (1989) wrote as follows: "theories . . . will help us analyze, predict, or otherwise explain the conditions of human existence, and, in doing so, provide the means and the understanding necessary to direct the processes of change" (p. 83)

relevant to the professionalism in advanced civil society and to the issue of the relationship between theory and practice in design. A variety of methodologies, particularly citizen-participation, has been pursued for learning the wisdom of people who will inhabit the environment that is designed. We need to elaborate our knowledge about the appropriateness of procedures such as who, when, and how to participate in design (Winkel, 1993). Of course the participation or collaboration itself does not mean any unconditional good. "One community" perspective presented by Min (1988), which intends to conquer the pervasive planner-user dichotomy, is interesting and suggestive in terms of the worldview of design.

REFERENCES

Adachi, T. (1959). *A study on extension of "function" in design/planning.* Dissertation for Degree of Engineering, Kyoto University.

Altman, I. (1992). A transactional perspective on transition to new environments. *Environment and Behavior, 24*(2), 269–280.

Altman, I. (1990). Toward a transactional perspective: A personal journey. In I. Altman & K. Christensen (Eds.), *Environment and behavior studies: Emergence of intellectual traditions* (pp. 225–255). New York: Plenum.

Altman, I., & Rogoff, B. (1987). World views in psychology: Trait, interactional, organismic and transactional perspectives. In D. Stokols & I. Altman (Eds.), *Handbook of environmental psychology* (pp. 7–40). New York: John Wiley & Sons.

Altman, I., Werner, C. M., Oxley, D., & Haggard, L. M. (1987). "Christmas Street" as an example of transactionally oriented research. *Environment and Behavior, 19*(4), 501–524.

Farbstein J., & Kantrowitz, M. (1991). Design research in the swamp: Toward a new paradigm. In E. H. Zube & G. T. Moore (Eds.), *Advances in environment, behavior, and design, Vol. 3* (pp. 297–318). New York: Plenum.

Gee, M. (1994). Questioning the concept of the 'user.' *Journal of Environmental Psychology. 14*, 113–124.

Ittelson, W. H. (1989). Notes on theory in environment and behavior research. In E. Zube & G. T. Moore (Eds.), *Advances in environment, behavior, and design, Vol.2.* (pp. 71–83). New York: Plenum.

Kaplan, R. (1987). Validity in environment/behavior research: Some cross-paradigm concerns. *Environment and Behavior, 19*(4), 495–500.

Klein, A. (1927). Neues Verfahren zur Untersuchung von Kleinwohnungs-grundrissen. *Stadtbau, 23*, 16–21.

Lynch, K. (1981). *A theory of good city form.* Cambridge, MA: MIT Press.

Min, B. (1988). *Research utilization in environment-behavior studies: A case study analysis of the interaction of utilization models, content, and success.* Dissertation for Ph.D., The University of Wisconsin-Milwaukee.

Nishiyama, U. (1946). A *study on commoners' house.* Dissertation for Degree of Engineering, Kyoto University.

Rapoport, A. (1990). *History and precedent in environmental design.* New York: Plenum.

Saegert, S. (1993). Charged context: Difference, emotion and power in environmental design research. *Architecture & Comportement/Architecture & Behaviour, 9*(1), 69–84.

Saegert, S., & Winkel, G. H. (1990). Environmental psychology. In *Annual review of psychology* (pp. 441–477). Polo Alto, CA: Annual Review.

Stokols, D. (1988). Instrumental and spiritual views of people-environment relations. In H. V. Hoogdam, N. L. Prak, T. J. M. Voordt, & H. B. R. Wegen (Eds.), *Looking back to the future, The Proceedings of IAPS 10* (pp. 29–43).

Wapner, S. (1987). A holistic, developmental, systems-oriented environmental psychology: Some beginnings. In D. Stokols & I. Altman (Eds.), *Handbook of environmental psychology* (pp. 1433–1465). New York: John Wiley & Sons.

Winkel, G. H. (1993). Environmental design evaluation as a change oriented research process. *Architecture & Comportement/Architecture & Behaviour, 9*(1), 85–98.

Yoshitake, Y. (1956). *A study on* tsukaware-kata [what space is used for]. Dissertation for Degree of Engineering, Univesity of Tokyo.

Chapter **25**

SOME ARGUMENTS FOR A COMPARATIVE DEVELOPMENTAL ENVIRONMENTAL PSYCHOLOGY WITH A LONG-TERM VIEW OF HISTORY AND CULTURAL PSYCHOLOGY

George Rand

Paraphrasing poet Paul Valery, "Europe is but a small peninsula of Asia."

In a review of the recent Mondrian exhibit, *New York Times* critic Herbert Muschamp expressed amazement that designer Martha Stewart, known for her lush and decorous lifestyle, bought for herself to live in a white, stripped-down, sparsely decorated house designed by ultramodern architect Gordon Bunschaft. Why would a person committed to popularizing diverse folk motifs, exploring the limits of colorful, emotional presentation of food and social style, take personal comfort in the mental space of a generalized Platonic abstraction, that is, a modernist home environment?

A similar logical conundrum applied in the birth of modernism at the turn of the twentieth century. Flat roofs and ribbon windows in architecture, global progressivism in education, positivism in philosophy, all introduced at the same time,

George Rand • Department of Architecture and Urban Design, School of Arts and Architecture, University of California-Los Angeles, Los Angeles, California 90024.

Handbook of Japan–United States Environment–Behavior Research: Toward a Transactional Approach, edited by Seymour Wapner, Jack Demick, Takiji Yamamoto, and Takashi Takahashi. Plenum Press, New York, 1997.

seemed to provide relief from the stifling, doctrinaire views of history, mixed with conjectural explanations about ethics and morality, that had come to dominate the end of the nineteenth century. Modernism also brought with it other dividends in a sense of personal freedom, an ethos of creativity and personal agency, and a reflexive view of the self derived from the new institutions of modernity rather than what were perceived to be the outmoded and outworn institutions of traditional societies.

Western social sciences were part of this larger modernization process. They participated actively in the "retreat into the present" (see Elias, 1991). This retreat, as it turns out, also gave comfort by sidestepping the human existential uncertainty that comes with recognizing the diverse sociocultural environments in which meanings and social or mental "worlds" are constituted. In this "today-centered" philosophy, modern psychology has come to measure its achievements by calculating utilitarian usefulness in correcting the ills of modern society. The birth of functionalism in psychology involved studying human behavior as if it were composed of an abstractly designed system of "well-meshing parts." These atomic parts were viewed as if they could be understood apart from a community of persons that has a set of mental representations and beliefs . Both the behaviorists and the phenomenologists denied the embeddedness of the psyche in cultural and/or social history. Psychology implicitly adopted in this universal position a distinctly Western European outlook. It presumed that behavior of the individual or a group could be understood without reference to the collective past, or the comparative development of other cultural traditions.

It is only in recent years, with the advent of poststructural theories, that interest has returned regarding the parallel paths taken by different cultures in constructing intentional worlds. Processes of functional adaptation and selection take on differential significance within a view of psychology that recognizes the potential significance of multicultural sources. I contend that part of the mission of the field of environmental psychology—as a contextual discipline—could be to rewrite psychological theories on top of a background of the long-term history of the person in his or her social, cultural, and geographic environment, including the comparative analysis of non-Western environments. This involves going from Eurocentric "functional laws" to a comparative developmental framework that examines the long-term evolution of person–environment relations within and across diverse cultural settings. Since there is no requirement that the identity of things remain universal across intentional worlds, the hope to decipher universal laws of person–environment interactions may be misleading and even distort the evidence.

Studies in environmental psychology in the seventies were culturally disengaged. They were primarily concerned with amassing generalizable functional data about person–environment relations on topics like crowding, personal space, environmental cognition, and environmental stress. Emphasis was given to multidisciplinary studies combining urban planning, architecture, sociology, and personality and social psychology. Even the landmark publication of the *Handbook of Environmental Psychology* fell short of examining cultural differences; although the volumes included surveys representing the international scope of the field, they merely conveyed data from researchers in different nations rather than probing these contexts in a comparative manner, as parallel paths in the evolution and history of person–environment relations.

It is ironic that these studies followed the model set out by "General Psychology." They engaged in a search for universal processing mechanisms for person–environment relations, in effect minimizing the impact of the cultural environment on behavior. In other words, the work perpetuated the idea of the division between intrinsic psychological mechanisms and extrinsic environmental conditions, based on the metaphor of "heredity" versus "environment."

Some contextualist theories came close to seeing the challenge to consider in a fresh way the issue of how institutional structure, culture, and context shape the person–environment relationship. For example, when Wicker's (1987) extension of Barker's behavior setting research tracked settings, exploring their life course as they develop, he revealed something about the interactive process in which person–environment relations are constructed, at least in a Western context. In the end, his augmented behavior setting theory added little to understanding the comparative evolution of settings and somewhat perpetuated the idea that all cultures are similarly constituted. It cast results in functional terms, and for this reason, lacked both a poetic vision of the arc of history, and an understanding of the play of culture in formulating and evolving settings. Also, his examination of settings presumes normative conditions of participation. There is little emphasis on "otherness" in these studies, or on the "politics of difference."

A few studies regarding shared uses of space under conditions of conflict are cited in the literature (e.g., the negotiation of territorial boundaries among warring groups in Belfast or South Africa). In these cases there is little reference to historic examples in the distant past (e.g., cosmopolitan cultures and the history of empires). One of the few studies to ground sociology in history was Lyn Lofland's (1971) study of the history of cities in which she compares the "appearential order" of the traditional European city to the "spatial order" of the modern city. This had great promise for incorporating a long view of the developmental history culture into the study of social space, but her thesis remained entirely Western and European. Also, at the time it was written in 1971 few studies had begun to embrace the concept of "difference" as a central concern. Even transactional theories (cf. Irwin Altman's research), which emphasized person–environment relationships to be a "two-way" street, and were based on comparative anthropology, tended to opt for a broadly conceived functional solution, and imposed a structuralist (atemporal, universal) interpretation upon the data.

Only in older anthropological theories of person–environment relations such as E. T. Hall's (1987) seminal works, and the more recent explorations in "ethnospace" by David Stea, David Canter, and colleagues (Canter et al., 1988), is there a hint of interest in "how" different sociocultural environments actually become different by virtue of the way they are differently constituted by different peoples in different settings.

Amos Rapoport's work is unique in this regard and is the most significant attempt to outline a historical anthropology of person–environment relations (see Rapoport, 1990). He explores the "long-term" historic and developmental aspects of person–environment relations and form–function relationships across cultures.

For example, in one trenchant citation regarding causes for the curved shapes of streets in North African villages, he notes historic studies that derived the origins of these sinoidal paths and their lack of ninety-degree intersections as related to the perceptual and behavioral characteristics of camels. The original idea was derived from rounding off the hard, ninety-degree corners of intersections in order to avoid sudden encounters that frightened the animals. Social identification with movement of pack animals through dense urban environments may have produced a liking for these features by descendents, long after cities prohibited access by pack animals to the in-town spaces that made them necessary in the first place. Of course, the derivation of "typologies" of urban space is more complex than can be accounted for with a single factor. But the point is that "typological features" persist long after their functional basis has disappeared. This forces use of an "abductive logic," one that operates by functional and formal analogies, through use of poetic, subconscious processes, in addition to inductive and deductive logics. Using this approach, we can hypothesize that culturally sanctioned person–environment relationships may have been transmitted from the Middle East by traveling merchants, reproduced in medieval Venice, and come to life as an influence in the rest of Italy and Europe, contributing uniquely curved forms to medieval streets and piazzas. As these street shapes become part of the larger "habitus mentalis," they were incorporated into the culture in the form of more abstract values, beliefs, and classification systems, as well as related social habits.

To summarize, one aspect of contemporary environmental psychology that needs to be addressed is the presumption that there is an underlying person–environment order that can be understood without reference to their constitution in relation to culture, or to the long-term history and evolution of person–environment patterns in comparative cultural contexts. With institutions of modernity came a disembedding of the person from traditional connections to a set of social habits that had evolved over a long time, that is, in relation to a specific locale, cultural context, etc. Inasmuch as the modernist view takes as its basis Enlightenment science, it denies the importance of intuitions and social habits that remain linked to a particular history, e.g., a set of regional, local, or peasant traditions, that continued to play an unconscious role despite the imposition of a modernist ideology and the categories of international culture.

MODERN VERSUS TRADITIONAL CULTURE IN THE UNITED STATES, EUROPE, AND ASIA

It is odd to have to call attention to the common international culture of the world trilateral partners—the United States, Europe, and Japan. We have become comfortable with global stereotypes based on modernity that link them to one another as parties in a single world-system. In the more local perspective of a cultural psychology, differences between these cultures can be highlighted rather than always focusing on the common culture.

Within modern European nation-states, there are residues of earlier forms of authority and the control and institutional power and violence that formed them.

Political scientist Robert Putnam's (1988) work regarding the history of political institutions in modern Italy suggests that the living character of contemporary person–environment relations can be traced back to long-term historical interactions that occurred in cultural regions, in his case studies of regions in Italy. Historically, collective mental life differed from one part of the Italian peninsula to another. In the north, villages were organized as a series of decentralized city-states, relatively prosperous and autonomous of one another. This area created a culture with a sense of personal agency and powerful local institutions. it is probably the basis for the current economic successes of ("Third Italy") that region, which some analysts see as rivaling the Japanese miracle, but based on completely different cultural habits and forms of organization. Areas in the vicinity of Rome were historically under the hierarchical authority of the Papacy. This developed into a community with habits leading to the formation of an integrated national government in contrast to a decentralized system of authority. Finally, the area of greater Sicily to the south had been organized much in the model of the caliphate Arab state, that is, with a combination of strict controls from above and considerable autonomy from below, and as a result, sought to maintain a position that combined the independence of a nation as a state within Italy. These diverse historical influences persist in the present inasmuch as they shape the values, belief systems, cognitive categories, and other aspects of culture. They are related to the comfort and style with which citizens approach social and political participation.

In a general series of studies, political scientist Francis Fukuyama (1995) extends this argument by showing that lingering "social habits," that is, ways of relating to the social and institutional environment, differ among contemporary modern cultures. The modern Chinese culture, he suggests, evolved under the influence of the Mandarin system and behaves differently from Japanese culture, which was formed under the influence of Tokugawa constructs, and, finally, the European cultures, still emerging from their legacy in the feudal state, have developed their own modes of constituting their social world.

What this suggests is that data on person–environment relations may not be as generalizable as hoped for by normative social sciences. It may require more of a person-oriented, ideographic approach that takes into account the comparative history of the environment, as well as its politics and economics. The past impact of general systems theory has been to examine all systems through a single biological conception of feedback, control, equilibrium, and homeostasis. There does not seem to be room in these normative notions for operations of power, the applications of surveillance and control by an elite, and ultimately the uses of social control and violence as a means of maintaining social order.

Particularly in relation to racial and ethnic divisions, there does not seem to be any means for understanding how constructions of person–environment relations can be used to restrict access by one sector of society to resources, or used in the interest of preserving power of another group. At the very least, there is a need for a theory which applies differently in the case of traditional versus modern social systems (see Talcott Parsons' [1991] "systems theory" for the modernist framework regarding social systems, and Weber's sociology for traditional social systems based on religion prior to the evolution of the bureaucratic state).

In the late-modern context, all cultures have come to recognize the paradox of the coexistence of the modernist global culture based on universal rules with traditionally based local cultures with their own individual identity. The dynamic interaction between local and global conceptions of culture in psychology has become the distinctive feature of postmodernism. It is now commonplace to ac- knowledge that CNN engenders global world views that then transform local cul- tural conditions. For example, the urban elite in Bangkok or Bombay now requires settings to support a cosmopolitan international society that is identical to the parallel societies in New York or London. These global person–environment norms found in the context of the Third World operate side by side with those of from the traditional culture. Similarly, large sections of modern Western cities like New York and London have been set aside to support traditional lifestyles (as in ethnic enclaves or immigrant communities) that resemble settings in the Third World, and are integrated into the culture in close proximity to modern settings.

Cultural Ideals

Imagine how the underlying ordering suggested by environmental psychology is seen through non-Western European and/or non-North American eyes. It might help to explicitly try to understand how the background presence of the European nation-state shapes perception of the world system within the European nation- state and, further, to contrast this perspective with assumptions based on those of more traditional class-segmented societies. It would also be useful to see the Anglo-Saxon cultural ideal (i.e., that of a hierarchical monoculture) as a particular way of constructing a culture based on an abstraction from the classical ideal.

The problem is that because of the lack of historical and comparative analysis, there is rarely an explicit discussion of these cultural ideals and their relationship to underlying scientific premises of the field. Even though we may agree that Western culture has produced a remarkable series of achievements, it is important to recast understanding in a comparative context of the global history of civilization.

Some aspects of traditional Japanese society, although non-Western, may be much closer to European monocultures than the culture of the United States. North American culture was partly severed from its feudal history. The feudal form that dominated Western Europe is not conspicuous in the United States, although it remains part of the political debate regarding fear of "big government." The United States, in this sense, has no legacy of monarchy to fall back upon. After generations of immigration, the only thing that is truly common among the myriad ethnicities and cultures is a vaguely held symbolic image related to the spirit of freedom and conveyed by the Statue of Liberty. It is not surprising, then, that a cultural unbraid- ing may be taking place, threatening the appearances of national unity. No longer is the issue of nationalism and citizenship something that can be presumed to exist based on the "oath of allegiance." There are multiple layers and claims regarding membership and identity defined in terms of geography, history, ethnicity, and culture. The authority of law is what holds the current system together.

There are examples of this phenomenon in my own work on "border" phe- nomena, e.g., studies of environmental adaptation of Korean and other Asian migrants to Los Angeles, or long-standing patterns of cultural mixing across the broad U.S.–Mexican border. For example, the U.S.–Mexican border region extends deeply into both countries, and in order to understand person–environment relations under these "border conditions," reference has to be made to traditional systems in which an administrative orderings coexist with local orderings at the level of individual communities. National politics may have little bearing on daily life in these local communities; they operate according to local rules. For example, new subcultures have been formed along the border—perennial commuters, "winter residents"—and each develops its own strategies for construing and dealing with the local environment. These adaptations are person–environment "designs" developed as local–global adaptations. These designs may not endure for very long, and they are usually transformed through political, social, and environmental interaction. The laws, regulations, and institutions of the state, and their respective global norms, coexist with local mores and institutions.

What the example suggests is that social systems and geographies are always in a process of dynamic change. In the Southwest, the boundaries of the United States were not fully resolved until 100 years ago, and even today, the influence of the indigenous Chiapas rebels in southern Mexico is having impacts far beyond its region. The class-segmentation that structures these local systems is different in character from structures in the modern global social world. This mix of modern and traditional systems seems to be quite commonplace, especially in Southern California, where we have loosely defined cultural frontiers rather than precisely fixed international boundaries.

Of course, beyond lines in space that serve as international borders, there are other conceptual boundaries that are less visible to outsiders. For example, in Los Angeles' Fairfax district there is an extensive orthodox Jewish community with a well-defined spatial zone that is only partly visible to the nonorthodox Jewish occupants. These invisible, reinstitutionalized parts of the city are commonplace examples of the impact of welfare on spce.

Culturally Based Environmental Psychology

Much of human behavior in relation to the environment is a product of culture and history and may be refractory to analysis in terms of discrete functional "factors" or "variables." There is an interesting story told about Bosnia from a work by Alfred van Cleef (1994), *The Lost World of the Berberovic Family*. A village is out of the way of the war that is coming out of Belgrade and Zagreb. The Serbs, Croats, and Muslims have lived here in relative harmony for years. The war now moves from two valleys over to one valley away, but the community continues to live in harmony. Then one day a young man is blown away, shot by a militiaman in the public square. Those over 50 years old instantly realize that the person who has been shot is the son of a partisan who killed the uncle of the shooter back in World War II. Then in a series of regressive moves, history is invoked as a justification for gruesome retaliatory actions in which a lot of people are killed and the village is drawn into the larger war. As time passes, the logical justification for war reaches deeper and deeper into the past, and so the present comes to resemble a heroic myth rooted in ancient cultural world views more than in a simple modernist setting.

This "fable" might be applied to any environmental interaction in which an event causes a breach to occur with the overtly conscious, canonic character of the modern environment. We see this in clinical work all the time, where the night dream that reveals the unconscious can upset or transform conscious reality. Or in the desperate retaliatory actions involved in gang drive-by shootings based on some deeply tribal fear that is evoked. There is no shortage of observations of these phenomena operating in seemingly modern settings.

It seems to me something is missing that bears on person–environment relations in traditional versus modern settings. First, that psychology can be looked at differently in the context of traditional social environments requires adding to the neutral field of environmental psychology the perspective of a cultural psychology. This is similar to the work of Amos Rapoport, who views the environment in all its dimensions, across all world cultures, and at levels ranging from the vernacular to high culture. Theory needs to be framed in a way that goes beyond the restrictions of contemporary modernism. It is easier to convert modernist terms into psychological variables suitable for laboratory research, but they often distort the meaning of the person–environment relationship, especially when they lack understanding of the shaping influence of their historic context.

To develop some of these ideas, I refer in a superficial way to the theory by Seymour Wapner concerning the developmental approach to environmental psychology. Wapner's theory specifies physical, interpersonal, sociocultural levels of the person involved in forming a holistic understanding of person–environment relationships (see Stokols & Altman, 1987). Taking "person–environment" as a unit of analysis, he examines different intersections of the person (biological-physical, interpersonal, sociocultural levels) and the environment (physical, interpersonal, sociocultural aspects).* Developmentally, we can expect to see changes in the direction of increasing differentiation and hierarchic integration with respect to this person–environment relationship.

The theory presumes there are typical processes of adaptation in which person–environment relationships proceed toward increasing adjustment or adaptation. Adaptive to what? The problem with this assumption is that it is apolitical and ahistorical. It examines processes within the cultural background of European modernism, and takes this framework to be universal. It therefore assumes a benign sociocultural context, and that processes of development are free to differentiate and integrate without competition, force, violence, or oppression from above (see Lefebvre, 1991). For example, it makes no statements about propaganda processes involved in advertising, or forms of adaptation that are prescribed or achieved by restricting access to information. Finally, it makes no statements regarding conflicting systems, such as religious traditions which have their own agendas and use their own subconscious metaphors to sway their flock. The simplest examples today

---

*Compare with the types of classification used by Rapoport for organizing knowledge about typologies of person–environment settings.

regard the choice between the modern definition of marriage and family and the traditional role of filial piety and family commitment.

The theory has great value for studying situations in which there are a range adaptations of people in critical transitions, such as the foreign sojourner coming to the United States from abroad and adopting a new host culture. This is true as long as the coherence and completeness of the host culture are clear and a concern with "difference" is not an issue. If the host culture, say, becomes wrenched out of balance by the massive influx of guest workers, or then reacts repressively as in the case of California's Proposition 187, all bets are off. To understand this dynamic would require expanding the seemingly neutral theory in ways that incorporate a more political or historical interpretation, e.g., including a theory about institutional racism, competition for labor resources, positioning guest workers within certain sectors of the city, etc.

A related example about the history of culture regards the Jews in Eurasia during the late nineteenth century. They were perceived as pariahs, and many Russian Jewish peasants were subject to violent Cossack raids. Like other oppressed minorities throughout the world, they basically had three choices: (1) migrate elsewhere, (2) organize themselves to oppose these oppressive forces, or (3) assimilate themselves within the dominant culture. Any and all of these choices evidence developmental patterns of increasing differentiation and hierarchical integration; but they are likely to have very different consequences for the historic development of culture and identity, as well as for the relationship of the local culture with surrounding cultures.

SOME TENTATIVE CONCLUSIONS

Perhaps it would be useful to think of a developmental environmental psychology that takes as normative the potential complexity introduced by taking one road or another at major change points in history—and focuses on this choice process as part of cultural development—rather than resting the theory on a series of phases that lay out "history" as if it were progressively revealing a "cultural core," in a Platonic or Aristotelian sense. This would be a true environmental psychology inasmuch as it shows how critical choices in relation to the environment shape the long-term course of history.

What we would like to propose, then, is that developmental environmental theory acknowledge that inconsistency and conflicting person–environment norms are inherent in all person–environment relations. Interesting transitions occur when conflicting norms, or more trenchantly, when establishing person–environment equilibrium, involve changing the dominant society. Transforming the person–environment matrix involves more than optimizing person–environment equilibrium. It may mean creating emergent new structures. These new designs may involve democratic means (e.g., political actions taken through forming networks and coalitions related to the environment), as well as nontraditional means (e.g., the spread of religion, communication through trade, immigration, colonization, and military occupation).

Take the case of Sun Yat Sen, who was born in China, developed an interest in Protestant religion, later migrated to Hawaii where he received religious training, then returned to China as head of state and was able to create new blends of old and new. Or the case of Peter the Great, who single-handedly brought European structures of government and social life to the Russian elite. If person–environment theory focuses on past examples that evince these contradictions, it may be able to invent new ways of studying contemporary settings as exhibiting conflicting and overlapping norms. This would bring a new focus to the field, one that examines person–environment relations as processes of representing and defending contrasting ways of construing the world. This approach would contrast with the dominant view that sees the process as one of optimization of person–environment equilibrium as the goal.

As we approach the end of the millennium, the question is whether it is possible to build a multicultural view of psychology.

The theory should be expanded in two ways:

1. Following the work of Robert Putnam and others, it should recognize that different social groups, ethnicities, and nationalities may have different sets of person–environment norms and social habits, and, as a result, different ways of characterizing ideal developmental processes, particularly those related to person–environment relations.
2. Rather than restrict theory to a unilinear process of development, it may be useful to structure discourse around development as multicultural processes, for example, similar to the mix of cultures that existed between Spain and the New World, or in contemporary Mexico and the Southwest United States.

It is therefore most interesting to engage in a Japan–USA conference on Environmental Psychology in which there are contrasting ideas about person–environment relationships. Is it possible to treat these perspectives as a border condition, one in which there exist a variety of cross-translations and back-translations. Perhaps this interface is most pregnant with information about person–environment relationships based on the ability to think about them through "others."

Can we also take person–environment relationships into new understandings of history, including the dynamics of race and class (e.g., those of nation, class, gender, and race)? For example, would the issue of "adaptation" of African-Americans to the current physical, interpersonal, and sociocultural environment take the theory into unknown territory?

These factors themselves might be developmentally ordered. For example, perhaps organizing social systems on the basis of "group rights" (especially those based on race) as a means of claiming equity is arguably more primitive. Building a nation-state that includes diverse ethnic and racial groups may be more "advanced," and also rarely achievable; only cases of Mexico, Brazil, and Cuba* offer stable examples, but each with their own problems. Social systems can be looked at independently based on their complexity, adaptability, leadership and organization,

*See Angel Pacheco's (Lucca-Irizarry & Pacheco, 1992) idea of "newyoricans" as an example of "queer" status that is likely to lead to more complex categories, and be more globally adaptive than the traditional system of Puerto Rican family development.

values, politics, reciprocity, and affectivity.* To achieve advanced systems, such as those that rely on greater individual choice, may require more complex systems of organization than the hierarchical networks that are currently in place. New forms of distributed networks can be coordinated with centrally organized command and control systems.

Finally, are there instances in which cultures can be made sufficiently "foreign" so that participants can feel themselves to be on the border between the idealizing formulae of Anglo-Saxon cultural values and "outsider" views from other cultures? Can this hypothetical framework be used to study long-term reactions to cataclysmic events such as the Bosnian war, or traumatic reactions to the atomic bombing of Hiroshima, taken as a multicultural phenomenon. Such a multicultural characterization might focus on events as seen through the lens of different cultures, for example, the active organization of a collective response leading to empowerment, or a passive stance leading to acceptance. Empowerment leads to perception of still more opportunities for organization and instrumental change; acceptance leads to other forms of adjustment that are based on more metaphysical beliefs, values and cultural categories. Added attention to the poetics of developmental processes will make it possible to acknowledge the many paths taken by cultures during their lifespan.

REFERENCES

Betsky, A. (1990). *Violated perfection: Architecture and the fragmentation of the modern*. New York: Rizzoli.

Canter, D., Krampen, M., & Stea, D. (1988). *Ethnoscapes*. Brookfield, VT: Avebury.

Elias, N. (1991). *The society of individuals*. Oxford: Blackwell.

Fukuyama, F. (1995). *Trust: Social virtues and the creation of prosperity*. New York: Free Press.

Hall, E. T. (1987). *Hidden differences*. Garden City, New York: Anchor.

Lefebvre, H. (1991). *The production of space*. Cambridge, MA: Blackwell.

Lofland, L. (1971). *A world of strangers*. New York: Basic Books.

Lucca-Irizarry, N., & Pacheco, A. (1992). *Environment and Behavior, 24*(2).

Parsons, T. (1991). *The social system*. London: Routledge.

Putnam, R. (1988). Institutional performance and political culture. *Governance: An International Journal of Policy and Administration, 1*(3).

Rapoport, A. (1990). *History and precedent in environment design.*. New York: Plenum.

van Cleef, A. (1994). *De verboren wereld van de familie Berberovic,[The lost world of the Berberovic family.]* Amsterdam: Menlenholf.

Wapner, S. (1987). A holistic, developmental, systems-oriented environmental psychology: Some beginning. In D. Stokols & I. Altman (Eds.), *Handbook of Environmental Psychology* (pp. 1433–1465). New York: Wiley.

Wachsler, L. (1995, November 20). Vermeer painted when war had torn Europe apart. *New Yorker*, 71(37), 56.

Wicker, A. (1987). In D. Stokols & I. Altman (Eds.), *Handbook of environmental psychology*. New York: Wiley.

*Here we recommend a combination of the categories used by self-psychology psychanalytic theory on the mirror self. This is also related to issues of gender and projection in the sense used by Aaron Betsky (1990) in justifying that "queer" space be seen as a higher achievement than more well-ordered urban spaces.

Chapter **26**

PROSPECTS FOR ENVIRONMENTAL PSYCHOLOGY IN THE THIRD MILLENNIUM

Kenneth H. Craik

Stokols (Chapter 23, this volume) has recognized the approach of the 21st century as a useful occasion to take stock of our field and its condition. I will organize my comments along a temporal dimension as well, with brief observations on forecasting trends in environment–behavior relations, forecasting trends in environmental psychology as a field of inquiry, and experiencing the current status of our field.

FORECASTING TRENDS IN ENVIRONMENT BEHAVIOR RELATIONS

Technology as an Engine of Discontinuity in Environment Behavior Relations

Stokols (1995, and Chapter 23, this volume) is quite right in identifying technological innovations as a likely source of changes in future behavior environment relations. One way to gauge the role of technology is to locate ourselves at earlier points in time. Because we now approach a new century, we might imagine a conference on the prospects for our field having taken place in the late 1890s at

Kenneth H. Craik • Institute of Personality and Social Research, University of California-Berkeley, Berkeley, California 94720.

Handbook of Japan–United States Environment–Behavior Research: Toward a Transactional Approach, edited by Seymour Wapner, Jack Demick, Takiji Yamamoto, and Takashi Takahashi. Plenum Press, New York, 1997.

Clark University. But the year 2001 in the Western calendar will designate not only the advent of a new century but also of a new millennium. Consequently, we might also imaginatively situate ourselves in the 1990s, perhaps at a gathering in the cathedral school at Reims led by the great teacher Gerbert (946–1003) (soon to be Pope Silvester II in 1001) (Lyon, 1964).

What forecasts about technology would have been predictively accurate at those two hypothetical gatherings? In his *History of Invention*, Williams (1987) lists 159 major technological innovations for the entire 7,000-year period from 6,000 B.C. through 1000 A.D., and, with regard to our two eras of interest, he cites 162 innovations for the 900 year period from 1001 to 1900, and 170 for the most recent 85 year period from 1901 through 1985. We have no reason to expect this accelerated pace of technological change to diminish in the future or to lessen its role in generating research issues for environmental psychology. In the recent past, for example, the production and delivery of electricity has been a prime instigation for research on the assessment and prediction of scenic quality impacts due to surface coal mining, electric power transmission lines, and the like, and on public perceptions of technological risks associated with nuclear power, storage of liquified natural gas, and so on (Canter, & Craik, 1981).

As the year 1001 A.D. approached, alarmed anticipations of the end of the world were widespread in Western Europe, derived from Biblical sources, a severe climate change in the 900s featuring cold winters and poor crops, an eruption of Mount Vesuvius in 993, and a conflagration that burnt much of Rome soon thereafter (Lyon, 1964, pp. 34–39). Without doubt, research on contemporary worldviews concerning perceptions of environment, technology, and society are worthy of assessment and monitoring as we move into the 21st century (Buss, Craik, & Dake, 1986). For example, Biosphere 2, a project linked to the technological possibilities of human space travel and colonization, appears to be motivated by a worldview that foresees a probable desperate need for new human environments beyond the earth (Bechtel et al., Chapter 16, this volume).

Sources of Continuity in Environment-Behavior Relations

Stokols has wisely granted as much attention to continuities as to discontinuities in his forecasts of environment–behavior relations. He places his primary emphasis upon relatively short-term continuities, extrapolating from societal forces already underway and clearly discernible (e.g., environmental pollution, an aging population).

Another source of long-term continuity in environment-behavior relations derives from the evolutionary past of our species (Buss, 1995). Indeed, as Rapoport (Chapter 28, this volume) notes, the implications of evolutionary concepts for environmental psychology are important but have been neglected.

For example, although the reconstruction of hominid behavioral evolution remains a daunting task (Tooby & DeVore, 1987), humans certainly appear to have lived in small groups from the Pleistocene environment onward (Cosmides & Tooby, 1989; Tooby & Cosmides, 1989). Even within modern urban environments, persons continue to live within their own relatively small, potentially specifiable idiographic communities (Craik, 1985, 1996b; Emler, 1990).

Thus, our own relatively small lifelong social networks represent a *continuity* (Kochen, 1989). The discontinuity flows from societal and technological change. In medieval Europe, for example, peasants were tied to their manor and not allowed to leave their area to travel or relocate without explicit consent of the lord. During the subsequent rise of the new villages, peasants who managed to reside in a given village for a year and a day without being claimed by a lord as a serf won their freedom (Nicholas, 1992). Our rights to residential mobility and travel have expanded steadily, along with facilitating technological innovations, such as the automobile, train, and airplane, on the one hand, and the telegram, telephone, e-mail, and Internet on the other. Thus, the *discontinuity* is that we can now enjoy community without propinquity (Webber, 1963).

This particular theme of continuity and discontinuity can be found in the reports by Bechtel and Michelson. In Bechtel's analysis of Biosphere 2, the seemingly isolated Biospherians actually had access to telephone, FAX, and e-mail networks, which dramatically expanded their personal communities beyond the seven other mission members. Michelson (Chapter 11, this volume) applied time use analysis to examine the perceived risk to the personal security of workers, especially women, whose employment schedules have dramatically expanded the range of times and places of their daily travels, dispersing their social networks far beyond the familiar residential neighborhood.

FORECASTING TRENDS IN ENVIRONMENTAL PSYCHOLOGY

Its Intellectual Scientific Structure: A Core Paradigm

Two decades ago, the intellectual-scientific structure of environmental psychology struck me as easy to discern and chart (Craik, 1977). In brief, a number of long-established research paradigms from mainstream psychology had found interesting puzzles to address in the arena of environment–behavior studies, mobilizing their respective Kuhnian packages of concepts, units of analysis, exemplary achievements, and methods (Kuhn, 1962; Masterman, 1970). This invasion of the paradigms accounted for the quickly attained, high level of research productivity of the emerging environmental psychology, as well as some vocal disenchantment with its several mutually esoteric research styles and its lack of conceptual coherence. Two decades later, the current situation is much more complex, featuring cross-paradigmatic research, an enormous range of applied studies, an exciting international expansion, and many important impacts and reverberations back onto mainstream psychology (Canter, Craik, & Griffiths, 1984; Craik, 1984–85, 1987; Stokols, 1995; Wapner, 1995).

The overall research program of environmental psychology is without doubt remarkably rich, thick, and extensive, but what is the nature of its present and future intellectual-scientific structure? My own conclusion is that environmental psychology has now attained paradigmatic status *in its own right*, and furthermore, it is organized around one of the handful of *core paradigms* of contemporary scientific psychology (Craik, 1996a).

The core paradigms of psychology are centered upon key notions, such as the ideas of (1) human evolution; (2) the human biological organism; (3) cognitive, affective and other psychological processes; (4) life-span development; (5) the person as a self-reflexive agent of integrated action; (6) the individual in social interaction; and finally, environmental psychology's focal problem, (7) the person situated in and transacting with the environment. Most other areas of psychological research constitute applied, specialized, or technical aspects or spinoffs of these core paradigms, as perusal of the subject index of any recent program for an American Psychological Association meeting will attest.

From this perspective, a different implication is found for Stokols' (1995; this volume) concern about the transparency of environmental psychology, that is, the possibility that the incorporation of person–environment components by other fields of research will somehow diffuse environmental psychology's intellectual-scientific identity. On the contrary, every core paradigm manifests this transparency; it is a clue to environmental psychology's centrality rather than a hint of its imminent dispersal or demise.

Promoting and Teaching Person–Environment Theorizing

One basic implication of this perspective upon the current and future intellectual-scientific structure of environmental psychology is that theorizing about person environment transactions is the fundamental activity of our field. Our goal should not be simply to adjudicate or select among past or current conceptual solutions to this problem but rather to provide a continuing forum for seeking new and creative formulations (Rapoport, Chapter 28, this volume; Walsh, Craik, & Price, 1992). In addition, our graduate programs should not only thoroughly review available conceptual resources (e.g., Altman & Rogoff, 1987) but also provide practice and apprenticeship in person–environment theorizing as an active process (Maddi, 1980; Stokols, 1995; Wicker, 1985). The future vitality of environmental psychology depends the recognition of this distinction between a static approach to theories as products and a creative orientation to theorizing as an ongoing process.

From a broad temporal vantage point, such matters as institutional support, research funding, and applied opportunities are important but must be placed in proper perspective. These resources may wax and wane over the years, but the abiding core problem actively and persistently addressed by environmental psychologists will remain a basic and central scientific concern. For example, specific urgent theoretical challenges currently include the meshing of person environment concepts within a life-span developmental perspective (Caspi & Moffitt, 1993; Wapner, 1987) and the detailed articulation of person–environment transactions within a quotidian analysis of the lived days of individuals (Craik, 1993, 1994).

The era of the late 1930s and early 1940s enjoyed a creative conceptual focus on person–environment relations (e.g., Brunswik, 1943; Lewin, 1936; Murray, 1938) that failed, however, to generate a paradigmatic research program. The sustained accomplishments and lively current condition of environmental psychology are now much more encouraging with regard to its prospects for the next century and beyond.

EXPERIENCING PRESENT-DAY ENVIRONMENTAL PSYCHOLOGY

Experiencing environmental psychology here and now through its current journals, books, and conferences generates a vivid impression of its tremendous variety and range of topics, applications, and research approaches. All of them are not only heuristically valuable in enlightening our ongoing formulations of person environment transactions but also worthy of individual appreciation in their own right.

One example must suffice. The application of environmental simulation has been oriented primarily towards the future, either in forecasting responses to proposed environmental projects for planning, design, and research purposes (Bosselmann & Craik, 1987; Marans & Stokols, 1993; Craik, 1995) or in envisioning more imaginative scenarios for entertainment (Dykstra, 1977). However, Kubota (Chapter 13, this volume) presents an ingenious and promising use of environmental simulation in order to computer encode historical scenes from paintings and photographs. This clever approach will allow us to capture, manipulate, and apply some of the wisdom of past generations concerning the small-grain, detailed amenities that are often lacking in our contemporary urban and countryside settings.

Experiencing present-day environmental psychology also continues to be influenced by its status as an understaffed scientific behavior setting (Craik, 1973). On the one hand, discouragement can arise from the paucity of concerted, strong science inquiry with regard to any particular topic. On the other hand, the field persists in offering newcomers the benefits of heightened responsibility and quick assimilation, diversity of activities and central roles, and the sense of obligation, involvement, and self-worth characteristic of understaffed settings (Barker, 1968; Wicker, 1979). These attractions, one trusts, will continue to sustain environmental psychology as an essential core paradigmatic field of contemporary scientific psychology.

ACKNOWLEDGMENTS

Portions of this chapter were adapted from Craik (1996a).

REFERENCES

Altman, I., & Rogoff, B. (1987). World views in psychology and environmental psychology: Trait, interactional, organismic, and transactional perspectives. In D. Stokols, & I. Altman (Eds.), *Handbook of environmental psychology: Vol. 1* (pp. 7–40). New York: Wiley.

Barker, R. G. (1968). *Ecological psychology: Concepts and methods for studying the environment of human behavior.* Stanford, CA: Stanford University Press.

Bosselmann, P., & Craik, K. H. (1987). Perceptual simulations of environments. In R. B. Bechtel, Robert W. Marans, & W. Michelson (Eds.), *Methods in environmental and behavioral research* (pp. 162–190). New York: Van Nostrand Reinhold.

Brunswik, E. (1943). Organismic achievement and environmental probability. *Psychological Review, 50*, 255–272.

Buss, D. M. (1995). Evolutionary psychology: A new paradigm for psychological science. *Psychological Inquiry, 6*, 1–30.

Buss, D. M., Craik, K. H., & Dake, K. M. (1986). Contemporary worldviews and perception of the technological system. In V. T. Covello, J. Menkes, & J. Mumpower (Eds.), *Risk evaluation and management* (pp. 93–130). New York: Plenum.

Canter, D. V., & Craik, K. H. (1981). Environmental psychology. *Journal of Environmental Psychology, 1*, 1–12.

Canter, D., Craik, K. H., & Griffiths, I. (1984). Environmental bridge-building. *Journal of Environmental Psychology, 4*, 1–5.

Caspi, A., & Moffitt, T. E. (1993). When do individual differences matter? A paradoxical theory of personality coherence. *Psychological Inquiry, 4*, 247–271.

Cosmides, L., & Tooby, J. (1989). Evolutionary psychology and the generation of culture, Part II. Case study: A computational theory of social exchange. *Ethology and Sociobiology, 10*, 51–97.

Craik, K. H. (1973). Environmental psychology. *Annual Review of Psychology, 24*, 401–422.

Craik, K. H. (1977). Multiple scientific paradigms in environmental psychology. *International Journal of Psychology, 12*, 147–157.

Craik, K. H. (1984–85). The international context of environmental psychology: Comments on Professor Yamamoto's paper. *Hiroshima Forum for Psychology, 10*, 70–71.

Craik, K. H. (1987). Environmental psychology and its international character. *Psychologie Francaise, 32*, 17–21.

Craik, K. H. (1985). Multiple perceived personalities: A neglected consistency issue. In E. E. Roskam (Ed.), *Measurement and personality assessment* (pp. 333–338). New York: Elsevier Science.

Craik, K. H. (1993). Accentuated, revealed, and quotidian personalities. *Psychological Inquiry, 4*, 278–281.

Craik, K. H. (1994). Manifestations of individual differences in personality within everyday environments. In D. Bartussek & M. Amelang (Eds.), *Fortschritte der Differentiellen Psychologie und Psychologischen Diagnostik: Festschrift zum 60. Geburtstag von Kurt Pawlik* (pp. 19–26). Göttingen: Hogrefe.

Craik, K. H. (1995). Environmental simulation and 'The Real Thing,' review of R. W. Marans, & D. Stokols (Eds.), Environmental simulation: Research and policy. *Contemporary Psychology, 40*, 779–780.

Craik, K. H. (1996a). *Environmental psychology*: A core field within psychological science. *American Psychologist, 51*.

Craik, K. H. (1996b). The objectivity of persons and their lives: A noble dream for personality psychology? *Psychological Inquiry, 7*, 326–330.

Dykstra, J. (1977). Miniature and mechanical special effects for "Star Wars." *American Cinematographer, 58*, 702–705, 732, 742, 750–757.

Emler, N. (1990). A social psychology of reputation. In W. Stroebe (Ed.), *European review of social psychology: Vol. 1* (pp. 171–193). New York: Wiley.

Kochen, M. (Ed.). (1989). *The small world*. Norwood, NJ: Ablex.

Kuhn, T. S. (1962). *The structure of scientific revolutions*. Chicago, IL: University of Chicago Press.

Lewin, K. (1936). *Principles of topological psychology*. New York: McGraw.

Lyon, B. D. (Ed.). (1964). *The high middle ages: 1000–1200*. New York: Free Press of Glencoe.

Maddi, S. R. (1980). The uses of theorizing in personology. In E. Staub (Ed.), *Personality: Basic aspects and current research* (pp. 333–375). Englewood Cliffs, NJ: Prentice-Hall.

Marans, R. W., & Stokols, D. (Eds.). (1993). *Environmental simulation: Research and policy issues*. New York: Plenum.

Masterman, M. (1970). The nature of a paradigm. In I. Lakatos, & A. Musgrave (Eds.), *Criticism and the growth of knowledge* (pp. 59–89). Cambridge: Cambridge University Press.

Murray, H. A. (1938). Proposals for a theory of personality. In H.A. Murray, *Explorations in personality* (pp. 36–141). New York: Oxford University Press.

Nicholas, D. (1992). *The evolution of the medieval world: Society, government and thought, 312–1500*. London: Longman.

Stokols, D. (1995). The paradox of environmental psychology. *American Psychologist, 50*, 821–837.

Tooby, J., & Cosmides, L. (1989). Evolutionary psychology and the generation of culture, Part I: Theoretical considerations. *Ethology and Sociobiology, 10*, 29–49.

Tooby, J., & DeVore, I. (1987). The reconstruction of hominid behavioral evolution through strategic modeling. In W. G. Kinzey (Ed.), *The evolution of human behavior: Primate models*. Albany, NY: SUNY Press.

Walsh, W. B., Craik, K. H., & Price, R. H. (Eds.). (1992). *Person-environment psychology: Models and perspectives*. Hillsdale, NJ: Erlbaum.

Wapner, S. (1987). A holistic, developmental, systems-oriented environmental psychology: Some beginnings. In D. Stokols, & I. Altman, (Eds.), *Handbook of environmental psychology, Vol. 2* (pp. 1433–1465). New York: Wiley.

Wapner, S. (1995). Toward integration: Environmental psychology in relation to other subfields of psychology. *Environment and Behavior, 27*, 9–32.

Webber, M. (1963). Order in diversity: Community without propinquity. In L. Wingo (Ed.), *Cities and space: The future of land use: Essays from the Fourth RFF Forum* (pp. 23–56). Baltimore, MD: Johns Hopkins University Press.

Wicker, A. W. (1979). *An introduction to ecological psychology*. Monterey, CA: Books/Cole.

Wicker, A. W. (1985). Getting out of our conceptual ruts: Strategies for expanding conceptual frameworks. *American Psychologist, 40*, 1094–1103.

Williams, T. I. (1987). *The history of invention*. London: MacDonald & Co.

Chapter **27**

WHAT IS THE SITUATION?
A Comment on the Fourth Japan–USA Seminar on Environment–Behavior Research

Susan Saegert

INTRODUCTION

Environmental psychology more than other aspects of psychology takes the situation to be of fundamental interest. By situation I mean the actual physical specifications of a person in a place with other people or alone, and also the convergence of cultural, social, and economic possibilities and constraints that constitute the meaning of the situation. Bruner (1990) explicates the meaning saturated nature of situations when he argues that the subject matter of psychology concerns the narratives that make sense in a culture. This argument, however, poses challenges for the cross-cultural understanding of such narratives, as exemplified in the 20 years of collaboration and exchange among Japanese and U.S. environmental psychologists.

The first exchange in Tokyo, Japan, made the difficulty of such narratives clear by calling attention to differences of culture and place among us even as it attempted to discover conceptual groundings for shared development of research and theory. But much has changed since then, the differences among us seem to concern more diversity of topics studied, and the different theoretical and methodological decisions that guide our work. Several researchers compared Japanese

Susan Saegert • Department of Environmental Psychology, Graduate School and University Center of the City University of New York, New York, New York 10036.

Handbook of Japan–United States Environment–Behavior Research: Toward a Transactional Approach, edited by Seymour Wapner, Jack Demick, Takiji Yamamoto, and Takashi Takahashi. Plenum Press, New York, 1997.

and U.S. settings, finding both differences between cultures (Demick, Ishii, & Inoue, Chapter 7) and places (Wapner, Fujimoto, Imamichi, Inoue, & Towes, Chapter 20) and similarities (Asai & Hirata, Chapter 18). The important change was that both differences and similarities were framed in a common narrative tradition.

The convergence of narrative across culture and place coupled with substantive and methodological divergence raises questions about how both place and culture situate meaning and about the nature of the field that supposedly contains the various papers. Rapoport argues that at present the topical and methodological divergence leads to accumulation of research rather than to cumulative knowledge. He attributes this state of affairs in part to lack of unifying theory that would be *general*, *abstract*, and would *avoid folk terms*. By adopting Bruner's concept of a psychological theory *about* folk psychology, and even more so, Valsiner's (1987) assertion of the cultural/folk psychological roots of psychological theory, I suggest that the problems Rappaport alludes to are not to be solved by constructing generalized, abstract, and "nonfolk" narratives. Rather, theory should help us understand the properties of generalizing narratives vis-à-vis those that make increasingly specific the places, people, and interactions in which they have meaning.

My proposal is that in both cases we situate knowledge within the discourse among persons, and in places. For a discourse to be general, abstract, and "nonfolk," it must escape culture and place. To some extent, the settings and practices of international scientific exchanges provide a place to be without place and to speak with minimal reference to a particular culture's narrative. The very existence of texts, in this case the papers that make up the basis for discussion, free meaning from a specific culture and author.* The Japan–U.S. seminar has achieved a level of generality, abstraction, and freedom from culture, by virtue of the situation and because of the unified theoretical nature of the work. But unlike narratives grounded in particular places, the narratives that make sense within the shared knowledge base often lack content specificity, not because they are "abstract" or "nonfolk" but because they are removed from both the physical and cultural world in which the research takes place.

Rapoport also calls for theory to be grounded in a broad synthesis of scientific literature. The interdisciplinary nature of environment–behavior studies seems instead to have led to a rather narrow shared literature, complemented by each author's own particular base of disciplinary and topical knowledge. To some extent, these shared and specialized knowledge bases constitute the cultural basis for the generation of narratives. But the specialized knowledge each author brings to the creation of narrative is disjoined from much of the shared environment–behavior literature. This disjuncture results not just from the lack of an adequate literature

---

*The import of this observation has been developed within the hermeneutic tradition c.f. as analyzed by Ricour (1991). Ricour argues that discourse also frees itself from the situation when it is understood as a more serious project than mere conversation. My argument, however, is not fully grounded in the debate that Ricour enters. While Ricour speaks of the reading of the text as an opening of the self to the world, the world lacks much concrete reality. My concerns are with the very palpable, obdurate, as well as ephemeral, qualities of the experienced world, and with the situational nature of understanding these qualities.

review, or a proper synthesis, but from the fact that authors stand in relationship to different literatures that teach different forms and contents for acceptable narratives.

Accepting the idea that research papers are narratives but that in the field of environment and behavior studies we can not assume that these narratives have a shared cultural base allows us to understand the problems Rapaport addresses somewhat differently. The desire for a commonly interpretable narrative replaces the goals of making papers more theoretical, abstract, and nonfolk. The call for better and more synthetic literature reviews is replaced with the idea of situating ourselves in definitions of problems that both convey the contribution of different literatures and of the spatial/psychosocial/cultural context in which the research took place. In Rapaport's system, good research rests on and contributes to, on the one hand, a theoretical base, and on the other, a literature that can be definitively synthesized. In stressing the interpretive and communicative aspects of research, particular scientific literatures are seen as guides to the construction of narratives, that on the one hand, lend them coherence, and on the other, close them off from narratives based in other conventions.

Rapaport's project and mine have, however, rather common intentions behind them. He emphasizes theory generation and synthesis. My goals for the field would be to develop narrative that can comprehend a wide variety of settings, topics, and audiences. The limits of generality from my point of view are not simply deficits in the unifying narrative, but also an inescapable characteristic of the situatedness of human experience. Since it is situatedness itself that seems to be the topic of this field, it should not be ruled out but rather given a conceptual space in which to assert itself. The following section explores differences in situatedness of the seminar papers with the goal of understanding the generality and limitations of different theoretical and methodological approaches to research.

SOURCES OF GENERALITY AND SPECIFICITY

The canonical form of a research question is akin to the occasions that Bruner indicates are the most likely to give rise to narrative. That is, an expectation of how a culturally appropriate sequence of events would occur, stated in terms of theory and previous research, is disrupted by the introduction of an event for which expectations are not so strong. If the event is too close to the expected sequence, the study would seem to add nothing. If the event is too outrageous, it seems to lack credibility within the general narrative framework established. In this set of papers, those drawing on organismic developmental theory most often took this canonical form. Most participants were familiar with the theory, and even more specifically, with the theory as it is understood in English. Thus texts having this form tend to succeed in being general. Their limitations were more in an inability to be informative about the specific situatedness of phenomena.

For example, Asai and Hirota illustrated how similar institutional goals, structures, and practices in two different cultures could generate somewhat similar responses, from a constrained set of possible responses. The authors noted the fact that residents were in both cases unwilling or unable to conform to the norms of the culture and that the institutions had as their purposes controlling or correcting this unwillingness. This in itself is interesting from the point of view of culturally appropriate narratives. However, previous research had found differences between Japanese and U.S. Americans in the extent to which they were amenable to social control, especially along dimensions which from a U.S. point of view have to do with individuality and autonomy. So the possibility of cultural rather than institutional generality of response was also plausible. The results of the study confirmed that the constraints of these institutions were powerful enough to transcend culture, but not necessarily in the ways that my U.S. perspective would think most likely. Perceived autonomy was strongly related to incarceration. However, perceptions of being controlled by others for Japanese residents of correctional institutions fell midway between those of Japanese high school students and residents of U.S. institutions. Similarly, the finding that both groups of Japanese perceived more order and clarity than the U.S. sample made sense. These findings were not, however, the ones the Japanese authors highlighted. They emphasized the similarity of ratings by institutional residents in both cultures. They also commented on the fact that both Japanese samples scored higher on self-expressiveness than Americans, replicating a Japan–U.S. comparison of workers. The "made-in-the-USA" scales that had been employed did not further understanding beyond the horizon of culturally based expectations.

Japanese papers written from the point of view of nonsocial science disciplines were the least comprehensible to me. However, the discussions of such papers were the most enlightening. Funahashi's paper on "Architectural Planning Research" (APR) fell into this category, and in addition, employed some concepts rendered in German and applied in Japanese. The paper criticized the narrowly physical aspect of the Japanese practice of APR and referred to the U.S. transactional critique of this approach—familiar territory so far. However, the most, to me, intriguing aspect of the critique was not included in the draft paper, in large part because of its lack of a cognate concept in English, at least within the U.S. context. Funahashi explained that the paper was proposing that the idea of a functional architecture that fit the needs of the user be replaced with the concept of "architecture for nothing." The ensuing discussion pointed out how such a concept was not a negation of architecture but a way of building to open up the undefined possibilities of living in space. It was not "form follows function" minimalism. Rather it was a recognition of the architect's limitations in knowing the multitude of functions humans could need or desire. The use of slides and the question-and-answer format of discussion allowed the author to differentiate this concept from the familiar multipurpose room. Although a succinct verbal formulation of the idea remained illusive, the presentation and discussion expanded my awareness of a distinctly Japanese aesthetic and conception of self-expression.

Cultural conceptions of the relationship of people and environments seem to have been expressed in the measurement methods indigenously developed. Among these papers, Japanese-generated methodologies had a person environment physical specificity that contrasted with the more psychosocially centered verbal nature of U.S. originated methods. (Of course, the fact that the official language of the seminar was English may have limited the possibility of reporting verbal methods indigenous to Japan.) The construction of methods and their use seemed to have different aims. Indigenous Japanese approaches seemed to presume a continuity between person and environment. The aspects of experience that were marked were the differentiations of person-in-environment units. U.S.-generated approaches marked out individuals and groups as against other individuals and groups. The environment was represented either as another form of "not-me" or as a means of intergroup conflict. However, it is difficult to know whether the differences in research methods are more related to culture or to discipline since the design fields were represented only by Japanese participants.

Given the qualifications mentioned above, indigenous Japanese methods seemed more suited for discovering generalities based in physical resonances between people and settings (Adachi, Kubota, Ohno, Suzuki, Chapter 9; Takahashi). The generalities illustrated by the U.S. verbal methods marked person-centered generalities or highlighted disjuncture (cf. Zube). The one U.S. example of nonverbal measurement related personal movement to psychosocial dispositions without including the physical environment as an aspect of the situation (Demick, Ishii, & Inoue). U.S.-generated methods also tended to be used to reinforce categorical generality, that is, generality based on the ability to classify verbal responses in a replicable manner (Wapner, Fujimoto, Imamichi, Inoue, & Towes) that can be sorted into environments categorized by social definitions such as country (Wapner, Fujimoto, Imamichi, Inoue, & Towes) or country and institution (Asai & Hirata).

Several ethnographic studies were reported by both Japanese and U.S. investigators, combining attention to physical details and extensive verbal data. They focused on the ways individuals, households, or groups used and transformed their spaces over time. In these cases, the research account was rendered in a manner closer to a folk narrative than to a "scientific" study. Culture, physical space, community and/or family dynamics and biological and social life changes were analyzed in relation to each other (Howell; Kobayashi; Minami). A transactional perspective on person and environment, and individuals and culture, was achieved by focusing on the interplay of physical space, social and personal action, and interpretation. Howell's work showed a much greater attention to physical detail than most U.S. papers. The theoretical concepts introduced included culturally/physically specified concepts such as permeability, as well as psychosocial concepts concerning development and role configuration.

The narratives of person-in-world, rendered in English, were cross-culturally coherent. The approach seems conducive to developing a framework that can illuminate both generality and difference. Sources of generality include a shared culture, a shared community circumstance (cf. relocation after a natural disaster, response to urban renewal), and life stage definitions. The interpretations and actions of persons and group within these conditions show sources of difference and point out the open-ended nature of person–environment relationships. Yet these sources of difference can be contained within the analytic framework.

All of the ethnographies presented a complex interweaving of culturally specific narratives, transcultural narratives, and place. Two of the U.S. accounts can be seen as continuing the prominent U.S. themes of the quest for autonomy and control. For example, Low's analysis focused on the different interests of social classes as they engaged in contests to control the environment. While both of the parks she studied were culturally hybrid forms, the newer Plaza de la Cultura provided space for the symbolic emergence of groups and activities that had previously had no place in the public space of the Costa Rican city she studied. Bechtel's paper also told a tale of disjuncture, this time between the aspirations and interpretations of Biosphere 2 inhabitants and the limitations of the physical world. Bechtel convincingly argues that the story of Biosphere 2 was a variation on the conquest of nature myth, including the frustration of man's quest by his biological dependence on the physical world in ways that were ignored at his peril.

The studies that described common human experiences, the aftermath of traumatic displacement by a natural disaster (Kobayashi, Miwa, and Mahi, Chapter 15) or the meaning of the destruction of familiar residential environments for the elderly (Minami, Chapter 10), produced narratives of loss and change with cross-cultural resonance. Kobayashi and his colleagues demonstrate the close link between the restoration of intimate connections between people and their environment and the restoration of psychic well-being after a disaster. Those who participated in the creation of a new environment, either in small ways such as decorating and tending plants or by actually building a new home in the form desired, seemed to restore the stage set for their valued identities. Those who did not sat idly in a context that did not engage them, some expressing even a loss of memory in a setting they moved into all at once without making it or choosing it.

Minami extended the concept of social disengagement of the elderly, familiar in the English language social science literature, to describe the relationships of the elderly with the environment. His account of how elderly residents enacted an autumn festival in the face of the destruction of the landscape of their neighborhood weaves together the relationship between the existing and remembered landscape, personal identity, and the generational transfer of memories of place and cultural meaning. Within this narrative is both the historically and geographically specific experience of an aging doctor who cared for the sufferers of the bombing of Hiroshima, as he obsessively chronicled the details of the physical dismantling of the postbomb neighborhood of his clinic, and the rather more general tale of an elderly woman's distress, decline, and death upon the cutting down of a beloved pine tree and her subsequent relocation to transitory housing.

Minami explicitly recognizes that his research not only tells a story but is also about the telling of stories, that place is in part specified through narrative:

> Despite ... repeatedly heard expressions of intimate transactions of neighbors, we could not confirm actual occurrences of sharing household goods and dishes in daily life base(d) in our observation. At the moment, we formed a hypothesis that omnipresence of the narrative accounts on the neighboring relationship is not a reflection of actual community life in this area but rather an act of narrative creation, or a kind of a "myth" shared by the residents. It is a community as it exists in the collective representation of the residents. Although the expressions described above might have actual counterparts in the real transactions in this community in old days, they are not observable events in the present life of the residents. In this respect, they are stereotyped narratives which correspond to the "original landscape" of the long-term inhabitants in this area.

Even if the community life as expressed in these narratives are not (sic) alive in this area any longer, the very act of narrating in this way by the local elder residents' may constitute continuity in the collective representation, and thus creates context in which the distinctive "D town-ness" is passed to next generations. One of the rationale to resume the autumn festival described above was that without it people, especially young children would forget the good part of D district. These stereotyped narratives, in this sense, are functional in guiding social ecology of the community under radical changes in the physical environment induced by the renewal project (p. 21–20).

The cross-cultural meaningfulness of ethnographic accounts together with the use of visual material, description, and quotations to display specific content offers a useful path toward an understanding of situated generality. In contrast to the quantitatively measured outcomes of other studies, the results of ethnographies are never mute or uninterpretable. The careful tracking of the movements of visitors through a garden, generated by a logic of observation rather than a story of how people behave, stands at some distance from narrative. It may be made comprehensible by a narrative the author constructs, by accounts of visitors, or it may remain in some sense meaningless. While this "meaninglessness" can be seen as a sign of the triviality of the research, it can also be the occasion for the invention of new narrative that brings into meaningfulness some heretofore alien aspect of person–environment relationships. Perhaps one challenge for ethnographic narratives is to create a sharper edge between the verbally coherent and the observationally or experientially distinct.

COMPONENTS OF SITUATEDNESS

Situatedness has been introduced as a criterion of useful environment–behavior research. An environment (as opposed to *the* environment) is a unique, specific time–space locale that can be "reached" in a conceptual sense by the construction of narratives that both mark out a recognizably distinct domain and relate it to a general narrative structure that gives it meaning. Situating actions within folk narratives is a method of ruling in and ruling out acceptable explanations and meanings. An unremarkable action in one situation may be outlandish in another. In much of social science, the description of the situation is one of the most highly conventionalized aspects of research. This point is clearly illustrated by the psychology journal format where an obligatory but brief literature review is followed by a highly conventionalized description of what was done, where, and by whom. The most generalizable finding can be developed when one argues that only what is described occurred and anybody, anywhere could do it. Each discipline has its own set of conventions that make unproblematic how to define the situation. Yet by accepting these, many aspects of what is to be regarded as general and what may be distinct about the situation addressed have already been determined. The parameters for the predictable, the unexpected, and the unintelligible have been established.

For the interdisciplinary area of environment–behavior studies, the first requirement for situating research is to determine how a situation is defined. In other social sciences, the situation is the backdrop against which culture, the society, the social group, the family, or the individual is studied, if it is even noted. In environment–behavior studies, the situation is the topic. Accordingly, environment behavior theorists explicitly define the characteristics of a situation (cf. Altman & Rogoff, 1987; Barker, 1968; Bronfenbrenner, 1979; Ittelson, 1973; Proshansky, Ittelson, Rivlin, & Winkel, 1974; Wapner, 1987). The first thing these theorists find to be apparent is that " the situation" exists at many levels of analysis and is inherently unbounded. This definition of the situation departs from the cultural narratives perspective Bruner advocates, and in doing so raises questions about the adequacy of explanations of meaning based on the presumption of a relatively bounded cultural universe in which the level of analysis is unproblematic.

Levels of Analysis

The definition of a situation is then taken as the first, and perhaps most fundamental, aspect of situating phenomena to be studied. A comparative, or definitive, analysis of the ways environment–behavior research defines situation is beyond the scope of this chapter. I will take the organismic-developmental definition of levels of analysis as a starting point, both because it is useful and because a number of the chapters in this volume derive from this theoretical perspective. Wapner identifies what he calls three levels of organization:

1. The biological level that calls attention to the sensory, vegetative transactions of human beings (organism) with their environments (ambiance).
2. The psychological level, described variously as focusing on "perceiving, thinking, remembering" and as the level of organization in which the "human operates in an environment, termed habitat, where social (people) and nonsocial (things) objects are present. Operating at this level, the human is referred to as agent and is directed toward episodic incentives such as finding food or a mate, avoiding danger by use of such instrumentalities as tools and mechanisms that enhance the senses and extend the functions of the body parts" (p. 1443).
3. The person–world level in which the socially and culturally defined person seeks meaning and identity follows or breaks rules and norms, performs rituals, makes a living, and in all other ways lives in the historically, geographically, and autobiographically specified world.

By paying attention to the different levels of analysis, we can raise questions about the interconnections among levels that would otherwise be ignored. Herein lies one reason why environment–behavior studies remain theoretically unsatisfying. Our theories commit us to raising such questions, but we usually do not do so in our research. Thus the research itself is not situated by theory and murky with regard to disciplinary presuppositions that would render the definition of the situation once more unproblematic. For example, Wapner et al. use the different levels of organization to categorize the response they get in an interview setting. However, the situation and method of measurement are all at the final level of analysis. Within the narratives of the respondents and of the authors, the organism/ambiance and agent/habitat experiences have already been strained, shaped, and selected through the enculturated, historical, autobiographical self. In addition, the definition of the second level of organization is confusing, especially if one accepts the idea that the psychological self is a borderline case of the second and third levels. Apparently aimed at what constitutes direct experience of a particular individual versus socially and culturally determined experience, no clear grounds for separation of the psychological level are given. Except for the very general information that respondents are either Japanese students studying in the United States or U.S. students studying in Japan, the empirical information of the study consists of narrative accounts by respondents and then categorization of the narratives by the researchers. Two problems with the theory become apparent: (1) The relationships among levels of organization are not clearly specified or measured in ways that suggest their separable, even if interdependent, nature; (2) and the theory itself and the research take for granted a shared cultural narrative that makes statements specifying levels of analysis meaningful. I suggest that these problems would not be fully solved by more careful definitions, nor better measurement. Rather, the idea of levels of organization must be addressed from within a narrative structure that reverses the order of dependence from the one recognized in the theory. The theory states that the experience of being a person in the world depends on being an acceptably successful agent in a habitat, which in turn depends on being a viable organism in environment or ambiance. As Heider (1959) pointed out in his essay "Thing and Medium," sequences of dependence of the sort referred to are given in experience in reverse order. The statements themselves can only be made and make sense because of the experience of being a person in the world. Even if the physical level of experience had been measured by chemical analysis of the blood or differential metabolism of Japanese and American food, the instruments of measurement and the meaningfulness of their readings are cultural products.

We would know more about how to specify levels of analysis if we examined how and when they function in narrative. Bechtel's account of Biosphere 2 (Chapter 16) offers a promising example. He placed the significance of a mismatch of environmental properties with human biology within the context of the actions of a small groups of people to meet their needs which in turn were made problematic not only by the presence of too much iron in the soil, but also by the negotiated definition and construction of the setting within a broad social, cultural, and historical perspective. He at once demonstrates that the situation was a complex cultural construction negotiated in complex interpersonal transactions, but he also demonstrated a certain level of independence of the biological from the interpersonal, and of both from the original narrative frame. Of course the absence of presentation of biological, atmospheric, or agricultural analysis may oversimplify the problem and its possible configurations. If it could be found that growing a special kind of potato would transform the soil composition, then the problem of the narrative might not be the overreaching of the human condition by the pioneering founder and staff, but rather the low level of scientific knowledge about agriculture at the time of the research. The point of this explication is to show the narratively embedded nature of both the topic of environment–behavior research and the practice of scholarship.

Narrative Incoherence

One implication of the above analysis might be that the most fruitful outcome of specifying levels of analysis would be to generate data from different levels of analysis that do not make sense. The pursuit of incoherent aspects of narrative is of course a hallmark of logic. But from the narrative point of view, these are the occasions for the development of more adequate narratives rather than the opportunity to point out errors. Setha Low (Chapter 21) explicitly uses the idea of looking for contradictions between the culturally received narrative and observations concerning use and personal experiences of people of different ages and segments of society. These contradictions are then reconciled within the academic narrative of global economic forces. This usurpation has the advantage of making sense to a culturally and disciplinarily diverse academic audience. Thus it succeeds in making the general specific, as well as in finding a general explanation for a specific occurrence. A further test of this approach would be to share the reconstructed narrative within the segmented culture of the folk narrative, or to attempt to act within that culture on the basis of the analysis. These additional approaches would deal with the tension between a more cogent argument for a wider audience and the possible loss of place–time specificity. To a Costa Rican skateboarder, the salient narrative about the Plaza de la Cultura would likely be very different than that which engages a potential purchaser of adjacent real estate or a political challenger to the Minister of Culture in 1976.

Situating the Narrator

Bruner's (1990) conception of meaning as culturally embedded narrative appears inadequate to deal with the actual practice. The weakness lies in the absence of serious discussion of what is meant by culture. Bruner uses the term to refer to the ways in which the members of a particular family understand and explain their lives and choices. He also uses it to refer to the way mothers talk to their children in a black ghetto in Baltimore. Bruner is most successful in arguing that fundamental psychological processes such as identity formation, decision-making, and learning are more properly understood as ways individuals participate in larger social processes, in what he calls "a cultural-historical situation" (p. 107). However, the definition of situation is left as something we know because we and he participate in a common "cultural-historical situation." Unfortunately for us, this is the introduction to environmental psychology rather than the conclusion. His wide-ranging and insightful exposition of the relationship between received narratives and individual experience and invention helps us understand the dynamic and developmentally significant nature of the bond between individual and context. We see more clearly how situation specifies the self. We must now work our way backward from that bond to explore the manifestations and specifications of situatedness.

CONCLUSIONS

If we as researchers understand our projects through the specific historical cultural situations within which our work has meaning, then it is reasonable to ask what the impact has been of the periodic dialogue and more sustained individual collaborations of participants in the Japan–U.S. Seminar on Environment-Behavior Research. Clearly the edge of cultural difference has been dulled. This change can be a sign of the increased power of our explanatory frameworks. The negative side may be that our research and theory obscure the situated in favor of the general. It might also be argued that historical cultural situations themselves are less distinct than previously, especially for members of international professions and users of electronic information technology. The transcultural multidisciplinary narratives of our texts are necessarily encountered from a specific vantage point. The physical space and time of a vantage point specify the meaning of the narrative by providing the stage set and choreography of action. The ethnographic approach incorporates folk narrative as an aspect of place meaning and place as an aspect of the identities and social action recounted. Yet the relationship of folk narrative to experience and action is proactively inexplicit and retrospectively an act of creativity. Scholarly narratives, and research methods, problematize events and make explicit the interpretive act. But they can also have the opposite effect of obscuring the difference between the situated, enacted, embedded narrative and the represented, recounted, analyzed narrative. The opportunity for discussion that the seminar allowed and that these written texts may not convey suggests methods of working outward.

Suggestions for Defining the Situation

1. We can situate our research better by making clear the specific literature and disciplinary traditions we are drawing on as well as the specific place and time characteristics of the work. Thus papers would situate the author in a particular view of what counts as knowledge, how it is obtained, and what counts as evidence, as well as a level of analysis, and definition of persons and situations. Such an approach would require some changes in format, although perhaps only a revival of long footnotes. As members of a community of scholars we should inquire about the "cultural-historical situation" of the author, the research process, and the resultant work as frequently as we ask for textual or statistical clarification.

2. Both terms of person–environment relationships should be accorded attention in both theory and measurement. For both social science and folk narrative, the obdurate, mute, and/or foreign aspects of the environment tend to evade incorporation in to a coherent explanatory framework. We should look for these jarring notes.

3. Situating research means admitting ignorance into academic discourse. Most of us at the seminar agreed that the information superhighway and the international, interdisciplinary nature of our field confounded our ability to deliberately define a topic, a theory, and an appropriate research method based on intellectual grounds rather than inability to keep up with the volume and flux of information.

4. More discussion of what can and cannot be said in different languages would help specify the way in which narrative is still culturally situated, even within globalized institutional settings and exchanges.

5. Inclusion of "specimens" in our research reports are more helpful than reports of only the analyzed or categorized data. Photographs and other nonverbal methods of documentation provide nonnarrative information. Direct quotations and untranslated phrases juxtaposed to the scholarly narrative can also provoke tension between the general and the specific.

As Kenneth Craik's (Chapter 26) comments made clear, we may not have a very good idea of what the situation is right now, at the end of the millennium, when the forms and boundaries of nations are shifting, religious and nationalistic tides are redefining cultures, and many social units that once seemed whole are fragmenting. But that then is the situation we understand ourselves to participate in. Although virtual reality and electronic media are sometimes said to make time and space irrelevant, they could better be understood as space–time forms that are socially and physically constructed and understood through these constructions though their potential for affecting experience may as yet be outside these constructions.

A CONCLUDING THOUGHT

By adopting an emphasis on narrative as the container of the meaning of a situation, the idea of the material world independent of human construction loses prominence. I think this is an overreaction, or perhaps wishful thinking. The material world is much a part of our most sophisticated social narratives, and should be. Humans have spent much intellectual effort on understanding, overcoming, or adapting to the material world as we know it. A variety of critiques of positivist social science point out the absence of a foundation for this knowledge that is separate from or more certain than the social world that creates it. However, within these social worlds, the interface of the physical and the psychosocial world, as we know them, is still a fruitful one. Habermas's (1972) distinction between technical and communicative knowledge seems most relevant here. Technical knowledge derives from an interest in changing the state of affairs, as distinct from communicative knowledge, which provides the grounds for understanding and democratic decision making about what should be done. According to Habermas, it is by bringing technical knowledge under the control of fair and democratic decision processes that human freedom is realized to a greater or lesser extent. The material world is a significant component of both technical and communicative discourses. Most cultural traditions, no matter how they are defined, have a place for the biological limits of life as well as for the existing but unknown universe. Environment–behavior research in its more interesting form examines both the coherent narratives that can be told within the livable and the knowable, and the edges of the unlivable and unknown.

REFERENCES

Altman, I., & Rogoff, B. (1987). World views in psychology: Trait, interactional, organismic, and transactional. In D. Stokols & I. Altman (Eds), *Handbook of environmental psychology, Vol. 1* (pp. 7–40). New York: Wiley.

Barker, R. G. (1968). *Ecological psychology: Concepts and methods for studying the environment of human behavior*. San Francisco: Jossey-Bass.

Bronfenbrenner, U. (1979). *The ecology of human development*. Cambridge, MA: Harvard University Press.

Bruner, J. (1990). *Acts of meaning*. Cambridge, MA: Harvard University Press.

Habermas, J. (translated by J. J. Shapiro) (1972). *Knowledge and human interests*. London: Heinemann.

Heider, F. (1959). Thing and medium. In F. Heider, *On perception and event structure of the psychological environment* (pp. 1 34).New York: International Universities Press.

Ittelson, W. H. (1973). Environment perception and contemporary perceptual theory. In W. H. Ittelson (Ed.), *Environment and cognition* (pp. 1–19). New York: Seminar.

Proshansky, H., Ittelson, W., Rivlin, L., & Winkel, G. (1974). *An introduction to environmental psychology*. New York: Holt, Rinehart & Winston.

Ricour, P. (translated by K. Blamey & J. B. Thompson) (1991). *From text to action: Essays in hermeneutics II*. Evanston, IL: Northwestern University Press.

Valsiner, J. (1987). *Culture and the development of children's action*. New York: Wiley.

Wapner, S. (1987). A holistic, developmental, systems-oriented environmental psychology: Some beginnings. In D. Stokols & I. Altman (Eds.), *Handbook of environmental psychology, Vol.2*, (pp 1433–1467). New York: Wiley.

Chapter **28**

THEORY IN ENVIRONMENT BEHAVIOR STUDIES
Transcending Times, Settings, and Groups

Amos Rapoport

INTRODUCTION

The title of this chapter resulted from the first (March, 1994) announcement of the fourth Japan–USA seminar, when I discovered that the general topic of theory crosscut and did not fit fully any of the proposed categories: environment–behavior Relations (EBR) in different sociohistoric times—the 21st century; EBR in different environments: general processes; EBR in different environments: the city; EBR in different subgroups: aging; and EBR in different subgroups: families. It best fit the second category, particularly "general processes," although the specific titles listed did not seem to match these.

Clearly, no criticism is implied of these papers, which I have not yet seen. I merely used these (as well as EDRA, IASTE, PAPER, Japan–U.S., and other meetings and journal articles) as stimuli to describe how I think environment behavior studies (EBS) should be approached. I contend that any explanatory theory of EBR must transcend groups, settings and times, and to be generally applicable when specifics are "plugged in."

The emphasis throughout is on the general nature of processes, partly because most of my research over the past 12 years has been on a book, *Theory in Environmental Design;* it has also been the thrust of my work generally (Rapoport, 1990a). I use this work as the "model system" and although (obviously) I think I am right,

Amos Rapoport • Department of Architecture, University of Wisconsin-Milwaukee, Milwaukee, Wisconsin 53201.

Handbook of Japan–United States Environment–Behavior Research: Toward a Transactional Approach, edited by Seymour Wapner, Jack Demick, Takiji Yamamoto, and Takashi Takahashi. Plenum Press, New York, 1997.

the argument more generally does not depend on that work (should it prove wrong).

In essence, this chapter is a metatheoretical sketch in outline form, emphasizing the need always to seek the most general formulations and to seek both linear and lateral connections—to work in EBS and as many other useful fields as possible. These are the attributes most lacking in EBS that make their current state and future prospects problematic.

Not only does the way EBS is subdivided raise questions and doubts. The *nature of the relationships among* these subdivisions and the larger framework is hardly discussed, let alone developed. My recent efforts are devoted to trying to construct and communicate precisely those relationships and that framework.

This deficiency leads to a set of problems with the state of the field, among them how it and its domain are defined and conceptualized, the lack of theory development, the nature of such theory, and other problems that follow. I will, therefore, briefly discuss some of the problems before I briefly sketch some possible responses.[1]

PART I: SOME PROBLEMS WITH EBS

If EBS is to be more than an ad hoc attempt to improve design, it must be considered as a new discipline. Among the attributes of a discipline are a clear and explicit definition of the domain and agreement about it (Rapoport, 1990b) and the existence of explanatory theory. Without such theory EBS has an *accumulation* of material *but is not cumulative*. The failure to become cumulative is also due to a tendency to ignore or neglect what has already been done.

"Reinventing the Wheel"

At the most basic level one is struck by the lack of knowledge among researchers of the literature. This has led me from being fairly optimistic about the field (Rapoport, 1977b, 1995c) to becoming ever more disillusioned and pessimistic; I am not even sure that EBS will make it to the 21st century unless it changes its ways and emphasizes, in a major way, synthesis and theory development.

One frequently finds "new" findings announced that have been known, sometimes for decades. Not only are thorough literature reviews neglected, but the often inadequate reviews often seem *pro forma* and are not really used. As a result, there is no continuity and hence no cumulativeness, no development and no progress. Cumulativeness is a minimum requirement for progress, although rapid progress requires theory, or even attempts to develop theory. Examples include archaeology (e.g. Rapoport, 1990b), ethology (e.g. Dewsbury, 1989), and modern biology. Another appropriate example is medicine, which has an applied side and the develop-

---

[1]These will be developed in the proposed book, *Theory in Environmental Design*.

ment of which is not only recent but very rapid. In effect, it has changed from an "art" to a "science-based profession" (Schön, 1983).

At EDRA 25 I discovered that others shared my view about the wheel (or even spokes) being reinvented. One response was a session at EDRA 26 that brought together three generations of EBS researchers.[2] Although an important initiative, it made clearer than ever the lack of cumulativeness or synthesis. It is instructive to compare it to a recent similar effort in biology, which also emphasized three generations—a symposium "Lizard ecology: The third generation" (Lang, 1995). Each of the four sections in the resulting volume is introduced by a contributor to previous volumes on the theme (in 1967 and 1983). Together, these three books span nearly three decades and the third emphasizes integration and a range of new developments based on the "solid foundation" of previous work. A *detailed* comparison with the EDRA 26 session would be most instructive, but the essential point, that EBS lacks cumulativeness, seems clear. Note, that although the book deals with lizards, it systematically covers numerous aspects of the topic over the three generations and integrates them. Even more significantly, the American Zoological Society is moving from the study of organisms to studies ranging from molecules to ecosystems; it is also changing its name to the Society for Integrative and Comparative Biology, and there have been university departments of integrative biology for some time. This is what is needed in EBS.

In fact, the accumulation that we do have (cf. Willer and Willer 1973, Willer 1987) can become counterproductive, since large numbers of isolated studies (e.g., about housing, neighborhoods, or environmental quality) cannot be read or used (e.g., Rapoport 1985, 1990e). Moreover, as long as synthesis and theory are lacking, it becomes necessary periodically to review earlier work, with new questions, concepts, or methods, and to expand such reviews across cultures, countries, and languages. Also, such studies on many topics not only repeat material known for some time but are often predictable even from the "theory" we do have, fit into such theory and provide data for theory—but are not so used.

Too Many Isolated Empirical Studies

At the moment most studies tend to be seen as specific, with no attempt to transcend their times, setting or groups, to be linked conceptually with other work, or to emphasize general EBRs and the mechanisms operating. Consider a single rather interesting study dealing with the important role of processions for an understanding of the form of a Sri-Lankan village (Bechhoefer, 1989). There is no reference to Johnson's (1965) work on "The processual element in architecture" or the work of Appleyard et al. (1964) and Carr and Schissler (1969) on the United States (all cited in Rapoport, 1977a). The study is not related to the important role of pilgrimage nationally and regionally in India and of processions in Indian cities (e.g. Madurai, which one of my students just studied (Vairavan, 1995), and many others). There is also no reference to the role of ritual movement among Australian

---

[2]This was organized by Nora Rubinstein and Herb Childress.

Aborigines, the Maya, Mayo Indians, and many other places and periods (cf. Rapoport, 1990b, c, d, 1993a, and references). Also neglected are the implications for EBS more generally, for environmental perception and cognition, relation to urban travel, to systems of activities in systems of settings (Rapoport, 1990c, or "behavior circuits" [Perin, 1970]), and so on. Some of these suggested relationships may appear rather far-fetched, but it is precisely such lateral linkages and the subsequent generalizations at much higher levels of abstraction that need to be emphasized, and it is striking how quickly they can develop.

Another example concerns 19th-century Cairene villas (Asfour, 1993) that can easily be related to the very large literature on syncretism in general (cf. Rapoport, 1983b), to comparable changes in India studied by A. D. King and others, and to the importance of modernity and its indicators more generally (Rapoport, 1990d).[3]

Inadequate Conscious Consideration of Other Fields

It seems strange that EBS has essentially restricted itself to the original fields with which it began, although even those are neither well enough known or adequately used. When other fields have been introduced, they have been the wrong ones from my point of view, e.g., phenomenology, hermeneutics, narrative, deconstruction, "critical theory" and the like. There has been no effort systematically to survey and introduce new often highly relevant *scientific* fields, although relevance only becomes apparent through relatively abstract concepts. Among such fields (and this is not an exhaustive list) are ethology and animal architecture, archaeology, cognitive science, evolutionary science, evolutionary psychology (e.g. Barkow et al., 1992), sociobiology, and hence the notion of "human nature," human ecology, biology and genetics, philosophy of science (especially biology), computational philosophy of science, history of science, metascience, as well as various sciences in general, not so much in terms of content as in terms of how they do things. Lacking is an active search for new sources of insight, interfield and intertheory support, and so on. As just one example, it seems quite clear that the persistent questions about constancy and change can be at least partly resolved by evolutionary science, evolutionary psychology, historical and cross-cultural analysis, and so on; there is no need merely to speculate.

Not Enough Emphasis on Synthesis

The neglect of synthesis in EBS is most striking. This contrasts dramatically with other (scientific) fields. Thus in one two-page article (Barinaga, 1995) there are two statements about things fitting together "very nicely" and things "coming together very rapidly," both regarding different studies of single neurons and their dendrites and at the level of the whole field. In the same issue of *Science*,[4] in

---

[3]The latter point has been developed in some detail in a paper (given at the conference on *Value in Tradition*, Tunis, Dec. 1994) entitled "Sustainability, Meaning and Traditional Environments" (Rapoport, 1994b).

connection with newly obtained ultraviolet data in astronomy, a first reaction quoted is about when these data, for which astronomers had been waiting for 20 years, would be ready for synthesis (Goldsmith, 1995). Note also the knowledge about *which data were needed*—on the basis of previous syntheses and theory.

There are also studies reviewing and synthesizing much work. One recent study on whiplash injuries reviewed 10,382 papers, concluding that only 346 were worthy of further study and finally using only 62 as valid. These were then synthesized (Altman, 1995).[5] In another example, Heinrich (1993) (cf. Holldobler & Wilson, 1990) critically digests, consolidates, and considers possible improvement in the literature (of over 1000 research papers) on thermoregulation in insects. In this way even a major, long paper can be reduced to a single line within the broader framework (cf. examples in Rapoport, 1990a). There is also the usual emphasis on *mechanisms* that is found constantly in scientific papers. One example (*Science,* 268[5208], 14 April 1995) devotes a major portion to mechanisms of signal transduction in cells, and other papers in almost every issue also deal with mechanisms.

Of course, for real synthesis one needs theory that can be seen as "abstracting and universalizing inquiry" (Kellert, 1993, p. 38, fn. 3). One cannot overestimate the importance of moving from unfocused, sporadic collections of papers to an integrated "cornucopian" body of literature (Clemens in Glen, 1994, p. 88; cf. Rapoport, 1973), i.e., moving from accumulation to cumulativeness, including links with other fields the relevance of which only becomes apparent when frameworks, concepts, and generalizations exist and are used.

We have, indeed, found out lots of things without, however, knowing how they apply or to which human groups; which settings have which effects, which mechanisms are involved, the scope and limitations of findings, or the validity of such findings (given the scarcity of replication). Nor are such findings linked and the relationships among them developed (or even discussed). As already suggested, too many findings can actually become counterproductive (Rapoport, 1973, 1985), since the many unrelated pieces, being accumulations, form a heap that cannot be used. This is made worse by the multidisciplinary nature of EBS, the variety of sources, and the presence of applied work that is often difficult to discover or to obtain (Rapoport, 1995b).

It is often easier to find things out than to know what to do with them. In EBS the emphasis is on finding out, and possibly thinking about, the relevance of such findings. More emphasis is needed on verifying and replicating findings, establishing the extent, scope, cross-cultural, cross-setting and cross-temporal validity, and limits, and to what extent, if any, scale changes or influences findings (Rapoport, 1995a, pp. 33–34). Such findings also need to be fitted into a larger framework and be used to establish linear and lateral connections within EBS and lateral connections with other fields.

---

[4]As in the above paper, I deliberately used only the issues of *Science* that arrived while I was writing this chapter.

[5]The possible role of meta-analysis in such syntheses is also worth considering.

I have randomly chosen one interesting chapter by Soma and Saito (Chapter 8), the title of which can generate a general series of questions that can serve as a simple example of such linkages. One could begin with the Whorfian hypothesis and the current view that it is too extreme, that all humans perceive the same colors although they may name colors differently and have many fewer names for them, although even then there are regularities in the sequence of name development (Berlin & Kay, 1969).[6] One can then consider color as a noticeable difference (e.g., references in Rapoport, 1977a, 1992a) and thus an important part of any repertoire of means ("how" as opposed to "what" and "why" things should be done). One could then ask about the cross-cultural use of color (references in Rapoport, 1990d) and its affective role (e.g., Philip, 1990/91), which may affect mood, leading to the psychological effects of color more generally, on which there is a large literature worth systematic exploration. One can then move to the meaning (high, middle, and low [e.g., Rapoport, 1988a, 1990d, g]), of color and hence its relation to culture and its role in judging environments (e.g., Foote, 1983). This, in turn can be related to other aspects of meaning both generally (Blanton, 1994; Rapoport, 1990d) and specifically (e.g., Cherulnik, 1991, who neglects color, and also Rapoport, 1990d, and Foote, 1983, who also discusses restaurants). This might allow studying the relative importance of colors versus other components of the repertoire in communicating various meanings, e.g., communicating identity, judgments of environmental quality, in the generation of ambiance, and so on. One can also test hypotheses about the relation of the use of color and other variables. A student of mine (S. Shaw) looked at patterns of color use vis-à-vis climate, landscape, and culture. His hypothesis that cold climates are correlated with bright colors was not supported, but needs more systematic work, e.g., plotting colors on a world climate map, and on maps of other features.

Color and color ranges can also act as indicators of identity generally, including ethnic identity, as has been found in Australia, Canada and the United States, e.g., colors in Little Italy versus Chinatown in New York (O'Donnell, 1995) or among Mexican Americans (Arreola, 1988). Colors can also be used as indicators of regional identity, as in São Paulo, Brazil (Monzeglio, personal communication), and some Swiss towns and regions actually specify permitted color ranges.

The link between color and culture allows color change to become an indicator of acculturation. It can then be linked to other aspects of acculturation, such as space use (Pader, 1993), form, materials, fences, planting, and so on. All these may also, and frequently do, work together, greatly increasing redundancy (e.g., Livingston, 1992). Such work then links acculturation in countries such as the United States, Canada, and Australia with work on developing countries (Rapoport, 1983a) and with culture change more generally. In turn, many studies of developing countries and culture change, related to the built environment, can be linked into very general frameworks, transcending specific locales, groups, settings, and times.

Color can become a sign of status (examples in Rapoport, 1990d) and as such can become a part of environmental quality. This allows color to be linked to a large

---

[6]This, in turn, provides a potential link to the whole question of taxonomies, whether of color terms or other attributes (see references in Rapoport, 1990d, h, and, e.g., Berlin, 1992, among others).

range of other variables and bodies of work (e.g., Rapoport, 1990e, 1995b), particularly if cultural landscapes are used as the unit of analysis (Rapoport, 1992 a, b). Such aspects of color can then lead to group conflict based on meaning (as in the case of Portuguese and Anglos in Toronto). The relation of color and status can also be studied historically, as in the case of Beijing (and ancient China generally) and other areas where color use was related to status by law, e.g., in terms of sumptuary laws.

All this allows color to be seen as part of both the perceptual and associational realms, and its latent aspects emerge as important. That, in turn, is a very general finding that links color to a large range of topics in EBS where latent aspects dominate. Moreover, there may also be constraints to theorizing from a range of other fields that bear on the topic (e.g., anthropology and ethnography, perception, cognitive science, linguistics, evolutionary psychology, etc.). There is also much other work that could be related to color, e.g. its role in the creation of ambiance (Rapoport, 1992a), which can also be inferred from observation (i.e., the "natural history stage") as well as from the use of "indirect sources" (novels, advertising, TV, movies, etc.) (Rapoport, 1990f). I could go on, since many other linkages suggest themselves, but enough has been said to show the process. This would have to be done *systematically*, aiming for maximum coverage, using all literature from relevant fields, locales, periods, and types of environments. Only then could it be fitted into a theoretical framework of EBR. This is the type of process that I believe to be essential for progress in EBS, and which is largely lacking.

A similar process could be applied to materials. It was predictable from general considerations of the importance of latent functions that materials are often chosen for their meaning (Rapoport, 1983a, 1990d). There are now at least several empirical confirmations from various locales (Kaitilla, 1991, 1994; Sadalla & Sheets, 1993). These, however, do not refer to each other, which would allow major generalization, particularly if linked more closely with my more theoretical work and the large amount of material scattered in many places, especially (although not entirely) from developing countries. Again, material from novels, paintings, TV, film, advertising, etc. can be useful. Similar topical linkages could be performed for almost any EBS paper, and these expanded frameworks could be linked into even larger networks and frameworks, en route to theory development as an answer to explanatory questions.

Inadequate Thought or Effort on Theory Building

Typically, once the process of establishing both linear and lateral linkages has occurred, certain general patterns are likely to emerge. These then need to be explained, mechanisms specified, and so on. A process of inference and conjecture to hypothetical best explanation takes place. These hypotheses need to be sufficiently clearly and explicitly stated so that predictions follow and testing can take place. The final result is explanatory theory, seen as *reliable* inference, well supported by (different lines of) evidence, the use of different methods, etc.

Such an emphasis on theory raises some potentially difficult questions, among which are the following: What is an explanatory theory (and is it generic or domain specific)? How do theories differ from models? What does a theory do? How does one go about constructing theory? What are the criteria for a good theory? How are theories tested? Given the extent of the literature on such topics, it is strange that remarkably little discussion about them has occurred in EBS, although it can begin relatively simply. For example, the attribute of testability (at least in principle) seems important in EBS because one hears much about the importance of *interpretation* in the social sciences. Yet, although one can consider interpretation as a form of explanation, *it is not testable* (Bell, 1994); moreover, one does not preclude the other (Henderson, 1993).

Models are also often mentioned in EBS. Again, a general review of types of models (of which there are many) and their relation to theory (on which there is a large literature) must be postponed, but it is useful here to distinguish between *empirical models*, designed to fit empirical data (which can be found in EBS), and *theoretical models*, derived from first principles (although, clearly, adequate data are needed to begin).

I have already made the point that EBS never passed through a "natural history" stage (Rapoport, 1986b). Moreover, since it developed starting with a number of different existing fields, what "theory" there has been has tended to be *fragmented*. This has been exacerbated by the fact that many of these particular fields have had no real explanatory theory (Rapoport, 1990a; Willer & Willer, 1973).

Lack of Agreement on the Meaning and Usage of Concepts and Terms

I have argued for quite some time that an additional problem in EBS has been a lack of common shared understanding and agreement about terminology and the meaning of concepts. I have even called (unsuccessfully) for conferences just on that topic. Note that Proshansky (1973) had already emphasized the need for the "theoretical analysis of concepts," pointing out that it is dangerous to take concepts for granted or as given. Although individuals have attempted this, the field as such has not. Note that although definitions and agreement are not *sufficient* for theory development, they are a *necessary* precondition. Moreover, theories have conceptual as well as empirical problems and the former have been neglected in EBS in favor of the latter.

Although my concern was with those concepts related to theory, there are simpler issues. Thus at the EDRA 26 "Three generations" session it became clear that whereas to me "cumulativeness" meant not "reinventing the wheel," synthesis and theory development, others using the same terms were really referring to accumulation of empirical data, which can be found in both EBS and other social sciences (e.g. Willer & Willer, 1973; Willer, 1987), although knowledge of these data is often inadequate.

Many concepts used in EBS have multiple meanings. If we cannot agree (like chemists did in the 18th century, physicists, molecular biologists, and others who often repeat the process at regular intervals), we need at least to *specify explicitly* how a concept is being used in any given case (although in the long run it will be necessary to reach agreement, because that is necessary not only for synthesis and theory construction, but in order to communicate and interpret various empirical studies).

Consider the very different use of "territory" by Sack (1986) and Taylor (1988) who differ profoundly about one of the attributes of this term—its scale or size (and others use it differently yet; e.g., Altman, 1975, Malmberg, 1980). It seems critical in all such cases to reach some common agreement after systematically reviewing the *whole* literature and identifying areas of both agreement and disagreement. Also, despite an early argument for distinguishing between the objective aspects of the environment and human subjective reactions to them (Wohlwill, 1973), these still tend to be confounded (e.g., Evans, 1995; Rapoport, 1994a, 1995c;). One needs objective description, characterization, and operationalization of the numerous components of the environment *as well as* human subjective responses to them. Only then can one begin to specify and deal with the degree of subjectivity involved, which, I hypothesize, increases as one moves from perception through cognition to evaluation, affect, and meaning and preference (Rapoport, 1977a, 1990d), although these then become empirical questions.

Inadequate Open, Explicit, Critical Debate

In EBS there is a tendency to shy away from criticism of others' work. The situation is, of course, very different in the sciences where it plays a major role (Rapoport, 1990b). Such debate can be about the meaning and usage of concepts, about the nature of the domain, about approaches to studying that domain, etc. I suspect that a thorough and systematic literature review would reveal remarkably little such debate relative to the number of studies, papers, books, and so on.

PART II: SOME POSSIBLE SOLUTIONS—A METATHEORETICAL SKETCH

In this part, I essentially discuss briefly some possible responses to the problems above and some possible attributes that an explanatory theory of EBR needs to have, such as minimal (epistemological only) assumptions, so that ontology becomes an empirical question; the nature of the domain; the issue of replication; the need for both linear and lateral interfield and intertheory linkages.

The conceptual and theoretical development of any field demands that sufficient data be available to begin. These are often gathered during the "natural history stage," which EBS never really had (e.g., references in Rapoport, 1986b, 1990b; cf. Goldsmith, 1995). At the same time such data must be justified, i.e., *reliable*. Although the quality of data improves with conceptual development, the process must begin somewhere (Rapoport, 1990b, and references). In addition to this lack, and the adoption and borrowing of bits of "theory" and concepts that has led to fragmentation and confusion, there is another potential problem. This concerns the use of "folk" theories and concepts (used in architecture, psychology, anthropology, etc.). There is much debate about this very complex and contentious topic, specifically whether folk theories and their terms can be improved or whether they must be given up entirely. Without discussing this here, it seems important at least to raise this issue explicitly because it will have to be faced eventually (cf. Rapoport, 1994a, 1995c).

Attributes of a Good Theory

Starting with a brief list from Kuhn (1970, p. 199) which he repeats in Horwich (1993, p. 338), a much longer list has been generated in our Ph.D. Theories class, which is also found and commonly used in the philosophy of science (e.g., McMullin, 1993). This includes simplicity, elegance or parsimony, fertility or fruitfulness, future potential, scope, consistency, empirical accuracy, explicities, generality, coherence and logical consistency, testability, explanatory power, unifying power, compatibility with other accepted knowledge claims (intertheory support), naturalness (lack of *ad hocness*), problem solving ability and so on. Thus guidelines exist not only on how to go about constructing theory, but on the attributes that a good theory should have.

Such a list raises questions about what some of these (rather vague) criteria mean; how they might be operationalized (or "measured") in theory appraisal; their relative importance; whether they remain constant or change over time in a given science or in different sciences. These and related topics and the literature on them (which are extensive) cannot be discussed here. However, one question needs to be at least mentioned—how one handles the retroactive nature of some of these criteria while working on theory. One way is through controlled analogy with other disciplines (e.g., Rapoport, 1990b), which provides both positive and negative historical examples (cf. Kuhn in Horwich, 1993). Even without further development here, this list, with all its faults, is useful.

There are also some other attributes that EBR theory should have: it should be *general* not specific; it should be *abstract* rather than concrete; it should *not* use everyday "folk" terms but rather clearly defined unambiguous concepts. The attributes would allow such theory to transcend groups, settings, and times, as well as many other specifics. In fact, one important test of such emergent theory is precisely its ability to transcend such divisions, to subsume many papers and studies, even those based on different approaches (e.g., observation, field work, surveys and experiments). Using such emergent theory, many empirical studies should become self-evident, moreover, it should help establish relationships among such studies and to specify likely mechanisms. It is important to emphasize that the development of such theory and the evolutionary development of knowledge are *not* contradictory. In fact, they are the same thing—theories develop and change, i.e., they evolve in ways which the philosophy and history of science and scientific practice itself make clear. In any case, the emphasis here is on *explanatory* theory, not on any "Grand Unified Theory" (although the more unified the better). Even such theories (called GUTs in physics) evolve over time.

Developing such theory essentially requires the opposite of the characteristics criticized in Part I, and both historical and contemporary evidence suggests that progress is rapid, i.e., fields "take off" when this is done.[7] Thus one needs knowledge of what has already been done and relating new research to that; replicating research; agreement on the domain, definitions, concepts and what they stand for;

---

[7]As one example, consider the following statement, "The results described at that meeting prove that the field has come a long way in just a few months..." (Pool, 1995, p. 499).

operationalization of these concepts; the study of mechanisms and how they work; synthesis and generalization leading to pattern identification; explaining these patterns (which is theory). Consider just a few of these briefly.

Replication of Studies

Replication involves more than repeating studies to check their validity (e.g., Sommer and Sommer, 1983) (assuming that is possible).[8] It also involves testing studies cross-culturally; identifying patterns of differences and similarities; identifying mechanisms; identifying the range(s) of conditions (to be specified explicitly) under which a given phenomenon is valid, remains the same, changes, or becomes invalid. Replication also involves resolving *disagreements*, e.g., about the maximum size of territories (Sack, 1986, Taylor, 1988), about whether street trees raise house values (e.g. Rapoport, 1977a and references; *Urban Forests*, 1994, p.7 vs. Orland et al., 1992),[9] or about the relative importance of house interiors or exteriors in communicating social meanings (Duncan et al., 1985 vs. Sadalla et al., 1987; cf. Rapoport, 1990d, p. 230).

It would be interesting to find out how many replications there are. My sense is that there have been very few, whichever of the many meanings of "replication" listed above is used, or in the sense of picking up interesting findings or ideas to see whether they apply (and if so—how, where, under what conditions). The suggestion that three-dimensionality of façades predicts preference (Westerman, 1976) has, as far as I know, never been studied further. Similarly with Wohlwill's (1983) (implied) suggestion that vernacular environments are liked because their irregularity resembles "natural" environments. There is also a potential lateral link between these two suggestions that may well be worth exploring (cf. also Schaur, 1991).

There can, of course, be too much replication (in the sense of repetition). Thus Heinrich (1993, p. 527) suggests that research can become redundant, with studies merely corroborating what is already known and reliably established, which implies knowing what has been done, and how well it is established.

Knowledge of the Literature(s)

It is essential, before doing any work, to familiarize oneself with all the literature(s) and, moreover, to *use* that knowledge explicitly in the study, so as to place it in context. This may often be difficult in terms of time, which suggests the urgency of forming adequate and complete data bases, to which task some researchers should devote themselves. Although some such attempts have been made in EBS,

---

[8]There are arguments in the philosophy of science literature about the impossibility of replication,that replication is a fiction or myth. But the fact is that it *is* being done all the time. In any case, I will neither review that literature here, nor argue the case, but accept (or assume) that replication or attempts at it in this sense are indeed possible, feasible, and done all the time in other fields.

[9]This assumption is also implicit in the Australian Green Streets project which, in effect, "disguises" higher density ("densification") partly through the use of vegetation (as discussed in more detail in the paper mentioned in fn. 3 above).

so far none have really succeeded. The task is admittedly made more difficult by the diversity and fragmentation of sources (Rapoport, 1995b). This also implies that ideally any empirical study would explicitly be embedded in a general context, whether a synthesizing framework or knowing what has already been achieved and identifying the gaps still remaining. This becomes more important as the speed of development increases, as it inevitably would. This implies getting away from practical problem orientations as well as disciplinary orientations and emphasizing *conceptual*, *abstract*, and *general* aspects even of specific problems.

Maximum Linear and Lateral Connections

The goal needs to be always to connect seemingly disparate items, bodies of literature, types of work, fields, approaches, and methods at all levels within EBS, as well as seeking intertheory and interfield linkages, support, and constraints. It would be a useful exercise to do this explicitly and systematically for all existing EBS work.

A major advantage would be that new material from whatever source immediately fits somewhere into that larger framework and sets up and suggests many relationships and linkages (seeking which should be a constant goal). These consequences make it almost inevitable that groups, settings, and times would be transcended.

A personal example. Writing this chapter coincided with my teaching an EBS course on urban design. All my courses are an attempt to provide and stimulate such EBS frameworks, of which any specific topic is the vehicle—whether vernacular design, urban design, housing or design for developing countries (all of which have many linkages among them). Many new materials, such as journals, books, and newspapers fit during the semester. Toward the end, within four days, three newspaper stories fit perfectly into the urban design framework and consequently also into the much broader EBS framework.

The first followed the very next morning a discussion of levels of meaning and my suggestion about the loss of high-level meanings in contemporary environments (Rapoport, 1988a, 1990d, 1990g, epilogue) in which I explicitly used churches as one example. The article on the design of "megachurches" (Goldberger, 1995) makes very clear that only middle- and low-level meanings are being expressed.[10] Moreover, because abstract concepts and a framework were being used, an article on *buildings* could illuminate a discussion of *urban design*.

The second piece (Kaufman, 1995) well illustrated the multisensory nature of environmental perception (Rapoport, 1977a, 1992a, b). It concerned smells (the "olfactory landscape," as it were, of New York City). The article ignores conflicts about, e.g., cooking smells (Rapoport, 1977a), and thus ignores culture, meaning, identity and stigma, environmental quality, and so on. Many other articles and portions of books (especially novels) would also fit (cf. Rapoport, 1990f).

Finally, discussions about the relative failure of pedestrian malls in the United States (e.g. Robertson, 1990) benefited from a story about the remotorization of a 20-year-old semipedestrian mall in Milwaukee (Mitchell street) (Nichols, 1995) and

---

[10]This was also clear from previous articles in this series, e.g., on combining mall-culture and the draw of entertainment with religious services.

also from an article in a general interest magazine that pointed out that as early as 1798 Americans did not like to walk (Garreau, 1994, p. 24). This also provides linkages to discussions of pedestrian movement as involving aspects of both culture and perception (e.g., Rapoport, 1981b, 1986a, 1987).

A framework able to use such material not only enlarges the body of evidence available, but also has methodological implications regarding the use of newspapers, advertising, novels, films, TV and so on (Rapoport, 1977a, 1990d, f) and allows the use of material based on very different approaches. Synthesis and theory allow data and concepts to be integrated despite varied approaches, to the extent that they are "scientific" or can be interpreted in such a way (for example, semiotics and meaning [Rapoport, 1990d, epilogue] or Bonnes et al.'s [1990] transactional approaches to urban "places" [cf. Rapoport, 1994a], which fits my "systems of settings" [Rapoport, 1990c]. Even phenomenology (as "raw data") and introspection may fit—as long as they are the starting point, not the end.

In this general process of synthesis and theory building, intertheory and interfield linkages and support are critical, i.e., linkages with fruitful fields, and disciplines. It is important to realize that "fruitfulness" changes and is established at conceptual, fairly abstract levels. This also applies to linkages that only become possible (or even *visible*) at such levels. This is particularly important in EBS, which itself is multidisciplinary. The value of such intertheory links is that no findings or theory can violate well-established findings, mechanisms, and theories in relevant fields. Such intertheory links thus also provide major constraints on theorizing. This can be visualized as shown in Figure 28.1; it can also be applied to the relation between EBS and design (Figure 28.2)—another example of maximum generality.

I had intended to use the two announcements of this symposium, EDRA 25 and 26, IAPS 13, and a few papers or topics from journals as examples. Lack of space makes this impossible and it will have to be postponed. I would, however, again emphasize the need for systematic, international surveys to synthesize findings, to ask specific questions, and to form hypotheses for testing, and to link the findings to larger EBS concepts. It is important that the survey be complete and systematic; that synthesis become a major effort so as to reveal general patterns, agreement, and disagreement (or anomalies), so that attempts can be made to explain all those. This leads to theory.

This general need to be able to synthesize and integrate different approaches and methods can create epistemological problems (e.g., Sheridan, 1994) yet it is essential and, I believe, possible. The only requirement for this to happen is the single assumption (or decision) to accept a "scientific" approach. Many data using different approaches can become usable as *data* (although they still need explanation) (cf. Rapoport, 1990d, epilogue). In fact, a multiplicity of methods and lines of evidence may strengthen theory.

Minimal Assumptions

In general, too much is assumed in EBS, including matters that should be seen as empirical questions within the framework of explanatory theory that itself assumes the *least* possible. Even ontology should be seen as an empirical question. Historically, many aspects of philosophy have become sciences and this process

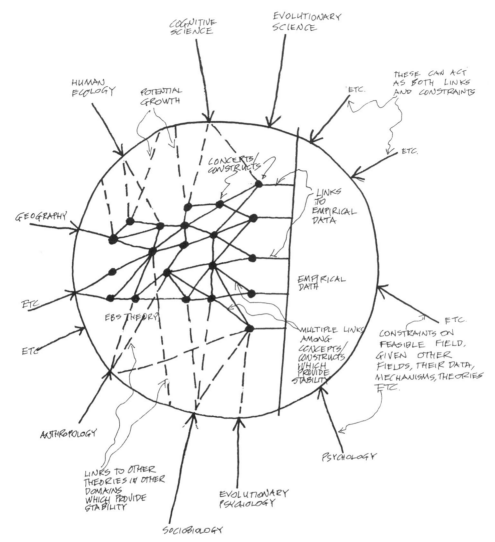

Figure 28.1. A diagramatic model of a theory and the intertheory linkages that constrain it. Based, in part, on Fig 3.2, Rapoport, 1990b, p. 69.

seems to be continuing (e.g., Himsworth, 1986). Although this is also beginning to happen to epistemology (as it is "naturalized"), at the moment only a single epistemological assumption needs to be made—that science is the only way of *knowing* the world, and hence of understanding, explaining, predicting it, and applying the knowledge gained. There are obviously other ways of *interacting* with the world (aesthetic, mystical, affective, conative, etc.) (e.g., Rapoport, 1990a, b). But it then also follows that even those other ways are knowable (and hence explainable) through scientific study—that *anything* is. This may prove wrong, but any other position becomes a self-fulfilling prophecy. Another consequence is that humans can be studied in the same way, and it has been suggested that rather than contrast-

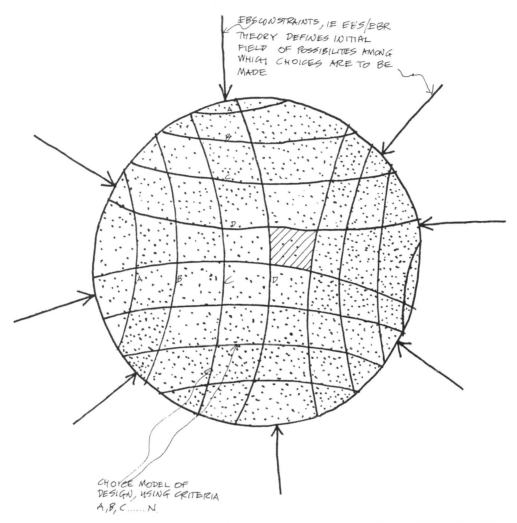

FIGURE **28.2.** The choice model of design with EBS constraining the initial field of possibilities. Based, in part, on Fig 1.7, Rapoport 1977a, p. 17.

ing "hard" and "soft" sciences, it is more useful to ask how to move from well-established, reliable knowledge to less-well established ones (e.g., Boulding, 1990).

Clearly, this raises a basic and highly contentious questions: What is science? Even accepting that the demarcation criteria are not as self-evident as previously thought; intuitively (and in practice) there exists a taxonomy of fields, and that distinction is made: We seem to know science when we encounter it. Although an adequate discussion is impossible here, some minimal sets of criteria can be given (e.g. Rapoport, 1990b, short list pp. 58–66, longer list pp.119–120). Moreover, the specifics regarding EBS can be developed, as in archaeology (partly) and in biology (more fully).

In any case, this single epistemological assumption suffices. Everything else follows, and no other major assumptions are needed, many becoming topics for empirical research, i.e., at most hypotheses to be tested, based on the best available knowledge from a variety of relevant fields. Rather than assuming the role of culture, for example (or speaking of "cultural theories"), the question becomes "What is the role of culture in any given case, vis-à-vis other variables (e.g., climate)?" In addition, as I have long argued that terms such as "culture," "environment," "tradition," "meaning," and so on need to be made more operational by dismantling (Rapoport, 1977a, 1988a, 1989a, 1990d, epilogue, 1990e, g, h, 1992a, 1993b, 1995c). This, in turn, becomes a *general* approach, emergent from the one assumption, and applicable to any topic (e.g., privacy) Altman, 1975; (Rapoport, 1977a). Dismantling, analysis, characterization, the study of mechanisms, imposing constraints, then synthesis and theory building is a typical sequence in science that typically asks: What is this? How does it work? How did we find out what we know? How can we find out more?

Maximum Generality

From the beginning, and at all times, one must strive for maximum generality, whether this be regarding philosophical assumptions general enough for EBS/EBR, formulations of the domain and parts of it, models, concepts, etc. Thus, as I argued at EDRA 26, "environmental aesthetics" is a perceptual part of environmental quality, which itself compresses and generalizes large amounts of information about physical, social, natural, and built environments; fixed, semifixed and nonfixed features, etc.; many studies (some of which may not even appear at first glance to be related) (e.g., Rapoport, 1990e, 1995c). The concept of environmental quality also leads to prediction and further material that fits that framework.[11] Moreover, the two can be linked by the three basic questions of EBS, my very general model of environmental evaluation and preference (Rapoport, 1977a, 1992b, pp. 37, which also introduces constraints), and so on. That also allows one to study how important perceptual aspects of environmental quality are vis-à-vis other components, for example, meaning ("symbolic aesthetics) (Rapoport, 1990e; cf. Khattab, 1993, 1994 esp. Chs. 5 and 6).

This generally also characterizes what I call the "choice model of design," which, like the others, applies to almost any situation. Here, however, I will only discuss the most basic and general formulation of the field in terms of what I call the three basic questions of EBS.

1. What are the characteristics of people, as members of a species, as individuals, and as members of various groups, ranging from families to societies, that shape the environment and, in design, should shape the environment so that it is congruent with these characteristics and supportive of them?

2. In what ways do which attributes of which environments affect what groups, in which ways, under what sets of circumstances, why and how?

---

[11]Thus environmental quality and meaning are central in my Tunis paper (see fn. 3), which has, since December 1994, been "confirmed" by several pieces of literature (both scientific and from newspapers).

3. Given this two-way interaction between people and environment, they must be linked in some ways: What are the mechanisms that link them and what are their characteristics (e.g., Rapoport, 1977a, 1983b, 1990b, 1995c).

The "environment" can then be conceptualized in two very general ways that are complementary: As the organization of space, time, meaning and communication, which leads to systems of settings/cultural landscapes made up of fixed, semifixed and nonfixed features, linked to behavior via systems of activities (including their latent aspects), and communicated by cues. This very general conceptualization (and dismantling it) once again accommodates a large amount of material (e.g., Rapoport, 1977a, 1990b).

This formulation seems to allow a maximum of potential lateral connections and also the derivation of other formulations (e.g., Moore, Tuttle, & Howell, 1985). Thus user groups and socio-behavioral phenomena are aspects of question 1; settings are a component of question 2. Question 3 introduces another important aspect: The emphasis on *mechanisms* generally, so that the identification, specification and characterization of mechanisms increase confidence, and scientific journals (and the history of science) provide many examples of their importance. Given their limited number the explicit specification and gradual characterization of mechanisms should not be too difficult (e.g., perception, cognition, evaluation, and preference—and hence choice and action)—meaning, supportiveness, and congruence). They are hence also very general and broadly applicable and also follow from the single epistemological assumption. Again, the mass of data we already have can help in starting this process.

The environmental quality and nature of settings (combining Barker et al., Goffman, Kaminsky, Kruse, Minsky, Abler, and Schank, etc.) (Rapoport, 1990d, esp. epilogue), and their cues, all become part of question 2. As usual, its use requires precision and explicitness. An environment must be shown to affect *defined* behavior in specific ways and these behaviors must be those people normally carry out (i.e., they are part of human ethology—except, possibly, in experiments). These behaviors must be capable of consistent identification, precise definition, and description, both in its original and changed state (cf. the "experiments" described in Rapoport, 1989b, 1990a). Also, the characteristics of any specific environment (system of settings) can be specified for any given case, again unifying the field (as does the use of cultural landscapes).

User groups can be derived from question 1; they are people with certain characteristics that can be specified in any given case, and their relative importance discovered; whereas "special user groups" tend to fragment the field, this formulation tends to unify it. This also applies to "culture," "subculture," "class," or whatever. Question 1 also allows new findings from other fields to be incorporated. One example is the recent discovery of the slow pace of human evolution (Gibbons, 1995), which makes evolutionary science, evolutionary psychology (e.g. Barkow et al., 1992), sociobiology, and related topics more important and part of the conceptualization of the field. Note that fields such as these do not neglect the role of culture, but rather study both possible limits of cultural variability and the evolution of culture itself.

In all these cases, *general principles* can be developed relatively easily and applied to people and settings with (specified) characteristics; the latter need to do certain (specified) things through certain (specified) mechanisms. Clearly specifics need to be researched, so that research will continue to be specialized concerning specified groups, settings and times, and so on. However, such research will not occur in isolation but in the broadest context or framework in both its point of departure and at least one of its goals. By not being limited to specifics, it will tend to coalesce, become cumulative, and help develop the broadest, most general theory. Yet, apparently my work has been criticized as being too generalized (e.g., Napier, 1994, p. 40).

Thus, while people will still specialize (as they must) there is also need for specialization in theory building. Moreover, any specialized subdomain or study must always ask, and consider, how it fits the larger picture and leave as many "dangling ends" or "hooks" to allow for connections.

CONCLUSION

I have briefly discussed a rather minimal set of problems and possible responses, which cannot be developed more fully here. I have, however, tried to show that even such a minimal set of response tends to lead to transcending times, settings, and groups (as would the use of cultural landscapes and systems of settings as units of analysis [Rapoport, 1990c, 1992b]). There is a vast amount of undigested material that needs to be studied, synthesized, and absorbed before (or at the same time as) many more empirical studies are undertaken. Only in this way, and by following the types of suggestions made, will it become possible to transcend settings, groups, times, and many other (often arbitrary) divisions.

I have already discussed settings. One needs to specify their spatial and temporal boundaries, cross-cultural and temporal constancies and differences, the nature and repertoire of physical cues (fixed, semifixed and nonfixed), and the redundancy needed under various conditions, the behaviors elicited and through which mechanisms. The important aspect of this list is its generality and high level of abstraction, so that it crosscuts and thereby transcends any specific settings, people, and their characteristics (and the constancy and change of those), all of which become empirical questions.

One could thus ask how important age is (whether of children or the elderly) vis-à-vis other variables. By specifying relevant human characteristics (pancultural, cultural, subcultural, age-related, etc.) (question 1), as well as attributes of settings (question 2) and mechanisms (question 3), one further transcends and links various divisions.

In addition to meetings devoted to some of the topics I have discussed, much more work is needed on organizing and synthesizing already existing material, including applied work (projects, reports, programs, POEs, etc.). The resulting synthesis should then be used in textbooks and conference organization (and classes) so that it develops, gets stronger, advances, and relates to conceptualizations of EBS.

This would produce a "cognitive map" of the field and domain, revealing concepts and their interactions as well as interfield and intertheory linkages. This might reveal an emergent framework on which more people might agree, particularly if it can be shown that many data and approaches fit, so that little is lost; in any case this can be improved with further work.

Several things are needed for any inquiry. Most obviously one needs to believe that it is worth doing. One then needs to specify a perspective or standpoint (in this case "science"). One then needs to examine the evidence for veracity and reliability. How well such evidence is established also depends on its relation to other fields and theories. Further work then follows by asking "what if" and generating hypotheses, counterfactuals, mental experiments, etc.

Clearly not everyone is interested in, or wants to do, theoretical work—or even basic empirical work. Some are committed to applied research, others are interested in research application—they want to *change* the world rather than to understand it. But as I have previously argued (Rapoport, 1986b, 1989b, 1990a, 1992c, 1995a, d) the only way to change the world in predictable ways (i.e., designing as applying knowledge) is through first acquiring knowledge through science.

In general, one begins with description (and we even lack descriptive languages, although attempts have been made by Cullen, Halprin, and Thiel). One then establishes patterns; explaining these is theory. Application follows. Thus these various aspects are complementary rather than conflicting. But the basis must always be an explanatory theory of EBR. Only then can predictable results be achieved, design seen as hypotheses validly and reliably proposed, and real evaluation done, because one cannot evaluate whether a thing does something well until one knows what it is supposed to do.

Explanatory theory is also essential to application for another reason—in order to generate *rules*. In the past these were generated through "selectionist" processes, whereas now they are "instructionist" and require reliable knowledge—but that is another topic.

REFERENCES

Altman, I. (1975). *The environment and social behavior*. Monterey, CA: Brooks/Cole.

Altman, L. K. (1995). Whiplash treatments found to be ineffective. *New York Times, (May 2)*.

Appleyard, D., Lynch, K., & Meyer, J. (1964). *The view from the road*. Cambridge, MA, MIT Press.

Arreola, D. A. (1988). Mexican American housescapes. *Geographical Review, 78*(3), 299–315.

Asfour, K. (1993). Cairene traditions inside Paladian villas. *Traditional Dwellings and Settlements Review, 4*(2), 39–50.

Barinaga, M. (1995). Dendrites shed their dull image. *Science, 268*(5208), 200–201.

Barkow J. H., Cosmides, L., & Tooby, J. (Eds.). (1992). *The adapted mind (Evolutionary psychology and the generation of culture)*. New York: Oxford University Press.

Bechhoefer, W. (1989). Procession and urban form in a Sri Lankan Village. *Traditional Dwellings and Settlements Review, 1*(1), 39–48.

Bell, J.A. (1994). Interpretation and testability in theories about prehistoric thinking. In C. Renfrew and E. B. W. Zubrow (Eds.), *The ancient mind (Elements of Cognitive Archaeology)* (pp.15–21). Cambridge: Cambridge University Press, .

Berlin, B. (1992). *Ethnobiological classification*. Princeton: Princeton University Press.

Berlin, B., & Kay, P. (1969). *Basic color terms: Their universality and evolution*. Berkeley: University of California Press.

Blanton, R. E. (1994). *Houses and Households: A comparative study*. New York: Plenum.

Bonnes, M., Mannett, L., Secchiaroli, G., & Tanucci, G. (1990). The city as a multi-place system: An analysis of people-urban environment transactions. *Journal of Environmental Psychology, 10*, p.37–65.

Boulding, K. E. (1990). Science: Our common heritage. *Science, 207*(4433), 831–836.

Carr, S., & Schissler, D. (1969). The city as trip: Perceptual selection and memory in the view from the road. *Environment and Behavior, 1*(1), 7–36.

Cherulnik, P. D. (1991). Reading restaurant façades: Environmental inferences in finding the right place to eat. *Environment and Behavior, 23*(2), 150–170.

Dewsbury, D. A. (Ed). (1989). *Studying animal behavior (Autobiographies of the founders)*. Chicago: University of Chicago Press.

Duncan, J. S. (Ed). (1981). *Housing and identity: Cross-cultural perspectives*. London: Croom Helm.

Duncan, J. S., Lindsey, S., & Buchan, R. (1985). Decoding a residence: Artifacts, social codes and the construction of the self. *Espaces et Sociétés, 47*, 29–43,.

Evans, G. W. (1995). What's wrong with self-reported measures in environment and behavior research? In A. D. Seidel (Ed.), *Banking on Design (EDRA 25)*, 83–91.

Foote, K. E. (1983). *Color in public spaces (toward a communication-based theory of the urban built environment)*. Chicago: University of Chicago, Dept. of Geography Research paper 205.

Garreau, J. (1994). Don't Walk. *The New Republic*, Issues 4157, p. 24 & 4158, p. 28.

Gibbons, A. (1995). When it comes to evolution, humans are in slow class. *Science, 267*(5206), 1907–1908.

Glen, W. (Ed). (1994). *The mass-extinction debates: How science works in a crisis*. Stanford, CA: Stanford University Press.

Goldberger, P. (1995). The Gospel of church architecture, revised. *New York Times, (April 20)*.

Goldsmith, D. (1995). Extreme ultraviolet satellites open new view of the sky. *Science, 268*(5208), 202–203.

Heinrich, B. (1993). *The hot-blooded insects (Strategies and mechanisms of thermoregulation)*. Cambridge, MA: Harvard University Press.

Henderson, D. K. (1993). *Interpretation and explanation in the human sciences*. Albany, SUNY Press.

Himsworth, H. (1986). *Scientific knowledge and philosophic thought*. Baltimore: Johns Hopkins University Press.

Holldobler, B., & Wilson, E. O. (1990). *The ants*. Cambridge, MA: Belknap Press of Harvard University.

Horwich, P. (Ed.). (1993). *World changes (Thomas Kuhn and the nature of science)*. Cambridge, MA: MIT Press/Bradford Books.

Johnson, P. (1965). Whence and whither: The processional element in architecture. *Perpecta*, 9/10, 167 178.

Kaitilla, S. (1991). The influence of environmental variables on building material choice: The role of low-cost building materials on housing improvement. *Architecture and Behavior, 7*(3), 205–221.

Kaitilla, S. (1994). Urban residence and housing improvement in a Lae squatter settlement, Papua New Guinea. *Environment and Behavior, 26*(5), 640–668.

Kaufman, M. T. (1995). Aroma experts agree: Smells are among the city's attractions. *New York Times, (April 22)*.

Kellert, S. H. (1993). *In the wake of chaos (unpredictable order in dynamical systems)*. Chicago: University of Chicago Press.

Khattab, O. (1993). Environmental quality assessment: An attempt to evaluate government housing projects. *Open House International, 18*(4), 41–47.

Khattab, O. (1994). *Lifestyle and environmental quality in the enabling settlement. Siwa Oasis, Egypt*. Ph.D. Dissertation, University of Newcastle upon Tyne (unpublished).

Kuhn, T. (1970). *The structure of scientific revolutions* (2nd edition). Chicago: University of Chicago Press.

Lang, J. W. (1995). Review of L. J. Vitt and E. R. Pianka (Eds.) *Lizard Ecology* (Historical and experimental perspectives). *Science, 267*(5204), 1668–1669.

Livingston, M. (1992). Bamboo housing and the mangrove of Guayaquil. *Traditional Dwellings and Settlements Review, 4*(1), 13 (abstract only).

Malmberg, T. (1980). *Human territoriality*. The Hague: Mouton.

McMullin, E. (1993). Rationality and paradigm change in science. In P. Horwich (Ed.), *World changes (Thomas Kuhn and the nature of science)* (pp. 55–78. Cambridge, MA: MIT Press/Bradford Books.

Moore, G. T., Tuttle, D. P., & Howell, S. (1985). *Environmental design directions*. New York: Praeger.

Napier, M. (1994). Review of M. Turan (Ed.) *Vernacular architecture: paradigms of environmental response*, Aldershot (UK) Avebury 1990. *Open House International, 19*(3), 39–41.

Nichols, M. (1995). Mitchell St. abandons 'mall,' tries to regain distinct character. *Milwaukee Journal-Sentinel, (April 23)*.

O'Donnell, S. (1995). *Towards urban frameworks: Accommodating change in urban cultural landscapes*. M. Arch. Thesis, Dept. of Architecture, University of Wisconsin-Milwaukee (May) (Unpublished).

Orland, B., Vining, J., & Ebreo, A. (1992). The effect of street trees on perceived values of residential property. *Environment and Behavior, 24*(3), 298–325.

Pader, E. J. (1993). Spatiality and social change: Domestic space use in Mexico and the United States. *American Ethnologist, 20*(1), 114–137.

Perin, C. (1970). *With man in mind*. Cambridge, MA: MIT Press.

Philip, D. (1990/91). Colour and affect: Three studies. *People and Physical Environment Research, 34–35*, 16–29.

Pool, R. (1995). A boom in plans for DNA computing. *Science, 268*, No.5210, 498–499.

Proshansky, H. (1973). Theoretical issues in environmental psychology. *Representative Research in Social Psychology, 4*, 93–107.

Rapoport, A. (1973). An approach to the construction of man-environment theory. In W. F. E. Preiser (Ed.), *Environmental Design and Research (EDRA 4)*(2), 124–135.

Rapoport, A. (1977a). *Human aspects of urban form*. Oxford: Pergamon.

Rapoport, A. (1977b). Our last (and first) ten years. *Man-Environment Systems, 7*(6).

Rapoport, A. (1981a). Identity and environment: A cross-cultural perspective. In J. S. Duncan (Ed.), *Housing and identity (Cross-cultural perspectives)* (pp. 6–35). London: Croom-Helm.

Rapoport, A. (1981b). On the perceptual separation of pedestrians and motorists. In H. C. Foot, A. J. Chapman, & F. M. Wade. (Eds.), *Road safety—research and practice* (pp. 161–167). Eastbourne (UK): Praeger.

Rapoport, A. (1983a). Development, culture change and supportive design. *Habitat International, 7*(5/6), 249–268.

Rapoport, A. (1983b). The effect of environment on behavior. In J. B. Calhoun (Ed.), *Environment and population (Perspectives on adaptation, environment and population)* (pp. 200–201). New York: Praeger.

Rapoport, A. (1985). Thinking about home environments: A conceptual framework. In I. Altman & C. M. Werner (Ed.), *Home environments (human behavior and environment*, Vol. 8) (pp. 255–286). New York: Plenum.

Rapoport, A. (1986a). The use and design of open spaces in urban neighborhoods. In D. Frick (Ed), *The quality of urban life: Social, psychological and physical conditions* (pp. 159–175). Berlin: de Gruyter.

Rapoport, A. (1986b). Culture and built form—a reconsideration. In D. G. Saile (Ed.), *Architecture in cultural change: Essays in built form and culture research* (pp. 157–175). Lawrence: University of Kansas.

Rapoport, A. (1987). Pedestrian street use: Culture and perception. In A. Vernez Moudon (Ed.), *Public streets for public use* (pp. 80–92). New York: Van Nostrand Reinhold.

Rapoport, A. (1988a). Levels of meaning in the built environment. In F. Poyatos (Ed.), *Cross-cultural perspectives in non verbal communication* (pp. 317–336). Toronto: Hogrefe.

Rapoport, A. (1988b). Spontaneous settlements as vernacular design. In C. V. Patton (Ed.), *Spontaneous shelter* (pp. 51–77). Philadelphia: Temple University Press.

Rapoport, A. (1989a). On the attributes of tradition. In J. Bourdier and N. Al Sayyad (Eds.), *Dwellings, settlements and tradition (Cross-cultural perspectives)* (pp. 77–105). Lanham, MD: University Press of America.

Rapoport, A. (1989b). A different view of design. *University of Tennessee Journal of Architecture, 11*, 28–32.

Rapoport, A. (1990a). Science and the failure of architecture. In I. Altman and K. Christensen (Eds.), *Environment and behavior studies—emergence of intellectual traditions (Vol.11 of Human behavior and environment)* (pp. 79–109). New York: Plenum.

Rapoport, A. (1990b). *History and precedent in environmental design*. New York: Plenum.

Rapoport, A. (1990c). Systems of activities and systems of settings. In S. Kent (Ed.), *Domestic architecture and use of space (An interdisciplinary cross-cultural study)* (pp. 9–20). Cambridge, UK: Cambridge University Press.

Rapoport, A. (1990d). *The meaning of the built environment*. Tucson: University of Arizona Press.

Rapoport, A. (1990e). Environmental quality and environmental quality profiles. In N. Wilkinson (Ed.), *Quality in the built environment* (pp. 75–83). Newcastle, UK: The Urban International Press.

Rapoport, A. (1990f). Indirect approaches to environment-behavior research. *The National Geographical Journal of India, 36* (1 & 2), 30–46. (Also in R. B. Singh (Ed.), *Literature and humanistic geography*, Varanasi (India) Banaras Hindu University, National Geog. Society of India, Research Publication Series (1990) 37, 30–46.

Rapoport, A. (1990g). Levels of meaning and types of environments. In Y. Yoshitake, R. B. Bechtel, T. Takahashi, & M. Asai (Eds.), *Current issues in environment-behavior research* (pp. 135–147). Tokyo: University of Tokyo.

Rapoport, A. (1990h). Defining vernacular design. In M. Turan (Ed.), *Vernacular architecture* (pp. 67–101). Aldershot (U.K.): Avebury.

Rapoport, A. (1992a). On regions and regionalism. In N. C. Markovich, W. F. E. Preiser (Eds.), *Pueblo style and regional architecture* (pp. 272–294). New York: Van Nostrand-Reinhold (paperback edition only).

Rapoport, A. (1992b). On cultural landscapes. *Traditional dwellings and settlements review, 3*(2), 33–47.

Rapoport, A. (1992c). Some thoughts on the future of environmental design. In S-K Kim (Ed.), *Tradition and creation* (Proceedings of 29th IFLA Congress), 223–233. Seoul, South Korea: KIFLA.

Rapoport, A. (1993a). On the nature of capitals and their physical expression. In J. Taylor, J. G. Lengette, & C. Andrew (Eds.), *Capital cities: International perspectives* (pp. 31–67). Ottawa: Carleton University Press.

Rapoport, A. (1993b). *Cross-cultural studies and urban form*. College Park, MD: University of Maryland, Urban Studies and Planning Program.

Rapoport, A. (1994a). A critical look at the concept 'place.' *National Geographical Journal of India, 40*, 31–45. (Also in R. B. Singh (Ed.), *The spirit and power of place: Human environment and sacrality. Varanasi: Banaras Hindu University*, National Geographical Society of India, Publication 41, 34–45.

Rapoport, A. (1994b). Sustainability, meaning and traditional environments, Berkeley, CA., Center for Environmental Design Research, *Traditional Dwellings and Settlements Working Paper, 75/IASTE* 75–94.

Rapoport, A. (1995a). On the nature of design, *Practices, 3/4*, 32–43.

Rapoport, A. (1995b). *Thirty three papers in environment-behavior research*. Newcastle, U.K.: Urban International Press.

Rapoport, A. (1995c). A critical look at the concept 'home.' In D. N. Benjamin and D. Stea (Eds.), *The home: Words, interpretations, meanings and environments* (pp. 25–52). Aldershot, U.K.: Avebury.

Rapoport, A. (1995d). Rethinking design (an environment-behavior perspective). In E. Schaur (Ed.), *Building with intelligence (Aspects of a new building culture)* Publication IL 41 (pp. 235–242). Germany: University of Stuttgart.

Rapoport, A. (in press). *On the relation between culture and environment*. Aris 3 (Carnegie Mellon University).

Robertson, K. A. (1990). The status of the pedestrian mall in American downtowns. *Urban Affairs Quarterly, 26*(2), 250–273.

Sack, R. D. (1986). *Human territoriality (Its theory and history)*. Cambridge: Cambridge University Press.

Sadalla, E. K. & Sheets, V. L. (1993). Symbolism in building materials: Self-presentational and cognitive components. *Environment and Behavior, 25*(2), 155–180.

Sadalla, E. K., Vershure, B., & Burroughs, J. (1987). Identity symbolism in housing. *Environment and Behavior, 19*(5), 569–587.

Schaur, E. (1991). *Non-planned settlements,* Institute for Lightweight Structures. Germany: University of Stuttgart, Publication IL 39.

Schön, D.A. (1983). *The reflective practitioner*. New York: Basic Books.

Science (1995). 267, No.5196 (Jan. 20), 330.

Sheridan, T. E. (1994). Review of J. D. Rogers & S. M. Wilson (Eds.), *Ethnohistory and archaeology* (Approaches to postcontact change in the Americas). *Science, 265*, (5169), 270.

Sommer, R., & Sommer, B. A. (1983). Mystery in Milwaukee: Early intervention, IQ and psychology textbooks. *American Psychologist, 38*, 982–985.

Taylor, R. B. (1988). *Human territorial functioning (An empirical, evolutionary perspective on individual and small group territorial cognitions, behaviors and consequences)*. Cambridge: Cambridge University Press.

Urban Forests, 14(4), August/Sept. 1994.

Vairavan, B. (1995). *Meaning in urban environment (A case study of Madurai, India)*. Term paper in Arch. 741, Dept. of Architecture, University of Wisconsin-Milwaukee (Spring) (unpublished).

Westerman, A. (1976). *On aesthetic judgment of building exteriors*. Doctoral Dissertation in Architecture, Royal Technical Institute, Stockholm (Sweden) (Swedish with English Abstract).

Willer, D. (1987). *Theory and the experimental investigation of social structures*. New York: Gordon and Breach Science Publishers.

Willer, D., & Willer, J. (1973). *Systematic empiricism: Critique of a pseudoscience*. Englewood Cliffs, NJ: Prentice-Hall.

Wohlwill, J. F. (1973). The environments is not in the head. In W. F. E. Preiser (Ed.), *Environment Design Research (EDRA 4), 2*, 166–181.

Wohlwill, J. F. (1983). The concept of nature: A psychologist's view. In I. Altman & J. F. Wohlwill (Eds.), *Behavior and the natural environment (Vol. 6 of Human behavior and environment)* (pp. 5–37). New York: Plenum.

Chapter **29**

ENVIRONMENT AND BEHAVIOR STUDIES
A Discipline? Not a Discipline? Becoming a Discipline?

Irwin Altman

INTRODUCTION

This chapter raises the question as to whether the now 25-year-old venture known as environment–behavior studies is or can ever be a distinctive "field" of study. In analyzing this issue, I will conclude that environment–behavior studies do not presently constitute a traditional "field" or "discipline." However, this does not mean that scholars and practitioners who have devoted themselves to years of environment–behavior work have wasted their time nor should it discourage newcomers from investing their energies in this topic. There is much exciting work to be done and there are enormous opportunities for practical and conceptual creativity. Most important, I will argue that status as a "field" or "discipline" is ultimately less important than understanding who we are, why we do what we do, and the connections or gaps between scholars and practitioners adopting diverse perspectives. That is, if we are to make progress in understanding the richness of environment–behavior phenomena, we must come face-to-face with some fundamental philosophical assumptions about our research and action. To date we have not done so and this has resulted in rather disjointed, noncumulative, and somewhat incoherent bodies of knowledge, approaches to understanding, and conceptualizations about environment–behavior phenomena.

Irwin Altman • Department of Psychology, University of Utah, Salt Lake City, Utah 84112.

Handbook of Japan–United States Environment–Behavior Research: Toward a Transactional Approach, edited by Seymour Wapner, Jack Demick, Takiji Yamamoto, and Takashi Takahashi. Plenum Press, New York, 1997.

The first sections of the chapter examine our status as a field. The following sections describe the assumptions we must explicitly state and examine if we are to magnify our future understanding of people and environments.

THE CONTEMPORARY SCENE

Several institutional indicators suggest that environment–behavior studies may be a "field" or "discipline." Thus, there are several established journals, some published for more than two decades. There are also many textbooks, edited volumes, and book series on selected topics, a handbook of environmental psychology, and so on. In addition, the Environmental Design Research Association (EDRA) has met regularly and published proceedings for more than 25 years; and the International Association for People-Environment Studies (IAPS) conducts conferences every 2 years; the Man-Environment Research Association (MERA) in Japan is an active organization. Moreover, other conferences occur frequently, several universities have established undergraduate and graduate programs or specialties in regular academic disciplines, and there are some new departments and schools.

Although these activities reflect institutional and sociological signs of a "discipline," I see the underlying organizational foundation of environment–behavior studies to be relatively fragile. For example, my impression is that the membership of organizations like EDRA consists of (1) a small core of participants who regularly attend meetings and who have been members of the organization for many years, (2) a larger cohort of participants who are active for a few years, but who eventually leave, never to return, or return and participate only sporadically; and (3) many one-time attendees and members. Thus the central core of participants and members is small, with relatively few newcomers who remain active for a long period of time. I also sense that the membership of EDRA and IAPS has reached a plateau or only grows slowly. There are probably many reasons for this pattern: slow growth of programs and positions in academic settings, limited research and grant funds, a small market for consultation and design by practitioners, a fragile worldwide economy, fewer travel dollars for professionals and scholars, and so on.

Far more important than these institutional signs, however, are deeper issues reflecting the present status of environment behavior work. I addressed some of these briefly in a recent paper (Altman, 1995). Rapoport (Chapter 28, this volume) offers a more comprehensive and penetrating analysis, citing several problems with research and scholarship in environment and behavior studies: (1) absence of cumulation of knowledge, repetition of research, and inattention to prior literature—yielding a "reinventing of the wheel" atmosphere; (2) too many isolated empirical studies that are not connected to one another; (3) inadequate attention to research and theory in other relevant fields; (4) insufficient attempts to synthesize and review research findings; (5) too little attention to theory construction; (6) lack of agreement or attention to the meaning and use of terms and concepts; (7) not enough open and regular debate of key issues.

On the other hand, Stokols (Chapter 23, this volume) points to a number of contributions of environment and behavior research over the years, including attention to significant environmental problems facing society, opening new lines of thought and research regarding people and environments, drawing together research and theory from different fields and subfields, and others. Consistent with Stokols's perspective, there has been some coalescing of knowledge in selective reviews of the literature in edited volumes and journals, in the 1987 *Handbook of Environmental Psychology* (Stokols & Altman, 1987), and in the *Annual Review of Psychology* and the *Annual Review of Anthropology*. In pointing to future promising possibilities, Stokols identifies a variety of social issues to which environment–behavior research can contribute in the coming decades, including health, global environmental change, information technology, aging, and others.

In spite of past contributions, I agree with much of Rapoport's analysis of the state of environment–behavior research. We do not regularly and systematically synthesize our knowledge, attend closely to others' research, or do the things necessary to build a cumulative body of theory, research, and practice. Moreover, I often sense that the subject matter, methodology, concepts, and questions posed by environment–behavior researchers are so diverse that the connection between ideas, people, and approaches are weak and often nonexistent, resulting in separate and isolated islands of activity. One has only to examine the proceedings of the Environmental Design Research Association conference, or of this volume, to see how varied are the interests of participants and how difficult it is to see links between studies—even when they deal with the same topic. Although individual projects are interesting and well conducted, one must struggle (and often fail) to see the interconnections between the work of different investigators and programs. To be sure, the diversity of research is rich, provocative, and interesting, but it simply does not hang together.

So, is there a solution to the problem? Along with his positive assessment of the contributions of environment–behavior studies of the past decades, Stokols (1996) calls for future efforts to develop middle-range theories of environment and behavior that bridge disciplinary perspectives, capitalize on the unique historical, cultural, geographical, and political contexts of regions of the world, develop community-based problem-solving strategies, and seek contextually oriented theories. Thus, Stokols and others view the environment–behavior field as having laid the necessary groundwork for the future, however imperfect it may be, and that the time is at hand to take a new "leap forward." Problems yes, but the products of the past and the potential for the future are very promising—if we move in the right directions.

Although Rapoport presents a more pessimistic view of the past, he too implies that environment–behavior studies have considerable potential if scholars focus on appropriate theory development, systematically replicate their research, pay attention to existing research literature, develop connections with related fields, and adopt a classic philosophy of science approach to methodology, research, and theory.

So, in spite of their seeming divergence about the current state of affairs, Stokols and Rapoport both foresee considerable potential in the future—*if* scholars, researchers, and practitioners adopt a more centripetal and unifying direction in the future, in contrast with the more centrifugal pattern of the past.

IS THERE A "FIELD" OF ENVIRONMENT–BEHAVIOR STUDIES?

Stokols and Rapoport, while differing somewhat in their assessments of the present state of affairs, not only point to the potentialities of the future, but also seem to share a belief about the existence and/or possibility of a "field" or "discipline" of environment and behavior studies. I too have long assumed that we were in the process of forming a new and innovative interdisciplinary field of environment and behavior studies—analogous in some ways to interdisciplinary fields in biochemistry, biophysics, materials science, genetics, neuroscience, and social ecology. We all believed that environment and behavior studies could become a new and exciting discipline, perhaps with its own unique philosophy of science, epistemology, theory, research, and practice.

For the first time, however, I now wonder whether it is possible for us to develop a unified and coherent discipline. Although I am presently an "agnostic," let me pose the extreme position that the environment–behavior area, *as presently constituted*, is not and may never become a viable discipline or field and that general and abstract calls for cumulation, synthesis, and theory development are not the underlying solution to our future. If we are to become a field or do productive work in the future without being a field, then we must proceed beyond these general calls to action. Instead, we must address and articulate to one another our fundamental philosophical, theoretical, and epistemological assumptions—matters that we usually do not verbalize and that are often held and acted on without awareness. Let me begin by stating some general propositions and assumptions underlying my agnostic position.

First, and a seemingly elementary point, environment–behavior phenomena necessarily involve linking some aspect of the physical environment to a *human process*. Any attempt to focus on the environment as environment has no meaning in the endeavor in which we are engaged.

Second, human processes—whether they involve individuals, groups, communities, or cultures—are preeminent. The physical environment is background context, milieu, and a factor affecting or affected by a human process. However, it is ultimately some human process that we are interested in understanding, changing, influencing, or enhancing. The physical environment is only an instrumentality in the service of understanding and bettering human functioning.

Third, there is an incredible array of human processes (even if we limit ourselves to the social sciences) that are linked with physical environments. These range from individual psychological to group to institutional to cultural and societal processes and to economic, political, and a variety of other human processes. These human processes vary in scale or scope and in level of functioning. Indeed, they are so multifaceted that we have historically defined disciplines such as psychology, sociology, anthropology, political science, economics, and others to address different aspects of human functioning. Although some scholars have faith in the ultimate unity of these disciplines and in a grand theory of human functioning, the fact is that we have not made strong strides in this direction over the past decades. Furthermore, we are also seeing the fracturing of traditional disciplines as the breadth and diversity *within* traditional fields is often greater than the differences *between* disciplines. It is likely, therefore, that we will see the evolution of new disciplines based on unique and emerging common interests. I welcome such changes because the intellectual and social issues of the times require new configurations of ideas and perspectives.

But can these new configurations include a discipline known as "environment–behavior studies"? Although I am an agnostic on the question, I will argue, for the moment, that there can be no discipline of environment and behavior studies. Why not? Because new fields or disciplines, like traditional ones, must be organized around selective human processes, with the physical environment only one of several factors linked to those human processes. Thus, I can conceive of new disciplines of family and interpersonal studies, life cycle processes, health, quality of life in communities and cities, cognitive science, and others, each focusing on different clusters of human processes. But it is inconceivable to me that there can be a discipline focused around the physical environment and *all* human processes. And that is what I think we have tried to do in environment and behavior studies—create a field that deals with *all* human activities in respect to the environment.

The reality is that our work as individuals and groups is necessarily delimited. Thus in my own long-term work on interpersonal relationships among friends, acquaintances, intimates, and in families, I have drawn on ideas, data, theory and methodology from psychology, sociology, anthropology, history, architecture, interior design, and other fields, to examine how close relationships are embedded in communities, cultures, and a variety of aspects of the physical environment. I am not interested in the physical environment per se, nor in all aspects of human functioning. Rather, I study interpersonal relationships in the social and physical environmental contexts in which they are embedded. Although I believe that there is no human process that is independent of its environmental context, this does not mean that *all* human processes are part of a unified field. The emerging reality for me, therefore, is that there is not now, and may never be, a single coherent field of environment–behavior studies. Rather, there are traditional and emerging disciplines, all defined by particular human processes with which they are concerned, and each dealing with environmental contexts within which their favorite human processes are embedded. But it is the human processes that define these fields, not their common interest in environmental issues.

WHY CONTINUE TO DO ENVIRONMENT–BEHAVIOR WORK?

Although the previous line of reasoning may be threatening and upsetting to those, like myself, who have devoted years of their professional lives to believing in and attempting to create a true and unified environment–behavior field, I say "Never fear." We have not wasted our time. Indeed, we should continue to pursue the directions proposed by Rapoport and Stokols and others, and we should continue to meet, exchange ideas, and investigate environment and behavior phenomena. Why should we do this in light of the possibility that we are not now and may not become a true "field" or "discipline"?

However much we may question the current state of environment–behavior research, I consider it to have been an enormously beneficial and successful venture. The opportunity to learn about a diversity of content, and to be confronted by and debate colleagues who view things differently has enormously affected my research and thinking (see Altman, 1990, for a description of my "intellectual history"). The same sort of mind expansion has occurred for many other social scientists and environmental designers—both academic and professional (Altman & Christensen, 1990).

Most important, rubbing shoulders with environment–behavior researchers and practitioners of all stripes opened my mind to the idea that there are many ways of "knowing" and many approaches to scholarly "evidence." Based on years attempting to incorporate holistically the physical environment in my studies of close relationships, Barbara Rogoff and I synthesized the writings of the philosophers John Dewey and Arthur Bentley (1949) and Stephen Pepper (1942, 1967) into a taxonomy of "worldviews" (Altman & Rogoff, 1987). In so doing, we described four worldviews—trait, interactional, organismic, and transactional—that are distinguished in terms of their units of study, treatment of time and change, and their philosophy of science (and, by implication, their derivative research methodologies). I came to see how the transactional worldview suited the work I had been doing and wanted to continue into the future. Indeed, the transactional perspective enabled my colleagues and I to treat temporal processes, psychological processes, and social and environmental contexts as an inseparable and holistic unit of analysis in cross-cultural studies of home environments (Altman & Gauvain, 1981, Gauvain, Altman, & Fahim, 1983), courtship, weddings and placemaking (Altman, Brown, Staples, & Werner, 1992), studies of family and community rituals (Werner, Brown, Altman, & Ginat, 1993), and the dynamics of polygamous families, including use of home environments (Altman & Ginat, 1996). Moreover, regular contacts with scholars from a variety of design and social science disciplines for more than two decades helped me to escape from the narrow bonds of my earlier thinking and training and opened my eyes to the possibility that there *are* alternative ways of knowing and doing useful research and theorizing.

My work has also been enhanced by discovering research methods used in other fields. I learned about archival analysis from historians and sociologists, qualitative analyses and participant observation from anthropologists, phenomenological approaches from philosophers and others, postoccupancy evaluation strategies from social scientists and environmental designers, and I even gained a bit of "an eye" about aesthetics and environmental design from colleagues in architecture, landscape architecture, interior design, and other disciplines. New forms of data, new sources of information, and new ways of looking at things gradually found their way into my research—especially qualitative analysis, ethnographic methods and content, and ways for describing communities and homes—all of which helped me gain a broader grasp of the environmental contexts of close relationships. I know many colleagues who experienced the same sort of "mind expansion" as a result of soaking in the ideas, ways of knowing, and perspectives of other disciplines.

So, although I acknowledge current problems of synthesis, cumulation of knowledge, theory, and other issues noted earlier, I also recognize that we have accomplished a great deal in the past decades. For some of us progress has been invaluable beyond measure. As a result, I firmly believe that we must continue our intellectual discourse in the indefinite future, whether or not there is now a "field," and whether or not we aspire to create a "discipline." Without continued exchanges, conferences, journals, books, and professional collaborations, we will drift back into the narrow confines of our traditional disciplines, and become insular and provincial in our approach to environment–behavior issues.

THE FUTURE: REVEALING TO ONE ANOTHER

But what should we do to advance our work and overcome present problems? Although general calls for synthesis, cumulation of knowledge, attention to theory, replication of results, and so on, are appropriate, they must be more specific and pointed. *We must penetrate to the intellectual core of our thinking, and reveal our philosophical assumptions, epistemology, theoretical underpinnings, and the bases of our decisions at critical choicepoints in research and theorizing.* If we do this systematically, then it will be possible to see the underlying rationale for our collectively diverse research and problem solving activities, better understand how our and others' approaches are similar and different, and discern why and how alternative research and problem solving choices yield different outcomes.

Revealing our assumptions and reasons for choosing alternative courses of action is essential whether or not we are or strive to become a field. It is the only way to establish links between bodies of knowledge and between conceptual approaches—especially in light of the incredible array of perspectives, assumptions, methodologies, concepts, and participants in environment–behavior studies. Following are the types of things that I believe we must thoroughly reveal to one another about each and every aspect of our work.

Units of Analysis

The four philosophical ways of "knowing" or studying psychological phenomena described by Altman and Rogoff (1987)—trait, interactional, organismic, and transactional worldviews—are distinguishable in terms of the unit of analysis they study, how they treat time and change, and their philosophy of science.

Regarding units of analysis: trait, interactional, and organismic worldviews focus on separate and independently defined psychological processes, environmental and social contexts. Although differing in specifics, these three worldviews assume that phenomena should be analyzed into discrete and independent elements, dimensions or subunits. In contrast, transactional approaches assume the inseparability of psychological, physical environmental, and social aspects of phenomena, and seek to understand unified wholes.

I believe that different disciplines, areas within disciplines, and individual researchers often adopt different worldviews—often without awareness or conscious choice—and make different assumptions about the unit of analysis they study. Although I do not claim that there is a "proper" or "correct" unit of analysis, it is essential for us to understand why a researcher chooses a particular unit of analysis in a project or program of research. Thus the meaning of research or theory on phenomena such as "residential satisfaction," "privacy," "place attachment," "crowding," or any other topic depends heavily on the unit of analysis to which the concept refers. In order to connect, compare, or differentiate research we must specify the unit of analysis being studied, the rationale for selecting a particular unit of study, and the philosophical worldview guiding our thinking. So doing will also force us to think carefully about our own and others' choices and decisions in studying a particular issue. And, as noted below, choices of units of analysis are related to philosophy of science and research methodology, yielding a powerful system guiding each piece of research and theoretical framework.

Stability and Change

Worldviews also differ in their assumptions about temporal aspects of phenomena. Whereas the trait worldview essentially assumes stability of psychological phenomena or change resulting from preestablished teleological mechanisms, the interactional worldview assumes that change comes from the interaction and influence of separate environmental and person/social entities, and/or the operation of underlying regulatory mechanisms such as homeostasis. Or stability and change occur in the organismic worldview because of directional and predetermined underlying teleological mechanisms. In contrast, transactional worldviews assume that stability and change are intrinsic aspects of psychological and social phenomena, and that change does not necessarily proceed in a predetermined direction or under the guidance of monolithic mechanisms. Rather, stability and change are emergent and more freewheeling in incidence and direction.

Here again it is important to understand how different scholars incorporate notions of stability and change in their work. Those who treat psychological and social systems as changing largely by virtue of external factors—a common approach in the interactional worldview—will adopt different theoretical and research strategies, and perhaps obtain noncomparable results, from those who treat stability and change as an integral aspect of phenomena—a perspective associated with a transactional worldview. Thus if we are to compare or link diverse research on environment–behavior phenomena, it is essential that we specify our approach to temporal processes. And in so doing we should also state *why* we chose a specific perspective on temporal processes, in order to more fully expose similarities and divergence among research approaches and programs.

Philosophy of Science

Contrary to the views of some scholars, Dewey and Bentley (1949), Pepper (1942, 1967), and our extension of their analyses (Altman & Rogoff, 1987) suggest that there are legitimate alternative approaches to science, inquiry, or "knowing." And these different approaches or worldviews may have different rules of evidence, scholarly goals, approaches to causation, and views about "objectivity." For example, trait worldviews emphasize Aristotelian *material causation*; that is, the cause of an event resides in the "essence" of a phenomenon and is internal to that phenomenon. Interactional worldviews, in contrast, emphasize *efficient causation*, or antecedent–consequent relationships between variables. Organismic worldviews focus on *final causation*, or the teleological underpinning of phenomena that "pull" them toward some ideal end state. Transactional approaches rely more on *formal causation,* or the description and understanding of patterns and organization of events involving confluences of people, space, and time.

Although comprehensive knowledge of a phenomenon ideally involves *all* forms of Aristotelian causation, there is a tendency for different scholars to highlight one or another approach to causation, depending upon their worldview. And it is easy to see how research and theory may take different paths for alternative approaches to causation. Thus an interactional researcher will primarily examine antecedent–consequent links between variables; an organismic researcher will study underlying dynamics that pull or propel a phenomenon toward a particular end state; a transactional scholar will attempt to piece together a coherent account of some event, marshaling principles and variables as necessary to develop a comprehensive profile of a particular occasion.

Once again, we consider it essential for scholars to articulate openly their assumptions of causation in every piece of research and/or theory they undertake. Although mapping between alternative ideas of causation is not always possible, a step toward linking diverse perspectives is to see how an environment–behavior phenomenon is "explained" and "understood" by different approaches to causation.

Other aspects of philosophy of science are also important, including worldviews that treat observer phenomena relations as objective, detached and separate versus observers being aspects of phenomena; approaches that seek universal and generalized principles of environment behavior phenomena versus those focusing on particular events that accept, but do not insist on, general principles as the goal of research, and so on.

I believe that it is essential for scholars to articulate to themselves and to others their assumptions and rationale for selecting a particular approach to causation (as well as other aspects of philosophy of science). Even though we may disagree about the validity, utility, or value of different perspectives, depending upon our training, disciplinary heritage, and personal predilections, we will at least know "from whence each of us comes," and therefore will have a constructive basis for debate, comparison, and synthesis of our work.

Research Methodology

There are many research methods within and across disciplines: experiments, quasi-experiments, surveys, questionnaires, interviews, observations, participant observations, phenomenological descriptions, archival analyses, historical analyses, hermeneutic techniques, and others. How do we decide to choose a procedure in a given project? This is a crucial question for researchers to articulate to themselves and to others, because methods are intimately linked to the questions addressed by research, determine the types of information gathered, and influence conceptions of phenomena. Although I personally believe that we should use whatever method best fits the problem, some researchers fit problems to methods, because they believe that certain methods are more "correct" ways to understand phenomena, and that it is not worth studying phenomena using less desirable procedures.

Ultimately, it is important to provide a rationale and be introspective about one's choice of research methodology because, as Altman and Rogoff (1987) suggested, there is somewhat of a relationship between worldview and research procedures. Thus adherents to an interactional worldview, which focuses on antecedent–consequent causation, are likely to use methods that incorporate an efficient causation perspective, e.g., experimental and quasi-experimental methods, correlational methods such as path analysis that provide sequential linkages between variables, etc. In contrast, transactional approaches tend to be methodologically eclectic, tailor procedures to phenomena and issues, and use methods enabling identification of patterns of relationships between variables.

Because environment–behavior researchers bring to bear so many different methodological perspectives, and a diversity of related underlying philosophical assumptions, research is not always comparable or translatable one to the other. But we can enhance connections with one another and better understand how environment behavior phenomena work if we clearly state our methodological assumptions. Why did we choose one method over another? What did we expect to learn about a phenomenon by the method we selected? What did we decide to forgo learning about by virtue of choosing a particular procedure? What are the strengths and limitations of the procedures we adopted in respect to the phenomena of interest? Once again, it is central that we reveal, be introspective, and articulate our underlying assumptions. It is through such revelations that we can benefit from diverse perspectives on environment–behavior phenomena and enhance our individual and collective growth as researchers and theoreticians.

Conceptual and Theoretical Approaches

Sometimes a research project derives from or is designed to test or address an aspect of a theory. Or a project may examine an issue that is not directly associated with a theory, concept, or hypothesis, but derives from a specific observation or question. Regardless of goals or origins, every research project or research program has some conceptual underpinnings or assumptions. Whether it be a full-blown theory, a general concept, or a vague question, all research is guided by some underlying principles, even if they are not explicitly stated by investigators. Because these underlying assumptions may vary widely across disciplines, areas within disciplines, or depend on the unique eye of a researcher, they are not often known by the consumers of research from other fields. In order to have any possibility of other than superficial communication between students of environment–behavior phenomena, it is once again necessary to uncover implicit conceptual ideas, and overtly explain concepts and theoretical guides to one another. Without doing so, our diverse concepts and theories will not be comprehended, our mutual consumption of one another's research will be fragmentary, and our cumulative progress will be crippled by absent or faulty appreciation of our respective scholarship.

A FINAL THOUGHT

I have called for "full and open disclosure" by environment–behavior researchers of assumptions underlying their research, including specification of unit of analysis, approach to temporal processes of stability and change, rationale for methodological decisions, and articulation of theoretical perspectives. I stated it as necessary for environment and behavior scholars to be mutually introspective and revealing, given the diversity of their viewpoints, training, and experience. It is, I believe, the only way for us to progress as an endeavor, or to eventually achieve status as a discipline or field of study. But even if we do not become a formal field, future progress is dependent on us understanding one another's assumptions about theory, philosophy, and epistemology. And it is by revealing who we are and what undergirds our research that we will accomplish more than superficial understanding of one another. We will also more constructively influence and be influenced by one another's underlying assumptions, and will see our own work enhanced by new and innovative approaches used by others. And forcing ourselves to expose our underlying assumptions will be self-informative. We will be better able to articulate why we think as we do, which may reinforce our beliefs and actions, reveal the limitations of our approach, or open up new possibilities that we hadn't thought of previously. Taken together, therefore, the difficult task of articulating our underlying assumptions may enhance our own work by allowing better comparison and connection with the work of others, and by laying a foundation for cumulation and synthesis of knowledge, theoretical advances, and a more integrated and systematic approach to environment–behavior research.

And I believe that we should engage in mutual self disclosure about our assumptions whether or not we ever become a bona fide "field" or "discipline." For me the issue of "field" or "not-field" is irrelevant. The central question is what we now need to do in order to conduct important environment–behavior research in the next 25 years. The path to achieving significant connections among us and with generations to come is through revealing and articulating the philosophical, theoretical, and epistemological assumptions and choicepoints that underlie our study of environment and behavior phenomena. A difficult and challenging task indeed. But we must do it, else we see the end of more than two decades of collaborative efforts by social scientists and environmental designers to understand environment–behavior phenomena and to create better places for people.

REFERENCES

Altman, I. (1990) Toward a transactional perspective: A personal journey. In I. Altman & K. Christensen (Eds.), *Environment and behavior studies: Emergence of intellectual traditions. Human behavior and environment: Advances in theory and research, Vol. 11* (pp. 225–256). New York: Plenum.

Altman, I. (1995). *Whither EDRA*. Presented at a symposium on Founders, stalwarts and heirs: Three intellectual generations of EDRA address its future. Environmental Design Research Association, Boston, MA.

Altman, I., & Christensen, K. (Eds.). (1990). *Environment and behavior studies: Emergence of intellectual traditions. Human behavior and environment: Advances in theory and research, Vol. 11*. New York: Plenum.

Altman, I., & Gauvain, M. (1981). A cross-cultural and dialectic analysis of homes. In L. S. Liben, A. H. Patterson, & N. Newcombe (Eds.), *Spatial representation and behavior across the life span* (pp. 283–320). New York: Academic Press.

Altman, I., & Ginat, J. (1996). *Polygamous families in contemporary society*. New York: Cambridge University Press.

Altman, I., & Rogoff, B. (1987). worldviews in psychology: Trait, interactional, organismic, and transactional perspectives. In D. Stokols & I. Altman (Eds.), *Handbook of environmental psychology: Vol. 1* (pp. 1–40). New York: Wiley.

Altman, I., Brown, B. B., Staples, B., & Werner, C. M. (1992) A transactional approach to close relationships: Courtship, weddings and placemaking. In B. Walsh, K. Craik, & R. Price (Eds.), *Person-environment psychology: Contemporary models and perspectives* (pp. 193- 241). Hillsdale, NJ: Erlbaum.

Gauvain, M., Altman, I., & Fahim, H. (1983). Homes and social change: A cross-cultural analysis. In N. R. Feimer & E. S. Geller (Eds.), *Environmental psychology: Directions & perspectives* (pp. 80–118). New York: Praeger.

Dewey, J., & Bentley, A. (1949). *Knowing and the known*. Boston: Beacon.

Pepper, S. C. (1942). *World hypotheses: A study in evidence*. Berkeley, CA: University of California Press.

Pepper, S. C. (1967). *Concept and quality: A world hypothesis*. La Salle, IL: Open Court.

Stokols, D., & Altman, I. (1987). *Handbook of environmental psychology: Vols. 1 and 2*. New York: Wiley.

Werner, C. M., Altman, I., Brown, B. B., & Ginat, J. (1993) Celebrations in personal relationships: A transactional/dialectic perspective. In S. Duck (Ed.), *Social context and relationships* (pp. 109- 138). Newbury Park, CA: Sage.

APPENDIX
LIST OF PRESENTATIONS OF PREVIOUS JAPAN–U.S. SEMINARS ON ENVIRONMENT–BEHAVIOR RESEARCH

(1) FIRST JAPAN–U.S. SEMINAR (1980, TOKYO, JAPAN)

Reference: Hagino, G., and Ittelson, W. H. (Eds.). (1980). *Interaction processes between human behavior and environment*. Tokyo: Bunsei.

Contents:

Shotaro Tsumakura, Welcome Address

Genichi Hagino, Opening Address: Trends in Studies of Environmental Psychology in Japan

Session One: Human Behavior in Various Physical and Social Environments

1. William Michelson, Basic Dimensions for the Analysis of Behavioral Potential in the Urban Environment
2. Susan Saegert, Residential Density and Psychological Development
3. Kitao Abe, Panic Potential as a Predictor of Human Behavior in Case of Disaster in Metropolitan Area
4. John Archea, Architectural Factors Affecting Behavior in Accidents, Crime, and Emergency Evacuation
5. Yoji Niitani, Urbanization and Pattern of Person Trip in Urban Areas in Japan
6. Kunio Tanaka, On Japanese Attitudes Towards Their Communities
7. Kaoru Noguchi, Perceptual Behavior in Traffic Environment

Session Two: Designing Human Habitats

8. Sandra Howell, Habit and Habitability

Handbook of Japan–United States Environment–Behavior Research: Toward a Transactional Approach, edited by Seymour Wapner, Jack Demick, Takiji Yamamoto, and Takashi Takahashi. Plenum Press, New York, 1997.

9. Robert Bechtel, Contributions of Ecological Psychology to Environmental Design Research

10. Mamoru Mochizuki, Japanese Art as Interactive Media—Where Dwelling Meets Nature?

11. Yoshio Nakamura, Landscape Perception and Man's Esthetic Intervention to Environment

12. Takashi Takahashi, Notes on the Concept of Space in the Japanese House

13. Koichi Tonuma, Human Scale in Metropolitan Area

Session Three: Environmental Perception

14. Seymour Wapner, Transactions of Person-in-Environments: Some Issues, Problems and Methods from the Organismic-Developmental Viewpoint

15. Seymour Wapner and William Ittelson, Environmental Perception and Action

16. Ichiro Souma, Cognition and Behavior in the School Environment

17. Takiji Yamamoto, Microgenetic Development of Environmental Cognition

18. Masaaki Asai, Affective Components of Environmental Cognition

Session Four: Synthesis, Three Trends of Discussion, and Future Perspectives
Three Trends of Discussion
William Ittelson, Closing Address—Environmental Psychology: Past Accomplishments and Future Prospects

(2) SECOND JAPAN–U.S. SEMINAR (1985, TUCSON, ARIZONA)

Reference: Ittelson, W. H., Asai, M., and Ker, M. (Eds.). (1986). *Crosscultural research in environment and behavior.* Tucson, AZ: University of Arizona Press.

Contents:

William H. Ittelson, Preface
Genichi Hagino, Message to the U.S.–Japan Seminar

1. John Archea, Behavior during Earthquakes: Coping with the Unexpected in Destabilizing Environments

2. Masaaki Asai, Driver's Perception of Road Settings—Semantic Evaluation, Its Relation to Physical Properties, and Recall of Surroundings

3. Robert B. Bechtel, Choice, Control, and Japanese and American Responses to the Environment

4. Clare Cooper Marcus, Design Guidelines: A Bridge between Research and Decision-Making

5. Kunio Funahashi, A Study of Pedestrian Path Choice

6. Kazuo Hara, A Transdisciplinary Model for the Concepts of "Environment" and Survey Studies of College Campus Atmosphere

7. Sandra C. Howell, The Psychoenvironmental Implications of Aging

8. William H. Ittelson, The Psychology of Technology

9. Masami Kobayashi and John Archea, Occupant Behavior during an Earthquake and Its Implication for Architectural Planning

10. Yoichi Kubota, Some Observations on the Imaginal Structure and Evaluation of Places in Terms of Townscape
11. William Michelson, Basic Dimensions for the Analysis of Behavioral Potential in the Urban Environment II: An Update on Methodological and Substantive Results
12. Ryuzo Ohno, Notion of Duality in Visual System and Its Implication for Environmental Design
13. Amos Rapoport, Settlements and Energy: Historical Precedents
14. Ichiro Souma, The Evaluation of Environment
15. Daniel Stokols, Transformational Perspectives on Environment and Behavior: An Agenda for Future Research
16. Takashi Takahashi, Polygon of Living Territory
17. Seymour Wapner, Jack Demick, Wataru Inoue, Shinji Ishii, and Takiji Yamamoto, Relations between Experience and Action: Automobile Seat Belt Usage in Japan and the United States
18. Ervin H. Zube, Advances in Applied Environmental Perception Research

(3) THIRD JAPAN–U.S. SEMINAR (1990, KYOTO, JAPAN)

Reference: Yoshitake, Y., Bechtel, R. B., Takahashi, T., & Asai, M. (1990). *Current issues in environment-behavior research*. Tokyo: University of Tokyo.

Contents:

Takashi Takahashi, Preface
Genichi Hagino, Opening Message
Yasumi Yoshitake, Opening Address

Work and Learning Environments
1. Masaaki Asai and Robert B. Bechtel, A Comparison of Some Japanese and U.S. Workers on the WES
2. Daniel Levi and Charles Slem, Comparison of Beliefs about the Effects of Technological Change between Japanese and U.S. Employees
3. Seymour Wapner, Fulvia Quilici-Matteucci, Takiji Yamamoto, and Takatoshi Ando, Cross-Cultural Comparison of the Concept of Necessity, Amenity and Luxury
4. Masao Inui, Relationship between Architectural Style and Office Interior Environment
5. Ichiro Souma and Takatoshi Ando, Environmental Cognition at Elementary Schools

Behavior and Safety
6. Satoshi Kose, Safety Standards Research in Japan: Development of Safety Recommendations for Domestic Stairs
7. Kunio Funahashi, Addressing System: Spatial Structure and Wayfinding in Japanese Towns
8. John Archea, Two Earthquakes: Three Human Conditions

Public Space and Landscape
9. Setha M. Low, Cross-Cultural Place Attachment: A Preliminary Typology

10. Toshihiko Sako, Cognitive Mapping Studies in Japan
11. Ervin H. Zube, A Cross-Cultural Exploration of Human-Landscape Relation-ships
12. Yoichi Kubota, Landscape Perception in Japan: A Note on the Transient Field of Landscape Experience

Theory Building

13. Amos Rapoport, Levels of Meaning and Types of Environments

Housing and Family

14. Irwin Altman, Barbara B. Brown, Carol M. Werner, and Brenda Staples, Placemaking in Social Relationships
15. Sandra C. Howell, Family Life: Habit and Habitability
16. William Michelson, Measuring Behavioral Quality in Experimental Housing
17. Tadashi Toyama, Identity and Milieu: A Study of Relocation Focusing on Reciprocal Changes in Elderly People and Their Environment
18. Takashi Takahashi and Kazuhiko Nishide, Behind a Mask: Personal Territorial-ity and Spatial Articulation
19. Masami Kobayashi, The Meaning of Kyoto Townscape
20. Robert B. Bechtel, Closing Remarks

Name Index

Abe, K., 435
Abler, 415
Adachi, K., 5, 15, 23, 389
Adachi, T., 359, 364
Adams, F.M., 101, 112
Agar, M., 317, 324
Ahrentzen, S., 151, 158,
    338, 342, 345, 346,
    347, 349
Akoi, J., 45f
Alexander, C., 128, 347
Allen, T.J., 340, 347
Altabe, M.N., 85, 98
Altman, D.G., 339, 353
Altman, I., 10, 89, 96, 134,
    147, 148, 335, 342,
    343, 345, 346, 347,
    361, 364, 367, 372,
    380, 381, 392, 397,
    407, 414, 417, 423,
    424, 425, 428, 429,
    430, 432, 433, 434, 438
Altman, L.K., 403, 417
Amabile, T.M., 344, 347
Amadeo, M., 15, 23
Amano, H., 52, 56
Anda, T., 52, 56
Ando, H., 190f, 191f, 192f
Ando, T., 437
Angel, S., 124, 128, 347
Appleton, J., 184, 198
Appleyard, D., 60, 80, 342,
    347, 401, 417
Araki, H., 23
Archea, J., 340, 347, 435,
    436, 437
Arias, O., 316
Arnold, H.F., 184, 198
Arreola, D.A., 404, 417

Asai, M., 2, 3, 8, 10, 11,
    261, 271, 386,
    387-388, 389, 436, 437
Asakawa, T., 55, 56
Asfour, K., 402, 417
Ashton, J., 336, 343, 347

Bachelard, C., 242, 244
Baird, L.L., 274, 281
Baker, A.H., 86t, 98
Baldassare, M., 336, 337,
    347
Barinaga, M., 402, 417
Barker, R., 113, 128, 150,
    158, 236, 239, 244,
    261, 273, 274, 281,
    344, 347, 367, 381,
    392, 397, 415
Barkow, J.H., 402, 415, 417
Barnard, K., 343, 347
Baron, A., Jr., 286t, 311
Baron, R.A., 337, 347
Barton, M.I., 86t, 96
Baum, A., 345, 347
Baumeister, R.F., 85, 96
Beatty, R.A., 183, 198
Bechhoefer, W., 401, 417
Bechtel, R.B., 8, 11, 59, 61,
    75, 79, 80, 149, 158,
    235, 236, 237, 239,
    240, 244, 271, 281,
    338, 341, 342, 347,
    348, 378, 379, 390,
    393, 436, 437, 438
Becker, F.D., 340, 341, 342,
    348
Bell, J.A., 406, 417
Belsky, J., 337, 348
Bem, S.L., 92, 96

Benedict, J.O., 342, 352
Ben-Porath, Y., 242, 244
Bentley, A., 428, 430, 434
Bergen, J.R., 164, 182
Berger, L., 351
Berlin, B., 404, 417, 418
Berren, M., 241, 242
Betsky, A., 375
Bezold, C., 338, 343, 348
Binder, A., 134, 148
Birdwhistell, R.L., 89, 96
Birren, F., 108, 112
Blanton, R.E., 249, 250,
    256, 257, 404, 418
Bochner, S., 286t, 310
Bond, M.H., 88, 96
Boneau, C.A., 348
Bonnes, M., 411, 418
Bosselmann, P., 381, 382
Boudon, P., 256, 257
Boulding, K.E., 413, 418
Brenner, G., 210, 234
Brink, S., 41
Bronfenbrenner, U., 342,
    348, 392, 397
Brown, B.B., 428, 434,
    438
Bruner, J., 135, 148, 385,
    386, 392, 394, 397
Brunswik, E., 380, 382
Buchan, R., 409, 418
Buhrmeister, D., 283,
    311
Bunschaft, G., 365
Burdine, J.N., 336, 349
Burroughs, J., 409, 421
Buss, D.M., 378, 382
Butcher, J., 241, 242, 244
Byrne, R.W., 60, 80

Campbell, A.C., 90, 96
Campbell, J.C., 25, 41
Canter, D., 113, 128, 342,
    348, 367, 375, 378,
    379, 382
Cargo, M.D., 339, 349
Carlson, R.J., 338, 348
Carpenter, E.H., 200, 208
Carr, S., 341, 348, 401, 418
Carrere, S., 242, 243, 244
Cartier-Bresson, H., 113,
    114f, 121f, 128
Casey, M.W., 60, 80
Cash, T.F., 85, 86t, 96
Caspi, A., 380, 382
Castenada, A., 311
Cazes, G., 242, 244
Chemers, M.M., 342, 345,
    346, 347
Chen, H.P., 110, 112
Cherulnik, P.D., 404, 418
Chihara, T., 85, 97
Childress, H., 401n
Chiswick, N.R., 89, 97
Chou, S.K., 110, 112
Choungourian, A., 101, 112
Christensen, K., 338, 340,
    348, 428, 434
Chung Tzu, 362n
Church, A.T., 283, 286t,
    286-288, 308, 310
Churchman, A., 342, 345,
    346, 347
Ciottone, R., 285, 311
Clark, S., 151, 159
Clark-Oropeza, B.A., 286t,
    311
Clearwater, Y.A., 347, 349
Clemens, 403
Cleveland, S.E., 86t, 87n, 97
Clitheroe, C., 344, 348
Cohen, S., 338, 344, 345,
    348
Comalli, P.E., 86t, 98
Cone, J.D., 336, 348
Conner, R., 336, 339, 343,
    348
Cooper, C., 345, 348
?Cooper-Marcus, C., 348
?Cooper-Markus, C., 339

Coopersmith, S., 92, 93, 96
Cormack, M., 286t, 311
Cosmides, L., 378, 382,
    383, 417
Cousins, S., 88-89, 91, 96
Craik, K.H., 9, 204, 205,
    208, 335, 343, 345,
    377, 378, 379, 380,
    381, 382, 383, 396
Crider, C., 87t, 96
Critchley, M., 86t, 87n, 96
Cronick, K., 334, 345, 351
Culhane, P.J., 204-205, 208
Cullen, 417
Cullen, G., 113, 128

Dake, K.M., 378, 382
Damon, W., 87t, 96
Dandonoli, P., 94, 96
Danko, S., 339, 342, 348
Day, K., 346, 348
de Ajuriaguerra, J., 87n, 96
DeFries, R.S., 336, 351
Demick, J., 1, 5, 6, 9, 10,
    11, 61, 80, 83, 84, 85,
    87n, 88, 89, 91, 93, 94,
    96, 97, 98, 343, 386,
    389, 437
DesLauriers, A., 87n, 97
DeVore, I., 378, 383
Dewey, J., 428, 430, 434
Dewsbury, D.A., 400, 418
Dr. M., 142
Dolce, J.J., 85, 98
Downs, R., 60, 80
Druckman, D., 336, 352
Duff, D., 91, 97
Duhl, L., 336, 339, 343, 348
Duncan, J.S., 409, 418
Dunlap, R.E., 336, 348
Dunlop, C., 343, 348
Dykstra, J., 381, 382
Dziuba-Leatherman, J.,
    337n, 349

Ebreo, A., 409, 419
Edney, J.J., 90, 97
Edwards, T.C., 336, 351
Ekman, P., 286t, 311
Elder, G., 342, 350

Elias, N., 366, 375
Elliott, D., 151, 159
Emler, N., 379, 382
Eshelman, P., 339, 348
Evans, G., 242, 244
Evans, G.W., 338, 345, 348,
    349, 407, 418
Everett, P.B., 336, 349
Eysenck, H.J., 101, 112

Fabian, A.K., 345, 351
Fahim, H., 428, 434
Fallon, A.E., 86t, 97
Farbstein, J., 341, 353, 363,
    364
Ffolliott, P.F., 199, 208
Fielding, J.E., 339, 352
Fiksdahl-King, I., 124, 128,
    347
Finkelhor, D., 337n, 349
Fisher, B., 337, 339, 350
Fisher, S., 85, 86t, 87n, 91,
    97, 283, 310, 311
Fiske, S.T., 344, 352
Fitzgibbons, M., 286t, 311
Fleming, I., 345, 347
Fletcher, J.F., 60, 80
Foote, K.E., 404, 418
Foust, T., 208
Francis, M., 341, 348
Francis, S., 339, 353
Franck, K., 151, 158, 338,
    342, 349
Frazier, W., 341, 353
Fried, M., 133, 143, 148
Friedan, B., 50
Friedman, S., 201, 208
Friedman, S.L., 342, 349
Fujihara, T., 302, 311
Fujii, 357
Fujimoto, J., 8, 283, 310,
    386, 389
Fukuyama, F., 369, 375
Funahashi, K., 9, 355, 388,
    436, 437
Furman, W., 283, 311

Galea, J., 90, 96
Gans, H.J., 133, 134, 145,
    148

Gardner, G.T., 336, 352
Garreau, J., 411, 418
Gauvain, M., 428, 434
Gee, M., 362, 364
Geertz, C., 134, 148, 317, 324
Gehl, J., 117, 128
Geller, E.S., 336, 349
Gellert, E., 87t, 91, 97
Gerbert (Pope Silvester), 378
Gibbons, A., 415, 418
Gibson, J.J., 173, 182, 249, 257
Ginat, J., 428, 434
Giuliano, G., 336, 349
Glass, D.C., 338, 349
Gleitman, H., 60, 80
Glen, W., 403, 418
Glick, J.A., 86t, 97
Goffman, 415
Goldberger, P., 410, 418
Goldsmith, D., 403, 407, 418
Goldstein, A.P., 337, 341, 345, 349
Goodman, R.M., 336, 349
Goodnow, J., 247, 257
Goto, S.G., 89, 98
Green, G.K., 85, 86t, 96
Green, L.W., 339, 340, 349
Greenbaum, P.E., 90, 97
Greenbaum, S.D., 90, 97
Greenberg, M.R., 340, 349
Grey, P., 343, 347
Grieneeks, J.K., 336, 348
Griffiths, I., 379, 382
Gump, P.V., 261, 271, 273, 274, 275, 281, 282, 344, 347
Gunderson, E., 243, 244

Habermas, J., 396, 397
Haga, H., 51, 52, 56, 57
Haggard, L.M., 361, 364
Hagino, G., 1, 3-4, 10, 334, 344, 346, 349, 435, 436, 437

Hall, E.T., 89, 90, 97, 113, 128, 163, 164, 182, 286t, 311, 367, 375
Hall, R., 344, 352
Halprin, 417
Hamaguchi, E., 88, 97
Hanazato, T., 6, 59, 340, 345
Handy, S.L., 338, 349
Hanson, J., 249, 251, 257
Hanson, P., 152, 159
Hanson, S., 152, 159
Hara, K., 436
Harrington, M.J., 336, 351
Harris, P.R., 347, 349
Harrison, A.A., 347, 349
Hart, D., 87t, 96
Hartig, T., 338, 349
Harvey, A., 151, 159
Harvey, D., 322, 324
Hasegawa, M., 55, 56
Hata, T., 7, 163
Havighurst, R.J., 52, 57
Hayashi, C., 88, 97
Hayes, S.C., 336, 348
Hedge, A., 339, 348
Heider, F., 393, 397
Heidmets, M., 345, 350
Heinrich, B., 403, 409, 418
Helmes, E., 241, 244
Helmreich, R., 240, 244
Henderson, D.K., 406, 418
Henderson, J.V., 325, 330
Hicks, I., 2
Higgins, E.T., 92, 97
Hillier, W., 249, 251, 257
Himsworth, H., 412, 418
Hirada, S., 198
Hirata, S., 8, 261-262, 263, 271, 386, 387-388, 389
Hiroshima, K., 3
Hiwatari, K., 18, 23
Hoffman, A., 94, 96
Holldobler, B., 403, 418
Holman, E.A., 337, 349
Holzman, P.S., 15, 23
Hood, B., 283, 310, 311
Hornstein, G.A., 285, 311
Horowitz, M., 91, 97
Horwich, P., 408, 418

Howell, M., 252f, 253f, 254f
Howell, S.C., 8, 247, 250, 257, 389, 415, 419, 435, 436, 438
Hubbard, P.J., 342, 349
Hull, W.F., 283, 286t, 311
Hunt, M.E., 340, 350
Hurt, S., 15, 23
Hwang, K., 336, 349

Ibarra, I., 207, 208
Imamichi, T., 8, 283, 310, 386, 389
Inoue, W., 6, 83, 88, 97, 386, 389, 437
Inoue, Y., 8, 283, 310, 386, 389
Inui, M., 3, 437
Ishihara, O., 52, 56
Ishii, S., 6, 83, 88, 89, 97, 98, 386, 389, 437
Ishikawa, S., 124, 128, 347
Ittelson, W.H., 2-4, 10, 337, 349, 362, 364, 392, 397, 435, 436
Iwawaki, S., 85, 97

Jacobsen, M., 124, 128
Jacobson, M., 347
Jahoda, M.A., 91, 97
Jodelet, D., 334, 349, 350
Johnson, P., 401, 418
Jordan-Edney, N.L., 90, 97
Jourard, S., 87t, 91, 92, 98
Julesz, B., 164, 182

Kahn, P., 242, 244
Kaitilla, S., 405, 418
Kaminoff, R., 345, 351
Kaminsky, 415
Kantrowitz, M., 363, 364
Kaplan, B., 284, 311, 312
Kaplan, E., 86t, 92, 97
Kaplan, R., 338, 349, 361, 364
Kaplan, S., 338, 349
Katz, C., 336, 347
Kaufman, M.T., 410, 418
Kawakami, N., 52, 57
Kay, P., 404, 418

Kellert, S.H., 403, 418
Kent, S., 248, 249, 257
Ker, M., 10, 436
Kertēsz, A., 113, 128
Kessel, F., 247, 257
Khattab, O., 414, 418
Kimura, I., 128f
King, A.C., 339, 353
King, A.D., 402
Kitayama, S., 85, 88, 97
Klein, A., 357, 364
Klineberg, O., 283, 286t,
    311
Kling, R., 343, 348
Knopf, R.C., 338, 349
Kobayashi, M., 7, 209, 213,
    233, 234, 334, 344,
    389, 390, 436, 438
Kochen, M., 379, 382
Kogo, C., 102, 112
Kohyama, T., 302, 311
Koizumi, A., 52, 57
Kondo, M., 7, 163, 173, 182
Korpela, K.M., 338, 349
Kose, S., 5, 25, 27, 28, 35,
    41, 42, 248, 257, 340,
    437
Koyano, P.W., 51, 52, 54, 56,
    57
Koyanoet, 52
Krampen, M., 375
Kruse, 415
Kruusvall, J., 345, 350
Kubota, Y., 7, 183, 198,
    381, 389, 437, 438
Kuhn, T., 379, 382, 408, 418
Kuller, R., 334, 344, 346,
    349
Kumagai, S., 52, 56

Lackney, J.A., 339, 342, 350
Lai, A.C., 112
Landau, B., 60, 80
Lang, J.T., 50
Lang, J.W., 401, 418
Lao Tzu, 362n
Larkin, P., 342, 348
Lasswell, H.D., 344, 349
Law, C.S., 200, 208

Lawton, M.P., 51, 52, 57,
    340, 341, 342, 346,
    350, 351
Leaf, A., 335, 336, 345, 350
Leavitt, J., 337, 350
Ledbetter, C., 240, 244
Lee, W., 129
Lefebvre, H., 319, 324, 372,
    375
Lefkowitz, J., 324
Leon, G., 242, 244
Lerner, R.M., 85, 87t, 90,
    94, 95, 97
Leveton, L.B., 60, 80
Levi, D., 437
Levy, D., 15, 23
Lewin, K., 380, 382
Lewis, M., 89, 90, 94, 97
Liang, J., 52, 57
Lindsey, S., 409, 418
Lister, F.C., 199, 208
Lister, R.H., 199, 208
Litton, R.B., Jr., 183, 198
Livingston, M., 404, 418
Lloyd, R., 60, 80
Lofland, L., 367, 375
Lollis, M., 90, 99
Low, S.M., 9, 313, 324, 334,
    335, 345, 390, 437
Lucca-Irizarry, N., 375
Luckmann, T., 134, 148
Luscher, K., 342, 350
Lynch, K., 344, 350, 363,
    364, 401, 417
Lyon, B.D., 378, 382

MacCallum, T., 235, 236,
    237, 239, 338
McDonogh, G., 324
McFarland, J.H., 83, 98
McKay, C.P., 347, 349
Mackworth, N., 15, 23
McLeroy, K.R., 336, 349
McMullin, E., 408, 419
McNally, C., 242, 244
McNally, R.J., 341, 350
McNeil, O., 285, 311
Maddi, S.R., 380, 382
Maeda, D., 52, 55, 57
Magnusson, D., 343, 350

Maki, N., 7, 209, 213, 233,
    234
Malmberg, T., 407, 419
Mang, M., 338, 349
Mann, R., 198
Mannett, L., 411, 418
Marans, R.W., 149, 158,
    340, 341, 348, 350,
    381, 382
Marcus, C., 436
Markus, H., 85, 88, 97
Maslow, A., 362
Masterman, M., 379, 382
Matsud, T., 55, 56
Matsukai, T., 52, 57
Mazumdar, S., 341, 350
Mead, M., 344, 350
Meehan, E., 336, 349
Meltzer, H., 15, 23
Merleau-Ponty, M., 86t, 97
Merton, R.K., 337, 341, 350
Meyer, J., 401, 417
Meyrowitz, J., 338, 342, 350
Michelson, W., 7, 149, 150,
    151, 152, 153, 154,
    156, 158, 159, 341,
    342, 345, 346, 348,
    350, 353, 379, 435,
    437, 438
Milanesi, L., 344, 351
Milgram, S., 334, 338, 344,
    350
Miller, P.J., 247, 257
Min, B., 364
Min, M.S., 89, 90, 97
Minami, H., 67, 88, 97,
    133, 134, 135, 148,
    299, 303, 311, 389,
    390-391
Minsky, 415
Miura, K., 7, 209, 233, 234
Miyagishi, Y., 163, 182
Miyatani, M., 89, 98
Mochizuki, M., 3, 334, 344,
    349, 436
Moen, P., 342, 350
Moffitt, T.E., 380, 382
Mokhtarian, P.L., 338, 349
Monzeglio, 404

Moore, G.T., 339, 342, 345, 350, 353, 415, 419
Moos, R., 261, 262, 263, 271, 339, 345, 346, 350
Morimoto, K., 52, 57
Morse, E.S., 248, 250, 257
Munce, S., 90, 96
Murray, H.A., 380, 382
Muschamp, H., 365

Nagai, H., 52, 56
Nagashima, K., 52, 56
Naglieri, J.A., 91, 97
Nahemow, L., 346, 350
Naito, K., 52, 56
Nakamura, K., 52, 57
Nakamura, Y., 184, 198, 436
Nakaohji, M., 25, 28, 41, 42
Nakatani, Y., 55, 57
Nakazato, K., 52, 57
Napier, M., 416, 419
Nasar, J.L., 89, 90, 97, 337, 339, 350
Neufert, 357
Neugarten, B.L., 52, 57
Newman, O., 342, 350
Newman, S.J., 340, 350
Nicholas, D., 379, 382
Nichols, M., 410, 419
Niino, N., 52, 57
Niit, T., 345, 350
Niitani, Y., 435
Nishisato, S., 107, 112
Nishiyama, S., 88, 97
Nishiyama, U., 357-358, 364
Noam, E., 338, 351
Noda, M., 232, 234
Noguchi, K., 435
Noguchi, Y., 55, 57
Nojima, E., 280, 282
Noschis, K., 342, 351
Novaco, R.W., 344, 351

Oberg, K., 286t, 311
O'Donnell, M.P., 339, 351
O'Donnell, S., 404, 419
Ohara, K., 209, 234
Ohno, R., 7, 163, 164, 173, 182, 389, 437
Ohta, A., 27, 42

Ohyama, Y., 198
Okamura, K., 55, 56
Okonogi, K., 232, 234
Okuda, S.M., 336, 351
Oliver, D., 242, 243, 244
Orland, B., 409, 419
Ornish, D., 243, 244
Ornstein, S., 339, 351
Osada, H., 6, 51, 340
Osaka, R., 15, 23
Osgood, C.E., 101, 112
Oskamp, S., 336, 351
Osmond, H., 134, 148
Otto, 15
Overton, D., 15, 23
Oxley, D., 361, 364

Pacheco, A., 285, 311, 375
Pader, E.J., 404, 419
Parkes, C.M., 134, 148, 232, 234
Parmelee, P.A., 340, 341, 342, 351
Parsons, T., 369, 375
Passini, R., 60, 73, 79, 80
Pastalan, L.A., 340, 351
Patterson, A.H., 89, 97
Peck, J.C., 338, 348
Pelletier, K., 339, 352
Pennebaker, J.W., 86t, 97
Penner, L.A., 85, 98
Pepper, S., 428, 430, 434
Perin, C., 402, 419
Perkins, D.D., 342, 351
Pessac, 256
Peter the Great, 374
Philip, D., 404, 419
Pinker, S., 60, 80
Pitt, D.G., 205, 208
Platt, J., 336, 351
Plimmer, K., 137, 148
Pol, E., 335, 345, 351
Pool, R., 408n, 419
Porter, R.E., 286t, 311
Porzemsky, J., 86t, 97
Poynter, J., 235, 236, 237, 239, 338
Preiser, W.F.E., 342, 351
Price, R.H., 380, 383
Proctor, L., 15, 23

Proshansky, H., 345, 351, 392, 397, 406, 419
Proulx, G., 60, 80
Putnam, R., 369, 374, 375

Quilici-Matteucci, F., 437
Quirk, M., 339, 351

Rainville, C., 60, 80
Ramirez, M., III, 285, 311
Rand, G., 59, 365
Ransberger, V.M., 337, 347
Raphael, B., 210, 234
Rapoport, A., 10, 89, 97, 251, 256, 257, 342, 343, 345, 351, 362, 364, 367-368, 372, 375, 378, 380, 386-387, 399, 400, 401, 402, 403, 404, 405, 406, 407, 408, 409, 410, 411, 412, 413, 414, 415, 416, 417 419, 420, 424, 425, 437, 438
Ratzka, A.D., 41, 42
Reddon, J., 241, 244
Reischauer, E.O., 88, 97
Reizenstein-Carpman, J., 345, 353
Resnick, M., 330
Rich, R.C., 342, 351
Ricour, P., 386n, 397
Riger, S., 337, 351
Rivers, S., 83, 85, 96
Rivlin, L., 341, 348, 392, 397
Rivolier, J., 242, 244
Robertson, K.A., 410, 420
Rodin, J., 87n, 98
Rogers, C.R., 98
Rogoff, B., 134, 147, 148, 342, 347, 361, 364, 380, 381, 392, 397, 428, 429, 430, 432, 434
Rokeach, M., 336, 348
Rosen, R., 338, 351
Rosenberg, M., 92, 98
Rosenberger, N.R., 88, 98
Rosenblatt, B., 87t, 98

Rosin, K.T., 330
Rotenberg, B., 324
Rozin, P., 86t, 97
Rubinstein, N., 401n
Ryan, E.J., 134, 148

Sack, R.D., 407, 409, 420
Sadalla, E.J., 409, 421
Sadalla, E.K., 405, 420
Saegert, S., 10, 335, 337,
    342, 343, 345, 346,
    350, 351, 361, 363,
    364, 385, 435
Saito, K., 52, 57
Saito, M., 6, 101, 102, 108,
    109, 110, 112, 404
Sakata, S., 55, 57
Sakihara, S., 52, 56
Sako, T., 8, 273, 275, 282,
    438
Salter, E., 60, 80
Sanchez, E., 334, 345, 351
Sandler, J., 87t, 98
Santostefano, S., 86t, 90, 98
Santy, P., 242, 244
Sarkissian, W., 339, 348
Sato, S., 52, 57
Sato, Y., 52, 57
Saunders, J.O., 336, 351
Schank, 415
Schaur, E., 409, 421
Schilder, P., 86t, 98
Schissler, D., 401, 418
Schlater, J.A., 86t, 98
Schoggen, P., 236, 244, 274,
    282, 344, 347
Schön, D.A., 401, 421
Schouela, D.A., 60, 61, 80
Schulz, R., 234
Schutz, A., 134, 148, 286t,
    311
Secchiaroli, G., 411, 418
Secord, P., 87t, 91, 92, 98
Shagass, C., 15, 23
Shanmugam, A., 286t, 311
Shaver, P., 283, 310, 311
Shaw, S., 404
Sheets, V.L., 405, 420
Shen, N.C., 110, 112
Sheridan, T.E., 411, 421

Sherwood, D.L., 336, 351
Shibata, H., 51, 52, 56, 57
Shimonaka, Y., 52, 57
Shontz, F.C., 86t, 87t, 98
Silberstein, L.R., 87n, 98
Silver, C.S., 336, 351
Silverstein, M., 124, 128,
    347
Simcox, D.E., 201, 206, 208
Simmel, M.L., 86t, 98
Singer, J.E., 338, 349
Slaby, R., 337, 351
Slem, C., 437
Smith, H.W., 90, 98
Soma, I., 3, 6, 101, 404
Sommer, B.A., 409, 421
Sommer, R., 149, 159, 342,
    347, 352, 409, 421
Sorell, G.T., 85, 97
Sorensen, J., 183, 198
Souma, I., 436, 437
Spelke, E., 60, 80
Spivack, M., 240, 244
Staples, B., 428, 434, 438
Stea, D., 60, 80, 367, 375
Steinberg, L.M., 60, 80
Stern, P.C., 335, 336, 343,
    345, 352
Stevens, D.A., 61, 80
Stewart, M., 365
Stokols, D., 9, 134, 148,
    242, 244, 281, 333,
    335, 337, 338, 339,
    341, 342, 343, 344,
    345, 349, 351, 352,
    362, 364, 372, 377,
    378, 379, 380, 381,
    382, 424-425, 434, 437
Stone, A.M., 341, 348
Stone, P., 151, 159
Stratton, L., 91, 97
Striegel-Moore, R.H., 87n,
    98
Struyk, R., 340, 350
Sugiyama, Y., 52, 57
Sundstrom, E., 342, 352
Sun Yat Sen, 374
Susa, A.M., 342, 352
Suyama, Y., 51, 52, 57
Suzuki, A., 28, 42

Suzuki, S., 209, 234
Suzuki, T., 6, 52, 56, 113,
    115, 129, 389
Swanson, D.C., 336,
    351
Szalai, A., 151, 159

Taft, R., 286t, 311
Takahashi, T., 1, 2, 3, 4, 5,
    11, 43, 50, 116, 129,
    231, 234, 340, 343,
    345, 389, 436, 437, 438
Takai, J., 302, 311
Takekawa, T., 52, 57
Talbot, J.F., 338, 349
Tamano, K., 55, 57
Tanaka, K., 9, 42, 134, 135,
    148, 325, 435
Tanaka, T., 302, 311
Tanaka, Y., 27, 28, 42, 286t,
    311
Tanigawa, K., 9, 325
Tanucci, G., 411, 418
Taylor, R.B., 342, 351, 407,
    409, 421
Taylor, S.E., 344, 352
Tentokali, V., 250, 257
Tetlow, R.J., 183, 198
Thiel, P., 174, 182, 417
Thomas, V., 325, 330
Thompson, J.K., 85, 98
Thorne, R., 344, 352
Thorud, D.B., 199, 208
Tobin, S.S., 52, 57
Toews, K., 8, 283, 310, 386,
    389
Tolman, E.C., 60, 80
Tomita, M., 102, 112
Tonuma, K., 436
Tooby, J., 378, 382, 383, 417
Toshima, T., 89, 90, 94, 98
Toyama, T., 231, 234, 438
Trafimow, D., 89, 98
Triandis, H.C., 89, 91, 98,
    286t, 311
Tsumakura, S., 2, 435
Tuttle, D.P., 415, 419

Ulrich, R.S., 338, 339, 352
Underwood, P.R., 232, 234

Vairavan, B., 401, 421
Vakalo, K.L., 340, 350
Valery, P., 365
Valsiner, J., 386, 397
van Cleef, A., 371, 375
Vassiliou, V., 286t, 311
Verbrugge, L., 340, 352
Vershure, B., 409, 421
Vila, B., 337, 353
Vining, J., 409, 419
Vrasawa, K., 52, 57

Wachs, M., 336, 349
Wachs, T.D., 342, 349, 353
Wachsler, L., 375
Walmsley, K., 41, 42
Walsh, J., 339, 353
Walsh, W.B., 380, 383
Wandersman, A., 342, 351
Wantabe, S., 52, 56
Wapner, S., 1, 3, 5, 6, 8, 9, 10, 11, 59, 60, 61, 80, 83, 84, 86t, 87n, 89, 91, 93, 94, 96, 97, 98, 133, 134-135, 148, 283, 284, 285, 288, 299, 302, 303, 310, 311, 339, 341, 343, 345, 351, 353, 361, 364, 372, 375, 379,

380, 383, 386, 389, 392, 397, 436, 437
Watanabe, A., 18, 23
Watanabe, K., 27, 42
Webber, M., 379, 383
Weber, M., 369
Weiss, R.S., 232, 234
Wekerle, G., 152, 159
Wener, R., 341, 353
Werner, C.M., 361, 364, 428, 434, 438
Werner, H., 61, 80, 83, 84, 86t, 87n, 91, 94, 98, 99, 284, 311, 312
Westerman, A., 409, 421
Wicker, A., 149, 159, 274, 282, 343, 353, 367, 375, 380, 381, 383
Wiesenfeld, E., 334, 345, 351, 353
Willer, D., 401, 406, 421
Willer, J., 401, 406, 421
Williams, A.F., 339, 353
Williams, T.I., 378, 383
Wilson, E.O., 403, 418
Wilson, L., 324
Wineman, J., 342, 353
Winett, R.A., 336, 339, 349, 353
Winkel, G., 342, 351, 361, 364, 392, 397
Witkin, H.A., 87t, 99

Wohlwill, J.F., 407, 409, 421
Wolff, H.G., 286t, 312
Worchel, S., 90, 99
Wylie, R., 91, 99

Yamada, S., 28, 42
Yamamoto, T., 1, 2, 3, 6, 59, 61, 80, 88, 97, 334, 340, 343, 344, 345, 349, 436, 437
Yamashita, K., 102, 112
Yasillo, N., 15, 23
Yasoshima, Y., 198
Yasumura, S., 56
Yokayama, H., 56
Yoors, J., 89, 99
Yoshino, S., 46f
Yoshitake, Y., 3, 11, 358-359, 364, 437
Young, O.R., 336, 352
Yu, E.S.H., 88, 99

Zaino, H., 163, 182
Zais, J., 340, 350
Zeisel, J., 135, 148, 341, 353
Zimring, C.M., 341, 345, 353
Zube, E.H., 7, 199, 200, 201, 205, 206, 208, 336, 353, 389, 437, 438
Zukin, S., 322, 324

Subject Index

A-bombing, urban renewal following, *see* Urban renewal
Aborigines, 401–402
Acculturation, 285–286
Adaptation, 285, 304–305, 309–310
Adventurer profile, 242
Affective experiences, of exchange students, 290
Affordances, 173, 249, 257
Africa, 88, 368
Agriculture, in riparian areas, 202, 203–204
Akabanedai site, 44, 49
Amae, 88
Anchor point theory, 281, 345
Annual Review of Anthropology, 425
Annual Review of Psychology, 425
Antarctic expedition, 242, 243
Antecedent-consequent causation, 432
Apparent head size estimation task, 91, 94
Appearential order, 367
Appropriating position, in urban housing, 46–47, 48, 49
Architectural intimacy, 89
Architectural planning research (APR), 9, 355–364
    definition of design in, 361–362
    environment behavior studies compared with, 359–361
    history of development in, 356–359
    situation and goals of design in, 362–364
    three fields in, 356
    transactional perspective in, 361
Architecture, interface between other disciplines and, 341–342
Architecture of *mu*, 362
Asia, *see also* specific countries
    body and self experience in, 88
    modern versus traditional culture in, 368–373

Asia (*cont.*)
    urbanization and quality of life in, 325–330
Assumptions, 84, 284–285, 430–433
    minimizing, 411–414
Attentional overload, 338
Australia, 344, 404
Australian Green Streets project, 409n
Aztec Indians, 319

Barrier-free design, 36, 38, 43
Bathrooms, 32, 33f, 34f, 35f, 36f, 39f
Bauentwurfs Lehre, 357
Bedfast (Netakiri) elderly, 26–27
Behavioral focal points (BFPs), 59, 62, 79, 345
    applicability of, 74–77
    main requirements of, 61
Behavioral maps
    in nursing home relocation, 62, 75
    of Spanish American plaza, 316
Behavior settings, 236, 237, 239, 240, 367
Beijing, China, 405
Being in urban structure, 119
Bereavement, 232
Bioculturalism, 286
Biological aspects of person, *see* Physical aspects of person
Biological level of organization, 392–393
Biosphere 2, 8, 235–243, 390, 393
    ecological psychology of, 235–240
    layout of, 238f
    richness of environment in, 239
    selection factor in, 240–243
Birds, in riparian areas, 204
Blacks, 336
Blind elderly, relocation of, *see* Nursing home relocation
Body action, 84, 86t, 90–91, 93
Body as object, 92

Body boundaries, 91, 95
Body buffer-zone task, 91
Body-Cathexis Scale, 91, 92, 93, 94
Body concept/image/representation, 84, 85, 87t
Body esteem, 84
Body evaluation, 91, 94
Body experience, 6, 83–95
    conceptual level of, 6, 84, 87t
    materials used in study of, 90–92
    overview of studies on, 85
    participants in study of, 90
    perceptual level of, 6, 84, 86t, 90
    sensorimotor level of, 6, 84, 86t, 90, 93
    statistical analysis of findings on, 92
Body fantasy, 84
Body mobility, 84
Body-Part-As-Object Test, 92
Body parts, use of in space, 84
Body perception, 84, 85, 86t, 90, 91, 93
Body-related attitudes, 85
Body representation, 91
Border phenomena, 370–371
Bosnia, 337, 371–372, 375
Brazil, 374, 404
Bryant Park (New York), 116, 314, 323

Cairene villas, 402
Campus plan, 8, 273–281
Canada, 336, 404
Canonical boundary conditions, 249
Caution District, 212, 228
Centrality, 250
Centralization, 329–330
Cherry trees, 186, 187, 188f, 189, 192f, 193, 197, 198f
Child abuse, 336–337
Children, 250
China, 369, 405
Churches, 410
City personnel, 326, 328, 329–330
Cognition, sensorimotor, perceptual, and conceptual aspects, 6, 84, 290
Cognitive experiences, of exchange students, 290
Cognitive maps, 59, 62–74, 78–79
    curves of corridors portrayed on, 69–70, 71f/t, 73, 79
    errors of relational positions on, 68–69
    figural presentation compared with, 60

Cognitive maps (cont.)
    literature review on, 60–61
    number of places on, 68
    overall structure portrayed on, 70–72
Color preferences, 6, 101–111, 404–405
    general order of, 103
    mathematical analysis by dual scaling, 107
    relative frequencies by hue and tone, 103-106
    stimuli used to determine, 102–103
    for white, 102, 103, 109–111
Come and go (yuki-Kau), 120–122
Communication, in family housing, 249–250, 251–255
Communicative knowledge, 396
Community health, 339
Compilation of Design Data for Architecture, 357, 358
Conceptual aspects of body, 84, 90, 91, 93
Conceptual aspects of self, 84, 90, 91–92, 93
Conceptual level of body experience, 6, 84, 87t
Connectedness to others, 88, 94, 95
Contextualist theories, 367
Correctional Institution Environment Scale (CIES), 262–271
Correctional institutions, 8, 263–265, 270–271, 388
Corridors, curves of, 69–70, 71f/t, 73, 79
Costa Rica, plazas of, see Spanish American plaza
Crime, 340, 341
    preventing at regional and international levels, 336–337
    urbanization and, 326
Cross-cultural studies
    of body and self experience, see Body experience; Self experience
    of color preferences, see Color preferences
    opportunities for, 345–346
Cuba, 374
Cultural intimacy, 89
Culture, 365–375
    modern versus traditional, 368–373
    urban public space and, 123–124

Debate, inadequate, 407

Dedifferentiation, 284, 286, 288, 291, 303, 304

Deficiency theory, 60, 79

Dementia, visual perception and, *see* Visual perception

Design
definition of, 361–362
goals of, 362–363

Development, 8, 84, 95, 139–140, 255–256, 283–286, 430
Ecology of theory development, 334
Experience and action, 285
Interpersonal relationships, 298–302
Orthogeneteic principle, 284–285
See also Elderly

Differentiation and conflict, 286, 288, 291, 303, 304–305

Differentiation and hierarchic integration, 284, 286, 288, 291, 303, 305

Differentiation and isolation, 286, 288, 291, 303, 304

Dining position, in urban housing, 46–47, 48

Disaster Relief Act (Japan), 212

Documentary surveys, 185

Door sills, 30–31

Dual scaling, 107

Dwelling design guidelines, for elderly, 5, 25–41
examples of, 29–35
lessons from Hanshin earthquake, 35–36
proposals for, 28–29
survey of extended family living in, 27
survey of local community living in, 28
survey of special housing for aged in, 27–28

Each in one's own way (omoi-omoi), 119–120

Ecological psychology, of Biosphere 2, 235–240

Education, urbanization and, 326, 328, 329

Efficient causation, 431, 432

Elderly
dwelling design guidelines for, *see* Dwelling design guidelines, for elderly
environmental design and community planning for, 339–340
nursing home relocation for blind, *see* Nursing home relocation

Elderly (*cont.*)
quality of life in, 6, 51–56
urban housing for, *see* Urban housing, for elderly
urban renewal and, *see* Urban renewal
visual perception in demented, *see* Visual perception

Empirical models, 406

Empirical studies, isolated, 401–402

Employment
in riparian areas, 201–202
urbanization and, 326, 328
of women, 7, 151–158

England/United Kingdom, 35, 51

Entry (genkan), 248

Environment, physical, interpersonal and sociocultural aspects, 5, 284, 286, 287, 289, 293–294, 295, 305

Environmental aesthetics, 414

Environmental Design Research Association (EDRA), 401, 406, 411, 414, 424, 425

Environmental disengagement, 138 140

Environmental psychology, 9, 333–347
Biosphere 2 and, *see* Biosphere 2
community problem-solving strategies in, 342–343
crime and, 336–337, 340, 341
cultural and historical aspects of, 9, 365–375
elderly and, 339–340
experiencing present-day, 381
forecasting trends in, 379–381
green, 335
health promotion and, 339
intellectual scientific structure of, 379–380
interface between architecture, urban design and, 341–342
pollution and environmental change in, 335–336
technological change and, 337–338

Environmental salience, 344–345

Ethnographic approach, to urban renewal, 134

Ethnospace, 367

Europe, 368–373, *see also* specific countries

Exchange students, 8, 283–310, 393
    American culture as viewed by Japanese,
        288–291
    development of interpersonal
        relationships by Japanese, 297–302
    experience and action of American,
        302–308
    experience and action of Japanese,
        291–297, 298t
Experience and action, relations among, 8,
    285, 287, 302
Extended families, 27, 38

Fair Housing Amendment Act, U.S., 35
Family housing, 8, 247–257
    context and behavior in, 251–255
    exploring alternative approaches in,
        249–251
    privacy in, 251
    rethinking stereotypes of, 248–249
    space use in, 255–256
Family planning, 326, 328, 329
Festive activities, 141–144
Figural presentation, 60
Final causation, 431
First Japan-USA Seminar on
    Environment-Behavior Research,
    435–436
5-Year Plan of Housing Construction
    (Japan), 28
Fixed seating, in urban housing, 46–47
Flooding, 29
Floor level, 29
Folk psychology, 386
Folk theories and concepts, 407
Formal causation, 431
Fort Huachucha, 202, 204
Fourth Japan-USA Seminar on
    Environment-Behavior Research
    (comment), 10, 385–396
France, 334
Fukae Town, Japan, 212, 214
Functionalism, 359, 361, 366
Fusuma (sliding doors), 248, 255, 256, 257
Futons, 47, 48, 248

Gender
    Biosphere 2 conditions and, 242
    body and self experience and, 85, 90,
        94

Gender (*cont.*)
    exchange students' experience and,
        297, 303
    quality of life and, 54, 55
    self-esteem and, 85
Generality, 387–391, 414–416
General systems theory, 369
Genkan (entry), 248
Geometric-technical properties, 84
Gila River, 201–204, 206, 207
Grand Central Terminal (New York), 122
Gran Hotel (Costa Rica), 323, 324
Granny annexes, 26
Green environmental psychology, 335

Handbook of Environmental Psychology,
    366, 425
Handrails, 32, 34f, 35f, 36f, 37f, 40f
Hanshin-Awaji earthquake, 35–36, 209
Happen to be present (iawaseru), 118–119
Hasegawa Dementia Scores, 16
Hawthorne Factory, 359
Health
    environmentally based strategies of
        promoting, 339
    urbanization and, 326, 328, 329
High school students, 8, 265–271, 388
Hikarigaoka site, 44, 49
Hiroshima bombing, urban renewal
    following, see Urban renewal *History
    of Invention* (Williams), 378
Hokkaido Southwest earthquake, 209
Holism, 8, 283
Holistic, developmental, systems-oriented
    perspective, 285, 287t, 288, 302,
    308
Homicide, 336, 337
Housing, 326, 328
Husband's domain, 44
Hypercoding, 151

Iawaseru (happen to be present), 118–
    119
I-kata, 115–116
Ina-gakuen Comprehensive Upper
    Secondary School, 273–281
India
    pilgrimages and processions in, 401
    urbanization and quality of life in, 326,
        327, 328, 329

Individuality, 88, 94

Indonesia, 326, 327, 328, 329

Industrial change, 326

Inkyo-beya, 43

Inquiry surveys, 185

Integrative techniques, *see* Time use
analysis

Interactional worldview, 428, 429–431,
432

Interdependence, 88

Intermediate maps, 71–72, 78

Internalized community, 145

International Association for
People-Environment Studies (IAPS),
411, 424

Interpersonal aspects of environment, 284,
285t
actual experience of, 295t, 296, 298t,
304, 305t
common problems in, 287
expectations concerning, 293t, 294
reports on, 289

Interpersonal Network Questionnaire, 299,
300

Interviews
of exchange students, 300–301
in Spanish American plaza studies, 316

Intrapersonal aspects of person, *see*
Psychological aspects of person

Italian-Americans, 250

Italy, 89, 368, 369

Japan, 334, 344, 346
architectural planning research in, *see*
Architectural planning research
body and self experience in, *see* Body
experience; Self experience
campus plan in, 273–281
correctional institutions in, 8, 263–265,
270–271, 388
dwelling design guidelines for elderly
in, *see* Dwelling design guidelines, for
elderly
elderly population in, 51, 340
exchange students in, *see* Exchange
students
family housing in, *see* Family housing
high school students in, 8, 265–271,
388
life expectancy in, 51

Japan (*cont.*)
modern versus traditional culture in,
368, 369, 370
quality of life for elderly in, 51–56
restoration housing in, *see* Restoration
housing
sociopsychological environments of
schools in, 261–271
urban housing for elderly in, *see* Urban
housing, for elderly
urbanization and quality of life in, 326,
327, 328, 329
urban riverfronts in, *see* Riverfronts,
trees on

Japan Color Research Institute, 110

Japan–United States Seminars, 1–5,
435–438

Japanese gardens, 7, 163–182
description of nonvisual information in,
174
description of visual information in,
173–174
environmental data creation of, 168–173
experimental design in study of, 165
procedures used in study of, 165–168
study site and participants in study of,
165

Jews, 371, 373

Jibun, 88

Kenchiku keikaku, 355, 357

Kibougaoka site, 44, 49

Kitchens, 250

Korea, 370
color preferences in, 102
urbanization and quality of life in, 326,
327, 328, 329

K-21 scale, 236, 237–239

Landscape values, in riparian areas, *see*
Riparian areas

Latin America, 88, 334, 345

Life expectancy, in Japan, 51

Life Satisfaction Index-K (LSIK), 52–56

Life-world, urban environment as,134

Literature, knowledge of, 409–410

Lost landscape, 141–144

Lost World of the Berberovic Family, The
(van Cleef), 371–372

Luxembourg garden (Paris), 116

Lyon station (Paris), 120f

Malaysia, 326, 327–328, 329
Man-Environment Research Association
    (MERA), 3, 4, 424
Mannen-doko, 48
Material causation, 431
Maya Indians, 319, 402
Mayo Indians, 402
Megachurches, 410
Meiji Restoration, 355n, 356
Mental retardation, visual perception and,
    see Visual perception
Mesoamerican plaza, 319
Mexican Americans, 404
Mexico, 374
Microgenetic process, in cognition, 61, 62,
    73, 78
Middle zone, 44
Migrants
    developmental analysis of, 286
    factors affecting adaptation of, 285t
Minnesota Multiphasic Personality
    Inventory 2 (MMPI-2), 241–243
Mizunashi River, 211
Modernism, 365–366
Modes of being in places, see Urban public
    space
Motivational salience, 344
Mount Unzen-Fugendake eruption,
    restoration housing for, see
    Restoration housing
Mount Vesuvius eruption, 378
M-P survey, 237–239
M town, 136, 139, 144
Multiculturalism, 288, 309, 374, 375
Multidimensional analysis, see Time use
    analysis

NAC Eye-Mark Recorders, 16
Nakaniwa (special places), 248
Nangai Study, 56
Narratives, 385–386, 387–388, 395
    incoherence in, 394
    situating the narrator in, 394
    urban renewal and, 135, 145
National Theater (Costa Rica), 322, 323
Natural disaster, 7, 35–36, 209–283, 334
Nedu site, 44, 49
Nepal, 326, 327, 329, 330

Netakiri (bedfast) elderly, 26–27
North zone, 44
Notre-Dame (Paris), 121f
Nuovo Cinema Paradiso (film), 120, 124f
Nursing home relocation, 6, 59–79
    behavioral focal points in, see
        Behavioral focal points
    behavioral maps in, 62, 75
    cognitive maps in, see Cognitive maps
    observation of movement in, 75
    questionnaire test in, 76–77

Occupying position, in urban housing,
    46–47
Olfactory landscape, 410
Omoi-omoi (each in one's own way),
    119–120
"Onagigawa Gohonmatsu" (wood-cut
    print), 190f
Optimum size of cities, 325
Orders for Discovery and Settlement, 319
Organismic worldview, 84, 428, 429–431
Organismic–developmental theory, 6
Orthogenetic principle, 61, 284, 286

Palais Royal (Paris), 115f, 116, 118f
Paris, France, 116, 118, 119, 123
Parque Central (Costa Rica), 316, 317–318,
    319–321, 324
Pedestrian malls, 410–411
Perceptual level of body experience, 6, 84,
    86t, 90
Perceptual salience, 344
Person, physical, psychological and
    sociocultural aspects, 5, 284, 286, 287,
    292–293, 294–295, 297
Permeability maps, 251, 252f, 253f, 254f
Personal memories, 50
Personal network, 301
Person-environment relations, 135,
    367–368, 369, 371, 372
    of exchange students, 284, 286
    nursing home relocation and, 59, 61
    promotion of theorizing in, 380–381
Person-in environment, 284–285
    category system, 289–296, 299–301,
        303, 308
Person-world level of organization, 392–
    393
Phenomenology, 366

Philadelphia Geriatric Center Morale Scale
    (PGC), 52
Philippines, 326, 327, 328, 329, 330
Physical aspects of environment, 284, 285t
    actual experience of, 294, 295t, 296,
        298t, 304, 305t
    common problems in, 287
    expectations concerning, 293t, 294
    reports on, 289
Physical aspects of person, 284, 285t
    actual experience of, 294, 295t, 296, 298t
    common problems in, 287t
    expectations concerning, 293
    reports on, 290
Physiognomic properties, 84
Pilgrimages, 401
Pine trees, 186, 187, 190f, 193
Placing oneself (personal world) in public,
    116–117
Plaza de la Cultura (Costa Rica), 316, 318,
    321–323, 390, 394
Pollution
    psychological and behavioral
        dimensions of, 335–336
    urbanization and, 326, 328
Pompidou Museum (Paris), 318
Postoccupancy evaluation (POE), 341, 359
Poststructuralism, 366
Post-traumatic stress disorder, 210
Predesign research (PDR) methods, 341
Privacy, 89–90, 251
Private zones, 44
Problem, theory, and method, relations
    among, 8
Processions, 401
Proposition 187, 373
Psychological aspects of person, 284, 285t
    actual experience of, 294, 295t, 296, 298t
    common problems in, 287t
    expectations concerning, 293
    reports on, 290–291
Psychological Distance Map (PDM),
    299–300, 303, 309
    for people, 305–308
    for places, 308
Psychological level of organization,
    392–393
Psychological part-processes, cognition,
    affect, and valuation, 8, 290
Public zones, 44

Quality of life
    in elderly, 6, 51–56
    urbanization and, 9, 325–330
Quantitative difference theory, 60, 79
Questionnaires
    for exchange students, 292
    in nursing home relocation, 76–77
    on expectation and experience, 92

Rain, protection from, 29
Replication, 409
Research methodology, 431–432
Restoration housing, 7–8, 209–233,
    390–391
    bereavement and, 232
    case studies of, 217–230
    description of volcanic eruption
        preceding, 211
    relocation to, 212–213
Revenues, 326, 328
Richness of environment, 239
Rillito River, 201
Riparian areas, 7, 199–208
    management position supported for,
        205–206
    perceptions of appropriate land uses,
        206–207
    perceptions on growth and change, 206
    related literature, 204–205
Riverfronts, trees on, 7, 183–198
    actual state of planting, 186
    documentary survey of, 185
    inquiry survey of, 185
    preferable visual images of, 189–193
    preferred species, 193
    spatial balance of tree arrangement,
        193–197
    traditional images of, 186–187
Rooms
    layout of in urban housing, 43–44, 48
    partitioning of, 248
Route maps, 60, 71–72, 78
Rwanda, 337

Safety, personal, 153–158
Safety belt use, 88
Safford, 201–204, 206
Saitama Prefecture, Japan, see Riverfronts,
    trees on
San Pedro River, 201–204, 205–206, 207

Scandinavia, 334, 344, 346
Schools
    campus plan in, 8, 273–281
    exchange students in, *see* Exchange
        students
    sociopsychological envirnoments of, 8,
        261–271
Scientific creativity, 344–345
Second Japan-USA Seminar on
    Environment-Behavior Research,
    436–437
Self-Cathexis Scale, 92, 93, 94
Self-concept/image/representation, 84, 87t,
    91
Self-drawings, 91, 93, 94
Self-esteem, 84, 85, 92, 93, 94
Self-Esteem Inventory, 92, 94
Self-evaluation, 92, 94
Self experience, 6, 83–95
    materials used in study of, 90–92
    overview of studies on, 85–89
    participants in study of, 90
    statistical analysis of findings on, 92
Self-fantasy, 84
Self-understanding, 84, 92
"Sendagi Dangozaka Hanayashiki"
    (wood-cut print), 192f
Sensorimotor level of body experience, 6,
    84, 86t, 90, 93
Sensorimotor palpation, 84
Setting deprivation, 240
Sex, *see* Gender
Sex-Role Inventory, 91, 92
Shimabara City, Japan, 212, 214
Shingashi River, 193f
Shintoism, 110n
Sicily, 369
Sierra Vista, 201–204, 206
Silver-housing schemes, 26, 27–28, 43
Sinlong-park (Taipei), 116, 123f
Situatedness, 391–394
Sketchmaps, 59
Sleeping position, in urban housing,
    46–47, 48
Sliding doors (fusuma), 248, 255, 256,
    257
Social-ecological perspective, 134
Social habits, 369
Social Logic of Space (Hillier and Hanson),
    249

Sociocultural aspects of environment, 284, 285t
    actual experience of, 295t, 296–297,
        298t, 304, 305t
    common problems in, 287
    expectations concerning, 293t, 294
    reports on, 289–290
Sociocultural aspects of person, 284, 285t
    actual experience of, 294, 295t, 296, 298t
    common problems in, 287t
    expectations concerning, 293
Sojourn, 283
Sojourners, 286–288, 308–309
    developmental analysis of, 286
    factors affecting adaptation of, 285t
Sonita Creek, 201
Soraku-en (Kobe city), *see* Japanese gardens
South zone, 44
Soviet Union, former, 337, 345
Space use, in family housing, 255–256
Spanish American plaza, 9, 313–324,
    390
    analysis of, 317
    interviews and historical documentation
        in, 316
    observation of, 316
    setting of, 317–318
    symbolism in, 319–320
    theoretical assumptions underlying,
        314–315
Spatial cognition, 60-61, 62, 73–74, 78–79
Spatial identity, 133
Spatial order, 367
Special places (Nakaniwa), 248
Specificity, 387-391
Square du Vert Galant et Pont des Arts
    (Paris), 114f
Staffing theory, 274, 281
Stairs, 27
Stand still (tatazumu), 122
Stereotypes, of family housing, *see* Family
    housing
Storage space, in urban housing, 48
S town, 136, 138, 143, 144
Stroop Color-Word Test, 89
Subject object distance, 91
Sumai-kata surveys, 357
Survey maps, 60, 71–72, 78
Sweden, 41, 51
Switzerland, 404
Symbolism, 319–320

Synthesis, lack of emphasis on, 402–405
Systems, 8

Taipei, Taiwan
    color preferences in, *see* Color
        preferences
    urban public space in, 116, 123
Taoism, 362n
Tatami rooms, 31, 38f, 47, 49, 248,
    256
Tatazumu (stand still), 122
Technical knowledge, 396
Technology
    as an engine of discontinuity, 377–378
    impact of change on individuals and
        groups, 337–338
Tekite studies, 240
Tempo regulation, 84, 90–91, 93, 94, 95
Tempo Regulation Test, 90
Testability, 406
Test of Self-Understanding, 92
Thailand, 326, 327, 328, 329
Theoretical models, 406
Theory, 432–433
    attributes of good, 408–409
    ecology of development, 343–346
    inadequate, 405–406
Theory in Environmental Design
    (Rapoport), 399
"Thing and Medium" (Heider), 393
Third Japan-USA Seminar on
    Environment-Behavior Research,
    437–438
Tianjin, China
    color preferences in, *see* Color
        preferences
    urban public space in, 116, 122
Times Square (New York), 314
Time use analysis, 7, 149–158
    as an integrative research method,
        150–151
    personal safety and travel contexts
        studied with, 153–158
TMIG Index of confidence, 54
Toilets, 32, 34f, 40f
Tokyo, Japan
    color preferences in, *see* Color
        preferences
    urban public space in, 116, 124

Tokyo Metropolitan Institute of
    Gerontology, 56
Trait worldview, 428, 429–431
Transactional approach, 284, 355, 361,
    367, 411, 428, 429–431, 432
    in architectural planning research, 361
    to exchange students, 284
Transportation, 326, 328
Travel, personal safety and, 153–158
Trees, on riverfronts, *see* Riverfronts, trees
    on
Tsukaware-kata, 358, 359–360
Tuileries garden (Paris), 116, 121f
Twenty Statements Test, 88
Twenty-first century, 333–335

United Kingdom/England, 35, 51
United States
    body and self experience in, *see* Body
        experience; Self experience
    color preferences in, 404
    correctional institutions in, 263–
        265
    crime in, 336–337
    elderly population of, 339–340
    environmental protection regulation in,
        336
    exchange students in, *see* Exchange
        students
    family housing in, *see* Family housing
    health care costs in, 339
    modern versus traditional culture in,
        368-373
Unit of analysis, 284–285, 429–430
Universal design housing, 29
Universities, exchange students in, *see*
    Exchange students
Urban design, interface between other
    disciplines and, 341–342
Urban housing, for elderly, 5, 43–50
    current trends in room layout, 43–44
    fixed seating and occupying position in,
        46–47
    overlapping and sequential separation
        of activities in, 48–49
    quality of space in, 49–50
    storage and room layout in, 48
    use of rooms in, 47–48
Urbanization, quality of life and, 9, 325–
    330

Urban public space, 6, 113–128
  being in urban structure, 119
  come and go (yuki-Kau), 120–122
  cultural factors and, 123–124
  each in one's own way (omoi-omoi),
    119–120
  happen to be present (iawaseru),
    118–119
  placing oneself (personal world) in
    public, 116–117
  stand still (tatazumu), 122
Urban renewal, 7, 133–148
  environmental disengagement of
    residents, 138–140
  ethnographic approach to, 134
  humanization via mediation in, 145–148
  lost landscape and festive activities in,
    141–144
  narrative acts and, 135, 145
  social-ecological perspective on, 134, 135
Urban riverfronts, see Riverfronts, trees on
Urban Villagers, The (Gans), 133
Utilities, 326, 328

Valuative experiences, of exchange
  students, 290–291

Verkehrsweg, 357
Violent crime, see Crime
Visual perception, 5, 15–23
  apparatus used to study, 16
  fixation behavior on stimuli, 18–23
  fixation time in, 18
  stimuli used to study, 16–17
  subjects studied, 16

Water barriers, 27, 32
Water resources, in riparian areas, 202–204
Way-finding tasks, 60–61
West End (Boston), 133, 145
Wheelchair accessibility, 26, 27, 29, 30–31,
  38-41
White, preference for, 102, 103, 109–111
Who Am I Test, 91
Wife's domain, 44
Willow trees, 186, 187, 189, 191f
Women, 7, 151–158, see also Gender
Work Environment Scale (WES), 271
World War II, 356

"Yatsumi-no-hashi" (wood-cut print), 191f
Yuki-Kau (come and go), 120–122